# Dispossession

## DISCRIMINATION
## AGAINST AFRICAN AMERICAN FARMERS
## IN THE AGE OF CIVIL RIGHTS

## Pete Daniel

THE UNIVERSITY OF NORTH CAROLINA PRESS / CHAPEL HILL

*This book was published with the assistance of the*
*Z. Smith Reynolds Fund of the University of North Carolina Press.*

© 2013 THE UNIVERSITY OF NORTH CAROLINA PRESS
ALL RIGHTS RESERVED
MANUFACTURED IN THE UNITED STATES OF AMERICA

Designed by Kimberly Bryant and set in Utopia by Tseng Information Systems, Inc. The paper in this book meets the guidelines for permanence and durability of the Committee on Production Guidelines for Book Longevity of the Council on Library Resources. The University of North Carolina Press has been a member of the Green Press Initiative since 2003.

Library of Congress Cataloging-in-Publication Data
Daniel, Pete.
Dispossession : discrimination against African American farmers in the age of civil rights / Pete Daniel.
p. cm.
Includes bibliographical references and index.
ISBN 978-1-4696-0201-1 (cloth : alk. paper) 1. African American farmers—Civil rights. 2. Racism—United States—History—20th century. 3. Farms, Small—Government policy—United States—History—20th century. 4. United States. Dept. of Agriculture—Evaluation. I. Title.
HD8039.F32U626 2013
630.89′96073—dc23   2012037475

17 16 15 14 13   5 4 3 2 1

*For Hallin*
*and the extended Daniel, Carbaugh, and Davis families*

# CONTENTS

# ILLUSTRATIONS

*It is one thing to write as a poet and another to write as a historian:*
*the poet can recount or sing about things not as they were,*
*but as they should have been, and the historian must write about them*
*not as they should have been, but as they were,*
*without adding or subtracting anything from the truth.*
—*Miguel de Cervantes,* Don Quixote

# PREFACE

*Dispossession* focuses on the third quarter of the twentieth century. Usually referred to as the civil rights era, this was a moment of extraordinary transformation in the rural United States. Science and technology applied to agriculture increased yields, made hand labor obsolete, and, combined with federal programs, drove 3.1 million farmers from the land. In the quarter century after 1950, over a half million African American farms went under, leaving only 45,000. In the 1960s alone, the black farm count in ten southern states (minus Florida, Texas, and Kentucky) fell from 132,000 to 16,000, an 88 percent decline. Whites also left southern farms during this decade, though the decrease was not as dramatic: 61,000 farms remained of the 145,000 a decade earlier, a 58 percent decline.

The U.S. Department of Agriculture (USDA) dismissed farm failures as the natural consequence of farmers' adoption of machines and chemicals, but in fact, the USDA shamelessly promoted capital-intensive operations and used every tool at its disposal to subsidize wealthy farmers and to encourage their devotion to science and technology. New was better; old was not meant to survive. By the 1950s, the intrusive tentacles of agrigovernment uncoiled from Washington through state and county offices and, paired with agribusiness, reconfigured the national farm structure. At the same time, the USDA erected high hurdles, often barriers, that discouraged or prevented minorities and women from securing acreage allotments, loans, and information. Paradoxically, the earlier increase in the number of black farms to a high of 925,000 by 1920 occurred during some of the nation's worst years of violence and discrimination since the Civil War, and the decrease intensified when government programs, civil rights laws, and science and technology promised prosperity and fair treatment. During this period, despite the profound implications of demographic chaos, the press seldom mentioned dispossession.

African American farmers stubbornly refused to go quietly from their farms and eloquently articulated and bravely resisted the discrimination that threatened them. They ran for county committee seats, confronted county executives, applied for loans, and brought suits to challenge discrimination. They were often unable to obtain credit, the sine qua non for modern agriculture, even for spring planting, much less for buying tractors, picking machines, and chemicals, nor were they favored by USDA personnel or policies. From its inception in 1862, the USDA was run by white men and, with the exception of the Negro Extension Service, excluded African Americans from decision-making positions.

The civil rights and equal opportunity laws of the mid-1960s prompted USDA bureaucrats to embrace equal rights rhetorically even as they intensified discrimination. This passive nullification, voicing agreement with equal rights while continuing or intensifying discrimination, did not rely on antebellum intellectual arguments or confrontations but instead thrived silently in the offices of biased employees. Phone calls and conversations at segregated meetings and conventions left no racist fingerprints, but the accretion of prejudice festered and ultimately grew into a plan to eliminate minority, women, and small farmers by preventing their sharing equally in federal programs. Despite overwhelming evidence of discrimination, incompetence, and falsehoods in many offices, the USDA never cut off funds to those offices, and apparently no white person was fired and few were even relocated or reprimanded.

An impressive number of organizations assisted black farmers in coping with bias in the USDA's Washington headquarters and in states and counties throughout the South, and young civil rights workers, African American farmers, and black extension agents made heroic contributions. This book focuses on the years prior to the *Pigford v. Glickman* decision in 1999, a class-action suit that won compensation for discrimination after 1981. *Pigford* exposed interminable USDA bias and established precedent for similar suits by Native Americans, women, and Hispanics. During the height of the civil rights movement in the 1960s, however, there was no check on USDA discrimination. Indeed, after the *Brown v. Board of Education* decision in 1954, southern USDA offices twisted programs to punish black farmers who were active in civil rights, and administrators in Washington acquiesced.

This book examines three USDA agencies: the Agricultural Stabilization and Conservation Service, the Federal Extension Service, and the Farmers Home Administration. These powerful pseudo-democratic agencies became repositories of prejudice and discriminatory practices as they hired office staffs, selected extension and home-demonstration agents, controlled information, adjusted acreage allotments, disbursed loans, adjudicated disputes, and, in many cases, looked after their own families and friends. The focus of this book seldom widens from farm issues, although the turbulent decade of the 1960s offered tempting diversions. The farmers and bureaucrats who appear in the pages that follow have seldom emerged as historical players, but they are an important part of the history of both the civil rights era and American agriculture.

# ABBREVIATIONS

| | |
|---|---|
| AAA | Agricultural Adjustment Administration |
| A&T | North Carolina Agricultural and Technical College |
| ACES | Alabama Cooperative Extension Service |
| ACP | Agricultural Conservation Program |
| ASCS (also ASC) | Agricultural Stabilization and Conservation Service |
| CCC | Commodity Credit Corporation |
| COFO | Council of Federated Organizations |
| CORE | Congress of Racial Equality |
| DASCO | deputy administrator of state and county operations |
| FBI | Federal Bureau of Investigation |
| FES | Federal Extension Service |
| FHA (later FmHA) | Farmers Home Administration |
| FSA | Farm Security Administration (also Farm Service Agency) |
| LCDC | Lawyers Constitutional Defense Committee |
| MCES | Mississippi Cooperative Extension Service |

| | |
|---|---|
| MFDP | Mississippi Freedom Democratic Party |
| NAACP | National Association for the Advancement of Colored People |
| NACAA | National Association of County Agricultural Agents |
| NASULGC | National Association of State Universities and Land-Grant Colleges |
| NSF | National Sharecroppers Fund |
| OEO | Office of Economic Opportunity |
| OIG | Office of the Inspector General |
| RAD | Rural Area Development |
| SCLC | Southern Christian Leadership Conference |
| SCS | Soil Conservation Service |
| SNCC | Student Nonviolent Coordinating Committee |
| SWAFCA | Southwest Alabama Farmers Cooperative Association |
| TAP | Technical Action Panels |
| USDA | U.S. Department of Agriculture |

# Dispossession

*I was born free,*
*and in order to live free I chose the solitude of the countryside.*
*—Miguel de Cervantes,* Don Quixote

# 1

# INTENDED CONSEQUENCES

On April 22, 1965, Secretary of Agriculture Orville Freeman encouraged the U.S. Department of Agriculture (USDA) staff to "put into effect with dispatch" comprehensive policies that would end discrimination. "The right of all of our citizens to participate with equal opportunity in both the administration and benefits of all programs of this Department is not only legally required but morally right," he stressed. Nearly every secretary of agriculture since Freeman has issued a similar plaintive decree.

The 1964 Civil Rights Act, of course, legally banned discrimination, but a scathing March 1965 report by the U.S. Commission on Civil Rights, the independent agency created by the Civil Rights Act of 1957 to investigate and report on a broad spectrum of discriminatory practices, alerted Freeman that racism infected every office in his department. *Equal Opportunity in Farm Programs: An Appraisal of Services Rendered by Agencies of the United States Department of Agriculture* revealed that blacks had no input in policy, had no representation on county agricultural committees, were refused loans and benefits, and suffered encompassing discrimination.[1] But civil rights laws and Freeman's memorandum, it seemed, only intensified the USDA's bureaucratic resolve to resist the concept of equal rights. Realizing that overt resistance would be futile, the staff perfected passive nullification, that is, pledging their support even as they purposefully undermined equal opportunity laws. By the 1970s, USDA leaders would claim full compliance with equal opportunity laws even as they subverted programs to deny benefits to African Americans, Na-

*Secretary of Agriculture Orville Freeman (left) with Federal Extension Service administrator Lloyd Davis. Courtesy of National Archives and Records Administration, Record Group 16, 2288 ST.*

tive Americans, Hispanics, and women. Despite support from President Lyndon B. Johnson, Secretary Freeman failed to control vindictive and parlous bureaucrats in Washington or to police state and county USDA offices throughout the South. At the moment that civil rights laws promised an end to discrimination, tens of thousands of black farmers lost their hold on the land, in part because of the impact of science and technology on rural life but also because they were denied loans, information, and access to programs essential to survival in a capital-intensive farm structure.

Orville Freeman came to the USDA after serving three terms as governor of Minnesota, and he succeeded Republican Ezra Taft Benson. Not even the dour and humorless Benson, who vowed to cut New Deal programs, could derail the runaway USDA train that delivered so much to so many. In January 1964, Rodney E. Leonard, one of Secretary Freeman's deputies, suggested to the secretary that the department had "developed

almost into a shadow government serving mainly rural America. We provide credit and power, we encourage conservation and recreation and
we support education." Leonard's vision was modest. "There were four or
five USDA programs in every one of the three thousand counties in this
country," Freeman recalled in a 1988 interview. "I mean, the magnitude
of the Department of Agriculture is very, very extensive and not realized
at all." By the mid-1960s, the USDA had become so vast, its constituency
so demanding, its programs so contradictory, and its lines of communication and responsibility so tangled that it often seemed at war with
itself. Paradoxically, the USDA's contradictions only strengthened it, for
it offered generous benefits to a vast constituency that wielded enormous
political power.[2]

Freeman had been reluctant to take on the USDA, with its two massive buildings (the main building located on the National Mall alongside
Smithsonian Institution museums), eight miles of corridors, and nearly
5,000 rooms. Under Secretary Benson, one room, it was reported, housed
a counterfeit-currency operation with a press and plates. When Freeman
arrived at the USDA in 1961, there were over 96,000 employees, some
12,000 stationed in Washington, and roughly 13 million farmers. In 2010, by
contrast, there were 113,000 employees and only some 2 million farmers.
Under Freeman, nearly all USDA employees were white, all supervisors
were white males, and, except in the Negro Extension Service, nearly all
blacks employed by the USDA were custodial workers. On April 5, 1963,
two years before Freeman issued his civil rights memorandum, his image
appeared on the cover of *Time* magazine. The *Time* story commented on
Freeman's World War II Marine experience, his terms as governor of Minnesota, his squash games with Secretary of Defense Robert McNamara,
and the complex issues facing U.S. farmers, but there was nothing about
discrimination in USDA programs or the drastic decline of farmers, African Americans in particular. "For some strange reason and coming in as
a liberal," he stated in a 1969 interview, "I worked well with Southerners; I had quite a number of Southerners in the department." He left the
implications of working well with southerners hanging, but his compat-
ibility with them may well have developed from his accommodation to
their prejudices. Freeman did not discuss the 1965 Commission on Civil
Rights report or comment on how the civil rights movement played out
in the South during his tenure. His interviews stress domestic and international agricultural policy, dealings with Congress, and his relationship
with presidents John F. Kennedy and Lyndon B. Johnson, but he ignored
discrimination and hardly mentioned African Americans.[3]

Freeman's pledge to end discrimination, like that of subsequent secretaries of agriculture, failed. New Frontier liberals such as Freeman, while offering support for civil rights, lacked grounding in southern history and culture, especially concerning how segregation and discrimination distorted relations between blacks and whites. By the time he came to the USDA, southern whites had demonstrated how viciously they would fight to preserve segregation, and as civil rights activity increased in the southern countryside, USDA officials manipulated government programs to punish activist farmers. Apparently, Freeman never realized the extent to which employees outside his executive staff, both in Washington and throughout the South, resisted implementing civil rights edicts or the fact that his aides protected him from most discrimination complaints. Without pressure from Freeman's office, discrimination would not only continue but also flourish.

Thirty years after Secretary Orville Freeman left office, Judge Paul L. Friedman handed down the landmark *Pigford v. Glickman* decision, the successful class-action suit brought by North Carolinian Timothy Pigford that found the USDA guilty of widespread discrimination. Judge Friedman began his 1999 decision with that familiar broken promise from the Civil War and Reconstruction eras: "Forty acres and a mule." The case concerned the USDA's sorry civil rights record and its denial of federal benefits to black farmers in the eighteen years since 1981. Judge Friedman suggested that General William T. Sherman's unfulfilled promise of land during Reconstruction resonated with black farmers' journey from slavery to freedom to sharecropping to ownership and, finally, to debt and dispossession. As the *Pigford* decision made clear, racism had continued to circulate through federal, state, and county USDA offices long after Secretary Freeman pledged to end it, and employees at every level twisted civil rights laws and subverted programs to the detriment of black farmers. Judge Friedman admitted that the *Pigford* case would "not undo all that has been done" but insisted that it was a "good first step."

By the time Congress finally appropriated the funds in 2010, many of the litigants had lost their farms or died, and there was no compensation for discrimination prior to 1981. When Judge Friedman handed down his decision only months before the end of the millennium, he observed that only some 18,000 black farms remained, and many of those were endangered. While black and white farmers throughout the country wrestled with mechanization, chemicals, and government programs, black farmers also confronted USDA discrimination. In the 1960s, the number of southern white farm owners decreased from 515,283 to 410,646, and the number of white tenants dropped from 144,773 to 55,650. Farms owned by blacks fell from 74,132 to 45,428, and black tenants declined from 132,011 to 16,113. Many tenants and sharecroppers, of course, became superfluous as tractors, combines, mechanical cotton harvesters, and herbicides reduced the demand for intensive hand labor. Adapting to capital-intensive operations threatened many farmers, but if African American farmers had left agriculture at the same rate as white farmers since 1920, former U.S. Commission on Civil Rights staffer William C. Payne Jr. calculated, there would still be 300,000 left. Underlying Judge Friedman's decision was a disturbing contradiction: black farmers suffered the most debilitating discrimination during the civil rights era, when laws supposedly protected them from bias. The increase in programs and the USDA's swelling bureaucracy had an inverse relationship to the number of farmers: the larger the department, the more programs it generated, and the more money it spent, the fewer farmers that survived.[4]

Black farmers who endured to the twenty-first century represented the remnants of former slaves who began the long march to ownership during the Civil War and Reconstruction. It was difficult to move from sharecropping, where the landlord sold the crop and paid the farmer, to tenancy, where the farmer sold the crop and paid rent to the landlord, to ownership. Black and white farmers fought stubbornly to maintain control of their crops and their labor only to watch crop-lien and labor laws erode their fortunes. Chained to a punishing annual work routine that in some ways resembled a stock car race—round and round with disaster likely at any moment—farmers battled nature, bankers, merchants, and landlords, and they were often poorer at settlement time in autumn than when they planted in spring. Their meager diets fostered rickets and pellagra, the lack of sanitation encouraged hookworm and dysentery, labor laws forced them to work to fulfill contracts or go to prison, and venal politicians ignored their education-starved children. Still, many escaped the cycle of debt and purchased land.

Slaves emerged into freedom with a keen understanding of farming that allowed many to navigate the boundary between exploitation and sufficiency. In rural areas, blacks and whites necessarily worked side by side, and despite white supremacy, friendly relationships developed across the color line. Industrious African American farmers deferred when necessary and earned the respect of their white neighbors, learning, as educator Booker T. Washington advised, how to cultivate white support. A combination of husbandry, diplomacy, and ambition allowed black farmers to secure land, and the fact that so many succeeded dur-

ing some of the darkest years of racist violence testifies to their character and determination. Demonstrating tremendous energy and sagacity,
they mediated a maze of law and custom and gained land and standing in southern communities. Despite their hard work, African Americans owned smaller farms and sold less than their white neighbors. In
1969, for example, nonwhite farms (mostly African American) averaged
78 acres compared to 310 for other farms, and only some 1,900 had sales
of $20,000 or more. Over half of all black farmers were over fifty-five years
old, and only 5 percent were thirty-five or younger.[5]

In the tense and troubled years after the 1954 *Brown v. Board of Education* decision, however, most of the advances made by black farmers
since the Civil War were erased. The decline of black farmers after World
War II contrasted starkly with their gains in the half century after Emancipation. By 1910, African Americans held title to some 16 million acres
of farmland, and by 1920, there were 925,000 black farms in the country. After peaking in these decades, however, the trajectory of black
farmers plunged downward. In a larger sense, there was an enormous
decline among all farmers at mid-century. Between 1940 and 1969, the
rural transformation, fueled largely by machines and chemicals and directed by the USDA, pushed some 3.4 million farmers and their families
off the land, including nearly 600,000 African Americans. From 1959 to
1969 alone, 185,000 black farmers left the land, and only 87,000 remained
when Richard Nixon entered office. Farm failures were endemic, and in
the 1950s, about 169,000 farms failed annually; between 1960 and 1965,
some 124,000 failed each year; and 94,000 per year failed between 1966
and 1968.[6]

What happened to African American farmers in the three decades from
1940 to 1974 can be measured both by their decline and by the fact that
their departure went largely unremarked. Although some scholars have
argued that the structural shift from labor-intensive to capital-intensive
operations explained the decline, others have argued that blacks fled the
countryside of their own volition, forcing large owners to use machines
and chemicals to replace them. Historians have rarely glanced offstage
at the vast USDA federal, state, and county apparatus that generated research, distributed information, assigned allotments, made loans, and
controlled funding that dictated the direction and pace of this rural transformation.

For a century and a half, the USDA has presided over monumental
changes in the U.S. countryside. Since its founding during the Civil War,
it has encouraged better farming methods, and over time, its staff has

swelled and its reach has extended to every crossroads and farm. In 1862, Congress provided for land-grant colleges in each state to focus on rural life. Because blacks were not admitted to these white schools, in 1892, Congress tardily established underfunded African American land-grant schools. The Hatch Act of 1887 created federal experiment stations that explored better farming methods and distributed information on helpful science and technology. Early in the twentieth century, the Federal Extension Service (FES), operating out of land-grant schools, became a conduit for feeding farmers advice on the latest science and technology from experiment stations, university research centers, and corporations. Some farmers welcomed and utilized research findings, others were skeptical of experts and outsiders, and still others never received information. USDA personnel, many educated at land-grant institutions, often denigrated farmers who did not accept their gospel of science and technology, echoing an enduring national tradition that pitted book learning against common sense and prized technology at the expense of husbandry. Knowledge handed down or gained by trial and error was devalued and forgotten while formulaic methodology and machines grew in importance. The staggering human cost that accompanied this transformation was eclipsed by the celebratory sheen of tractors and picking machines, insecticides and herbicides, and hybrids and genetically engineered seeds.

In popular memory, the conflicted history of rural life and the civil rights movement settled into a revised version that recast the South's segregated, impoverished, and backward history into a neoconservative success story. Distaste for the civil rights movement converted white Democrats into southern-strategy Republicans, and conversely, African Americans continued to leave the party of Lincoln for that of Roosevelt, Kennedy, and Johnson. Science and technology, the tale went, ended backbreaking work, freeing sharecroppers and tenants to move to lucrative urban jobs throughout the country. This heroic fable of capital-intensive agriculture demeaned its labor-intensive forebears by dismissing millions of farmers as inept and unable to adjust to science and technology. This sanitized version of rural life ran parallel to that of a successful civil rights movement in the 1950s and 1960s that brought equal rights to all Americans. Both tellings leveled mountains of conflict and ignored valleys of despair, for the transformation of the southern countryside in the mid-twentieth century painfully affected millions of people.

An ideology of progress infused the transformation of rural life. Early in the twentieth century, the spread of electricity, automobiles, powered flight, and automation stirred enthusiasm for modernization and rejection of older ways. In her significant analysis of rural industrialization, *Every Farm a Factory*, the historian Deborah Fitzgerald targets "economists, farm managers, employees of agricultural colleges, and particularly farm and home-demonstration agents, rural banks and insurance companies, and agricultural businesses such as those centered on farm machinery and seeds" as the agents of change. This collection of experts and entrepreneurs, many of whom had no practical farming experience, dreamed of large mechanized operations run on scientific principles by efficient managers who would replace small and less businesslike farmers tied to almanacs and labor-intensive work. While anthropologist James C. Scott's discussion of rural high modernism in *Seeing Like a State* does not exactly fit the pattern of transformation in the U.S. South, it is highly suggestive. He quotes Liberty Hyde Bailey, head of President Theodore Roosevelt's Country Life Commission, who identified a long list of experts who would enlighten benighted farmers. In the first decades of the twentieth century, agricultural economists and other specialists formed professional societies, and land-grant schools expanded curriculums to reflect modern ideas. Scott observed that "the rationalization of farming on a huge, even national, scale was part of a faith shared by social engineers and agricultural planners throughout the world." Both Scott and Fitzgerald analyzed how the Soviet Union in the 1920s began reshaping its rural landscape by erasing established farmers (and their know-how), awkwardly embracing mechanization, and enticing U.S. agricultural engineers and implement specialists to furnish expertise. By 1927, U.S. implement dealers had sold 27,000 tractors to the Soviet Union for its gigantic farming operations, but U.S. observers realized that without proper management, the Soviet experiment was flawed. Planners in both countries shared a vision of large mechanized farms utilizing the latest scientific ideas, and as Fitzgerald suggests, American advisers regarded Soviet farms as experiment stations where they could test ideas that would have alarmed American farmers and politicians. Scott argues that the huge, mechanized, but inefficient Soviet farms were the epitome of high-modernist theory, which dismissed farmers' culture, skills, and knowledge. Under Soviet direction, modernism was boldly, even bloodily, forced on the countryside.[7]

While Soviet collectivization sacrificed between 4 and 20 million rural people, the United States instituted the New Deal's Agricultural Adjustment Administration (AAA) and other rural programs that offered relief but also provided a platform to support mechanization and scientific agriculture. A crisis of overproduction and, with the exception of World War I, low prices prepared the way for sweeping change. By the time the New Deal arrived in the spring of 1933, farmers were desperate. Surpluses had driven prices far below the cost of production, voluntary efforts to reduce production had failed, and farmers realized that without government intrusion prices would continue to fall. The Great Depression and the New Deal thus arrived at a pivotal moment in the history of U.S. agriculture, and planners seized on the crisis to transform rural life.[8] Ironically, the USDA blueprint called for higher production through science and technology when overproduction and a mammoth surplus had caused the farm crisis in the first place. In the Soviet Union, the state forced tractors on large collectives with uneven results, but in the United States, it would not be the state that bought tractors, implements, and fertilizer but rather enthusiastic farmers subsidized by the USDA.

To make the rural countryside legible, the AAA compiled information on farming operations across the country, including farm acreage, crop history, ownership, and tenure. The section grids, large operations, and neat homesteads that epitomized Secretary of Agriculture Henry A. Wallace's rural Iowa and much of the Midwest offered an idealized blueprint for New Deal agricultural policy. The South, however, with its plantations, sharecropper plots, small farms, unkempt dwellings, weedy fields, and labor-intensive cotton and tobacco cultures suggested the backwardness so abhorrent to agricultural planners, and black farmers in particular were perceived as unsuited for the modernist agenda. To force the South into conformity with the modernist blueprint required not only trimming away sharecroppers, tenants, and small owners but also providing subsidies and tax advantages for larger farmers to invest in machines and chemicals. The New Deal agricultural program, much like the Soviet version, was boldly modernist, but it featured subsidized private rather than state ownership, the vote rather than the gun, and the semblance of democracy rather than the hammer of authoritarianism.

In the early New Deal years, programs such as the Federal Emergency Relief Administration, the Resettlement Administration, and the Farm Security Administration eased the hardship of some poor farmers, but by World War II, the USDA bureaucracy had become a silo that primarily fed substantial farmers. Scott stresses that modernist planners scorned experienced farmers and dismissed their accumulated knowledge as backward and valueless. In the United States, modernist planners dismissed the enduring tradition of yeomen farmers who delighted in annual crop cycles and were satisfied with earning enough to continue farming another year. As the noted poet, novelist, and commentator on rural life Wendell Berry put it, "The idea of a farm included the idea of a household; an integral and major part of a farm's economy was the economy of its own household; the family that owned and worked the farm lived from it."[9] As the population became more urban, the public lost touch with traditional farming. Farm Security Administration photographs show both prosperous and failed farmers, but the most poignant images are of dismayed down-and-out people on the road, caught between field and factory. Even the language was impoverished as the rich and varied rural vocabulary that had ruled horses and mules, elaborated work routines, and pondered weather gave way to the sterile scientific and technical terminology of machines and chemicals. At the same time, many farmers learned the intricacies of machines and gained an enduring fondness for tinkering.

Many African American farmers tilled small farms, in part because discrimination narrowed their path to prosperity and in part because farming on a small scale provided a respectable and satisfying lifestyle. The nearly 1 million black farm owners who tilled the land in the 1920s had no more of a voice in farm policy than did black farmers when USDA agencies later revolutionized farm structure. White men both formulated and executed agricultural policy. To carry out these plans, the USDA established offices in every county in the nation. When faced with voting on AAA acreage-reduction programs generated in Washington, farmers weighed whether to participate or risk market forces. President Franklin D. Roosevelt's secretary of agriculture, Henry A. Wallace, deemed the acreage-reduction plan the epitome of democracy inasmuch as farmers voted first for the program and then for the people who would execute it on the county level. Given farmers' legendary independence, the AAA used both carrot and stick. In exchange for approving the 1933 acreage-reduction plan, farmers received payments and benefited from higher prices. Most farmers approved the AAA's attempt to regulate the supply of commodities and thus keep prices stable, but New Deal agricultural programs made it impossible for farmers to escape federal rules that over the years became increasingly opaque. Early in the New Deal, a South Carolinian counted twenty-seven federal agencies addressing rural life in his county. Discrimination was also inscribed onto New Deal legislation, and after the *Brown v. Board of Education* decision in 1954, it took on a sharper edge.[10]

The Depression and the New Deal gave way to World War II, and change occurred so rapidly during the war that the sense of possibilities both at home and abroad shifted substantially. Veterans returned home with distressing war memories and a poignant awareness of human frailty. Enlistment dreams were retooled into discharge realities, and many veterans yearned for a conventional family, home, and job. Thousands of rural folk who moved to urban areas to work in defense plants during the war found that entertainment, job opportunities, and higher wages overpowered their intention to return to rural life. As African Americans observed these changed priorities, they also searched for incremental cracks in the segregation wall. The war experience generated tremendous energy and, especially among those too young to serve in the military or work in defense jobs, dreams of opportunities that would mark a generation. The 1950s witnessed a sweeping transformation in the South, and although segregation increasingly became an embarrassing relic that fit into neither national ideology nor international politics, it remained deeply rooted in southern soil and national policy.[11]

The New Deal greatly expanded the reach and power of the USDA. Increasingly, powerful farmers and pliant bureaucrats operated the machinery that disbursed federal funds and information. By 1960, directives relevant to USDA programs filled volumes of the Code of Federal Regulations, and swelling bureaucratic offices executed the rules. Such dense and legalistic regulations not only disguised USDA strategy but also presented a formidable challenge to farmers attempting to understand agricultural programs. In addition, federal tax policies favored larger farmers in numerous ways and encouraged nonfarmers with large incomes to enter farming and use accounting magic to profit from loss. Farmers generally used different accounting methods from nonfarmers, and the complexities of cash accounting, lower capital-gains rates, and the expensing of items favored larger operations. The tax laws favored lawyers, doctors, and others with large incomes who could write off a loss at farming against nonfarm income. The tax laws were wickedly complex, and efforts at reform usually expired in congressional committees.

Historians have yet to fully explore the importance of the tax codes in restructuring agriculture. USDA files hold numerous complaints from small farmers condemning doctors and lawyers for buying farms to lose money while, at the same time, increasing the supply of commodities and lowering prices. "We have farmers who get their total income from farming, as I do, trying to make a living, support a family, and educate my children," Tennessee farmer Herschel C. Ligon complained to Secretary Freeman in 1965, "competing with people who are farming to lose money, thereby being assisted by income tax deductions and/or participating in the ASC [Agricultural Stabilization and Conservation Service] program and the price support programs and making a most comfortable living from other sources." In late 1968, a Treasury Department official asked Secretary Freeman about the extent and effect of tax-code favoritism to nonfarmers. Freeman advocated getting rid of "'farm tax havens' for individuals and corporations with substantial nonfarm income" and even admitted that such tax favoritism posed "a new threat to family farmers." A few months later, Freeman left the department, and the code remained intact.[12]

As New Deal agricultural programs matured, small farmers were left with little besides Jeffersonian rhetoric, while large farmers invested USDA payments in machines and chemicals. USDA programs rewarded local elites, for extension agents fed information to prosperous farmers, the same class that dominated Agricultural Stabilization and Conservation Service (ASCS) committees and were most eager to expand acreage and adapt the latest methods. Bankers, farm-supply businesses, and implement dealers welcomed the infusion of federal funds. Through the Extension Service, land-grant universities, and experiment stations, county elites drew on science and technology and colluded with agribusiness and agrigovernment. John H. Davis, a Harvard professor and USDA administrator, coined the term "agribusiness" in 1955, defining it as "the sum of all farming operations, plus the manufacture and distribution of farm commodities." In its broadest context, it refers to the farms, firms, and lobby groups that thrive on the production, processing, storing, shipping, and marketing of food and fiber.[13] Its counterpart in the public sector, agrigovernment, complemented business interests and included the USDA headquarters bureaucracy, experiment stations, research facilities, regulation units, and acreage-policy divisions; land-grant universities; state agricultural offices; and county agricultural employees and committees. Agribusiness and agrigovernment cooperated, some might argue conspired, to replace labor-intensive with capital-intensive farming operations. Federal agricultural policy and laborsaving science and technology became weapons that ruthlessly eliminated sharecroppers, tenants, and small farmers. The human dislocation caused by this transformation was masked by the USDA's upbeat and sterile bureaucratic vocabulary, which focused on the tools of modern agriculture and justified USDA policies by denigrating those who resisted them as hopeless and backward obstructionists.

Agrigovernment was an amorphous conglomeration of federal, state, county, and university components. Captured by visions of large efficient farms, mindful that powerful farm organizations supported these goals, and aware of congressional pressure to aid wealthy farmers, agrigovernment offered programs that pleased a diverse constituency. Land-grant universities regenerated this ideology, and new USDA employees, most already favorably disposed, accepted the agrigovernment blueprint. At the county level, extension agents and program supervisors worked with successful farmers who could best take advantage of the latest scientific advancements. USDA employees understood that career advancement came from encouraging capital-intensive innovations, and they attached themselves to the vision of large mechanized operations that used chemicals for insect and weed control. Why bureaucrats who rarely profited from their sycophancy so willingly supported agribusiness seems inexplicable, although they no doubt identified with wealthy corporate interests that fueled the rural transformation.

The historian Paul Conkin placed this rural upheaval in national perspective and provided a vivid example of the shift to capital-intensive farming. In 1900, it took 248 hours of labor to produce a bale of cotton; in 1950, as tractors and mechanical pickers were entering the fields, it took 100 hours; in 1990, with a full complement of machines and chemicals, it took 5 hours. Yield per acre increased from approximately 270 pounds at the end of World War II to 513 by 1972. Similar changes tore through all of rural America, and the revolutionary shift from labor-intensive to capital-intensive operations had profound implications. In the twenty years after mid-century, the agricultural workforce was halved while the total value of products increased 40 percent.[14]

Most of the millions of disappeared farmers left quietly, and those remaining were seldom deemed newsworthy. During the 1950s and 1960s, black farmers occasionally shared press coverage with voting rights, school integration, and sit-ins, but rarely did reporters focus on their daily lives. They reported on southern blues, rhythm and blues, country, and rock 'n' roll, all embedded in southern rural culture, but they did not make the connection between the rural origins of music and its production and reproduction. The rural musical roots of such artists as Howlin' Wolf, B. B. King, and Carl Perkins, among others, were not considered pressworthy. Their careers flourished, of course, because they left the farm. In addition, the static of the Vietnam War, civil rights demonstrations, violence, and cultural changes obscured not only black farmers' tenuous survival but also the USDA's callous violation of their civil rights.

It was in this context that the rural South moved away from the labor system that, with only a few adjustments, had emerged after the Civil

War and that ended abruptly for many farmers in the mid-twentieth century, in the twinkling of an eye. It was a rural work culture that shifted from plowing with mules and horses to driving tractors, from chopping weeds to spraying herbicides, from hand-picking cotton to picking it by machine, and from neopaternalistic and exploitive sharecropping to lavish subsidies for wealthy farmers and paltry leavings for the poor. While some prescient leaders understood the revolutionary implications of this transformation, no political, business, or philanthropic visionary offered a plan either to keep farmers on the land or to accommodate the families that left. Indeed, farmers were often blamed for their own eviction, judged as lazy failures and welfare cheats and hounded by punitive bureaucrats.

Machines encouraged new skills and a vocabulary suitable for analyzing tractors: valves, magnetos, fuel pumps, spark plugs, lubrication, carburetors, transmissions, three-point hitches, gang plows, and, of course, horsepower. Implement dealers sprang up across the South offering jobs to mechanically inclined workers and introducing farmers to conversations that turned from what made a good mule to what made a good tractor, combine, or picking machine. To some farmers, machines and chemicals fractured vital links with the gratifying annual cycles dictated by their crops. It was not so much that farmers loved hard work as that work was part of a familiar routine. Flue-cured tobacco farmers, for example, cut firewood in winter to heat the curing barns in summer, sowed a plantbed, plowed their fields, transplanted the seedlings, hoed, cultivated, wormed, and suckered before barning began in July. Then for six weeks, they barned tobacco by priming, stringing, and hanging it in barns and curing it for the better part of a week. After the fields were bare, they graded the leaves, tied them into hands, and proudly took their tobacco to auction warehouses, where, in a matter of seconds, a line of buyers led by a chanting auctioneer bought their crop. By the 1960s, as allotments shrank, government regulations were tailored to encourage concentration of tobacco farms. Originally assigned to a particular farm, allotments increased land value and could be used as collateral, and they were further commodified when regulations allowed them to be rented in the 1960s and, finally, sold in 1982. A government policy thus metamorphosed into a commodity. Cotton farmers followed a less complex but equally ingrained routine of field preparation, planting, hoeing, cultivating, and, in the fall, picking and ginning. During World War II, International Harvester developed a successful mechanical picker on the Hopson Plantation near Clarksdale, Mississippi, and in two decades, machines had displaced hand pickers. Self-propelled combines developed during World War II doomed binders, shockers, and threshing machines belted to large steam engines in the prairie rice culture of Texas, Louisiana, Arkansas, and Mississippi. Herbicides ended the need for chopping weeds. Each commodity demanded its unique allotment policy, and contentious farmers brought suits and sought advantages. Science and technology plus government regulations had enormous consequences that forced millions of farmers to seek fortunes elsewhere.[15]

As capital-intensive farming evolved, there remained among rural people a nostalgic aftertaste of the old ways. In her 1980s interviews for the National Museum of American History's Oral History of Southern Agriculture project, the historian Lu Ann Jones listened to farmers' ambivalence about change. "The easiest work that I have ever done I believe on a farm," Darlington, South Carolina, farmer Tom B. Cunningham boasted in 1987, "was walking behind a mule, plowing. I could do it all day and ride a bicycle all night." The hardest work, he complained, was driving a tractor. "I ached and my muscles got sore." The skill and sweat of plowing with a mule first evolved into sitting on a hard metal seat on a tractor and then into being encased in a cab with modern comforts. "Today my men don't ride a tractor unless it's air-conditioned and everything else," Mamou, Louisiana, farmer Leslie Ardoin observed in 1988, adding with a note of condescension, "It's like doing housework, almost." Tractors on his farm plowed 150 to 200 acres a day. "It's got air-condition, it's got radio, it's got every convenience. A man don't even get hot working," he scoffed. Wendell Berry, the notable observer and critic of rural change, summarized working with animals. "In the first place, it requires more skill to use a team of horses or mules or oxen than to use a tractor," he explained. "It is more difficult to learn to manage an animal than a machine; it takes longer. Two minds and two wills are involved."[16]

Florentine Daniel was born in Franklin, Louisiana, in 1913. "See, I didn't get a learning—I went to work," she told Lu Ann Jones. When she was young, Daniel tended children, and as she matured, she worked in the fields planting and cutting cane by hand. She also did housework but preferred the field. "I loved to work in the field," she said over and over. "I loved to get out there and make a day's work and come on back to the house." The plantation wake-up call came at 4:00 or 4:30, and she would get some housework done and prepare her lunch before the second bell called her to the field. She would do more home chores after the field work. Whatever their thoughts on hard work, black farmers believed that civil rights laws offered the promise of equal rights that would give them parity with whites in obtaining acreage allotments, credit, information, and access to government largesse.[17]

The weight of agribusiness and agrigovernment crushed many farmers, and several letters to Georgia's U.S. senators, Herman Talmadge and Richard Russell, presumably from white farmers, suggest that class played a major role in the distribution of USDA payments. In 1969, Georgia farmer James R. Wimberly, owner of a 100-acre farm in Dodge County, complained to Talmadge that a county USDA staffer told him his farm was too small to bother with and that "all of the assistance needs to go to the large Farmers or large land owners because they have a lot invested in farming." Wimberly drew the conclusion that such a policy would "rob the Poor and make the Rich Richer." In 1970, Donald Wane wrote Talmadge from Statesboro, "I am becoming increasingly more upset over the manner in which FHA [Farmers Home Administration] appears to humiliate the farmer." Surveying the increasing concentration of landownership, C. E. Sheffield advised Senator Richard Russell, "We are living now much the same as the Europeans where they have Land Barons."[18] These farmers personified the alienation and dissatisfaction that were sweeping across rural America because of the USDA's focus on large farms.

USDA officials and state and local leaders rarely discussed civil rights policies with African Americans. Indeed, when *Brown* came down in 1954, prominent southern whites chose not to consult with African Americans and lead southerners away from segregation but to reinvigorate the racism of their fathers. A timorous President Dwight D. Eisenhower nervously avoided civil rights, pious Secretary of Agriculture Ezra Taft Benson ignored African Americans altogether, and the southern USDA machinery was retrofitted to deny benefits to blacks who advocated civil rights. Whatever goodwill had existed along the color line before the *Brown* decision vanished, and the white South and much of the white North girded for a neo-Reconstruction effort that would terrorize African Americans and nullify *Brown* and any laws meant to enforce equal rights.

Had brave white leaders emerged quickly after the *Brown v. Board of Education* decision in 1954, it is possible that the clergy, educators, and good-hearted whites could have met with African Americans to mutually construct a transformation of southern society. White male leadership failed. African Americans were never asked to participate in discussions about a post-*Brown* South. Instead, white politicians and many antilabor businessmen set a tone of aggressive resistance that echoed antebellum nullification rhetoric, and white citizens groups throughout the South enforced a code of noncompliance by shunning whites who did not support segregation and punishing African Americans who attempted to send their children to white schools or register to vote. Blacks and whites who did not conform to the segregation code were denied credit, harassed, shunned, and subjected to violence. Despite such repression, a series of major crises erupted along the color line in the mid-1950s that climaxed in Mississippi's Freedom Summer in 1964.[19]

Most white southerners, of course, were not eager to abandon a segregation system that gave them enormous advantages. Still, during the fifties, white southerners confronted unflattering newspaper and TV coverage that undermined their carefully constructed fiction of separate-but-equal bliss. Southerners often came across in the media as crass and unsympathetic because their accent, unapologetic use of the word "nigger," naive views of civil liberties, and history of violence offended many viewers, both north and south. Citizens' Councils, conscious of a public relations problem, produced TV and radio spots and seeded their segregationist message with state-sovereignty arguments. Gradually southern whites became adept at handling the media, shielding their fanged racism behind code words and smiling agreeably at critics while ignoring civil rights laws. Cunning USDA staff, meanwhile, became adept at voicing support for civil rights laws even as they ignored them.

The situation for rural blacks was bad enough before the *Brown* decision, but as USDA programs were sharpened into weapons to punish civil rights activity, the NAACP decided to help some of its rural members. Forty-two-year-old Presley Flakes, his wife, and his seven children lived outside Shaw, Mississippi, in a four-room frame house on a farm that he bought in 1946. He grew cotton, corn, peas, and beans that grossed about $5,000 a year, and he earned another $1,200 from seasonal employment. Flakes's farm had a barn, toolhouse, and smokehouse, and he owned a tractor and a car clear of debt. He also belonged to the NAACP. Citizens' Council member John T. Smith usually financed Flakes's spring planting, and he held a first mortgage on the farm for $720. When Smith learned of Flakes's NAACP membership, he refused to loan him money for the 1956 crop and added that "he wouldn't help any more 'n———.'" Flakes appealed to the NAACP in January 1956 to lend him $1,000 so he could continue farming. In a similar case, Jake Tanner had owned his forty-acre farm near Holly Bluff, Mississippi, since 1917, and from his six acres of beans, five of cotton, and four of corn he earned about $2,000 a year. He owned a mule and a cow. Because he received mail from the NAACP, the Greenville Production Credit Association and all other potential credit sources refused to lend him money. The NAACP's Mildred Bond and Medgar Evers reported that he was warned "that because of NAACP he'd better *not* even ask for loan or help." Looking back on those times, Henry Woodard, an African American farmer in Tunica County, Mississippi, told Lu Ann Jones in 1987, "When the civil rights push got to its peak, that's when they went tearing the houses down and moving us off the farm."[20]

Because of apprehension generated by the *Brown* decision, in many areas of the South whites became more repressive. The population of Terrell County, Georgia, was 67 percent African American, and 4,500 residents, 55 percent of them black, lived in Dawson, the county seat. Along Main Street, there was a bank, a sports shirt factory, a motel, several filling stations, and a farm-implement store. A red brick courthouse with a clock tower dominated the skyline. The police chief complained in 1958 that blacks watched TV news "telling what the Supreme Court has done and what the Federal Courts say and all about civil rights, and they begin thinking." Both the police chief and the sheriff suspected that communists were behind civil rights activity.[21]

A 1958 NAACP report noted that Dawson police had shot and beaten some dozen blacks over the past several years. In one case, the report observed, the victim's girlfriend was having an affair with one of the police officers, and another rumor circulated that the sheriff was "sending a message to some Negroes to stop fooling around with the girls with whom white men were going." This suggested, of course, the power that white men wielded over the lives and bodies of black women and men. Belonging to the NAACP attracted unwanted attention. Reverend Gibson, a respectable black farmer, the report continued, "has been subjected to an economic freeze-out and verbal threats in recent months." When he attempted to buy a tractor tire, Gibson was sent to talk to the store owner, who told him "he was one of those 'niggers' who was starting all that trouble and that he wasn't going to get a so and so thing unless it was a pistol ball in his head." The reverend owned an eighty-seven-acre farm but was unable to get any credit in Dawson. The county NAACP president owned an 800-acre farm and had also been threatened. The tense situation prompted the NAACP to ask for immediate action "before Negroes, who are said to be arming, start retaliating in kind."[22]

Rural blacks who had accumulated land and resources, such as Reverend Gibson and the head of the Dawson area NAACP, were vulnerable to economic reprisals, but many nevertheless risked their lives and property to fight for equal rights. Occasionally civil rights organizations would intervene to help threatened farmers. Francis Joseph Atlas, an

East Carroll Parish, Louisiana, farmer, had attended Tuskegee Institute and learned the brick mason trade. He was a Sunday school superintendent at the Progressive Missionary Baptist Church and a Mason. In 1948, he attempted to register to vote but was rejected. He tried again in 1950 and then again in 1960 as the civil rights movement gained momentum.[23] In September 1960, the fifty-five-year-old Atlas was subpoenaed to testify before the U.S. Commission on Civil Rights in New Orleans, and there he made clear the difficulties he faced when attempting to register. When he returned home, no gin would handle his cotton, and "seed, gasoline, and other firms" refused to deal with him. The black neighbor who usually custom-combined his soybean crop refused, citing fear of "what white people might do to him." Atlas customarily grossed between $7,000 and $10,000 a year, but in 1960, he lost his bean crop worth some $3,000. After he complained to the Justice Department, it brought action in federal court, and under duress a gin agreed to handle his crop and merchants decided to serve him, but these difficulties put him $4,700 in debt. An NAACP regional secretary visited Atlas and accompanied him to the FHA office, and after negotiating many bureaucratic hoops, Atlas secured a loan. Atlas's wife, relatives, and friends, the regional secretary learned, "have been extremely critical of him for agreeing to testify before the Civil Rights hearing," and he encouraged the NAACP "to support his commendable stand." The NAACP sent Atlas a check for $815.35, and on April 30, 1961, Atlas wrote that he was "deeply grateful." Because Atlas was a landowner, literate, and a brave and respectable citizen who testified before the Civil Rights Commission, the NAACP lent him enough money to continue farming.[24]

Lela Turner, her husband, and her stepson sharecropped near Sledge, Mississippi, on the J. T. Aldenson plantation, and she claimed her family was cheated on their cotton crop, though she had no way to prove it because she was not allowed to see Aldenson's books. Aldenson settled with his croppers in December when all of the cotton was picked and ginned, but Turner needed money in November 1962 to pay overdue bills and buy winter clothes. She asked for financial aid from the National Sharecroppers Fund (NSF). "We have no food in the house at all, and have been nearly starving all year mostly," she explained in November. "I started writing songs and poems," she continued. The traditional tasks of sharecroppers had changed since tractors, chemicals, and picking machines now did most of the work, so the usual chopping and picking income had dwindled. "Now," she despaired, "there isn't any work to do for the farmers." In January 1963, Turner applied for an FHA loan, but the office staff did not even allow her to finish stating her request before saying no.[25]

In the summer of 1961, Winson Hudson contacted the NAACP's Medgar Evers about starting a branch in Harmony, Mississippi, and when Evers spoke at the Galilee Church, the sheriff, intent on intimidation, sat in the front row. When Hudson later sent a twenty-two-page letter to African American congressmen Adam Clayton Powell and Charles Diggs complaining about FHA discrimination, she had to smuggle the letter out of Leake County since whites were monitoring her mail. In the fullness of time, the letter moved from congressional offices to the White House to Secretary Freeman's office and ultimately arrived back at the Leake County FHA office. There Supervisor Charlton Phillips called Winson Hudson and her husband Cleo into his office and rudely told them that he ran the county and that if they got FHA money, they would have to deal with him. Her letter had followed a well-beaten path, for many complaints to Washington about county USDA discrimination returned to the very office cited in the complaint. Winson Hudson later became head of the county NAACP and led a movement to integrate the schools, and her activism dried up her family's credit and provoked threats and bombings. Still, her integration activity, letters to Washington, and constant protest to the Washington FHA office eventually wore down Charlton Phillips. He spent an afternoon helping Winson and Cleo Hudson fill out FHA loan forms.[26]

To survive, many farmworkers relied on USDA surplus commodities such as meal, rice, butter, and dry milk. Through the Agricultural Marketing Service, the USDA distributed surplus food to state and local agencies that parceled it out to the unemployed, those unable to work, or those who earned too little to survive. While regulations explicitly forbade discrimination against anyone participating in voting rights or other civil rights activities, Mississippi officials blatantly ignored such rules. In 1962, Mary Burris went to the Ruleville city hall to sign up for commodities but was told to go to Indianola, the Sunflower County seat. The woman at the Indianola desk first asked why she had not signed up in Ruleville, then told her to take a seat. After a two-hour wait, the woman judged Burris's documents incomplete and instructed her to get statements of her earnings from every planter she had picked cotton for—by the next morning. Other sharecroppers often ran into obstinate clerks who invented trivial reasons to deny them commodities. When Gertrude Rogers went to the Ruleville office, the mayor announced that those who attempted to register to vote would not get commodities and "that he was going to mess up all of them." Civil rights leader James Forman asked USDA secretary Orville Freeman to give Sunflower County "top priority" and end the use of commodities as a "tool to discourage voter registration and to punish those who have attempted to register."[27]

Tennessee's Fayette and Haywood Counties epitomized the confluence of civil rights activity and mechanization. Planters evicted sharecroppers and tenants who attempted to register to vote and replaced them with machines and chemicals. In April 1960, the NAACP secured a Justice Department consent decree to halt economic reprisals, and a half dozen organizations provided relief for evicted farmers. The Justice Department prevented the eviction of 300 families in the two counties, but a dozen evicted Fayette County families were living in tents on a black farmer's land. While food and clothing gave respite, there seemed no future in agriculture for these families, for preventing eviction and providing food did not address the larger problems resulting from mechanization. Writing in March 1963, the Southern Conference Education Fund's Anne Braden observed that although African Americans were winning the vote, "people's life situation is really not changing at all, many still live on the edge of starvation—and people just get discouraged." Thinking strategically, Braden envisioned "a gigantic government program for redevelopment of this area." Fay Bennett, executive director of the NSF, also reflected on the situation: "I don't suppose anyone thinks it will be possible to keep all of the sharecroppers and tenant farmers on the land so that they can earn a good living." Bringing in industry, she suggested, would help as long as hiring practices were nondiscriminatory. "This is what we are trying to help to bring about in underdeveloped areas all over the world, and this is what we need in the backward areas of the rural South." Both Braden and Bennett grasped that issues far larger than isolated evictions were rumbling through the rural South. It was almost as if the earth was opening up and swallowing black farmers, for only in rare cases did an eviction receive press coverage.[28]

While many African American farmers worried about finding a few hundred dollars to start the planting season or fretted over whether there would be seasonal chopping and picking jobs, the Delta Council, an organization of white Mississippi planters and businessmen, confidently envisioned full mechanization. When the Delta Council Labor Committee met in March 1961, it focused on weed control and the "judicious use of herbicides, flame cultivation, and spot-hoeing as compared with hand-hoeing and cultivation." Only when herbicides did not completely clean the fields did they turn to day laborers to chop stubborn weeds. At the

Delta Council board of directors meeting in November 1962, members learned that machines picked 68 percent of the cotton crop. Even as the demand for labor diminished, the Delta Council went on record opposing a minimum wage. Between 1950 and 1960, newspaperman Hodding Carter III wrote, the eighteen Mississippi Delta counties lost 200,000 African American farmers.[29]

While the civil rights movement challenged a wide spectrum of discriminatory laws that demeaned African Americans, it inexplicitly ignored agricultural policy that struck directly at the livelihood of tens of thousands of black farmers. The landmark moments of the civil rights movement—*Brown v. Board of Education*, the Montgomery bus boycott, the desegregation of Little Rock Central High, the Freedom Rides, the March on Washington, the Civil Rights Act of 1964, and the Voting Rights Act of 1965—focused primarily on urban and national issues. Black farmers appeared in walk-on roles as they registered to vote, housed civil rights workers, or suffered violence. Their work, tenure, daily and seasonal routines, and hopes and dreams did not rise to national attention. In an urban and largely prosperous country, farmers mattered less and less. They were regarded abstractly as poor and brave, and both the discriminatory world they inhabited and the economic forces swallowing them faded into background noise as the large engines of agribusiness and agrigovernment vibrated through the countryside.

The Economic Opportunity Act of 1964 created the Office of Economic Opportunity (OEO), but in many southern states, administrators spent only a portion of the funds they were allotted and used Community Action Programs as weapons to discourage blacks. After the *Brown* decision, Citizens' Councils, the Delta Council, and politicians had hoped that mechanization would drive African Americans north and whiten southern states. Greenville newspaperman Hodding Carter, for example, saw mechanization as a way of encouraging blacks to move out of the South. Many black farmers, of course, were not interested in moving and stubbornly resisted state placement-service efforts to send them out of the state for work. As they analyzed the contradictions between their paltry living conditions and those of planters and witnessed federal programs corrupted by white administrators, many determined to remain in their communities and compete for federal funds. As federal programs proliferated under the Great Society in the mid-1960s, southern whites choked back their opposition to federal social programs and, attracted by good salaries, vied with African Americans for employment.[30]

Title VI of the Civil Rights Act of 1964 banned racial discrimination in federal programs and allowed funds to be cut off for noncompliance. *Christian Science Monitor* reporter Josephine Ripley observed in November 1964 that $2 billion flowed into the South for agricultural programs, including $1 billion for price support, $66 million for extension, and $26 million for the FHA. Ripley wrote optimistically, and as it turned out naively, that the USDA would end discrimination.[31] The enforcement tool, cutting off funds, was an idle threat, scoffed at and mocked by USDA bureaucrats.

Foreshadowing the more thorough report compiled by the U.S. Commission on Civil Rights in 1965, the NSF analyzed USDA programs in 1963 and sent its report to Secretary Freeman. It cited the case of African American farmer Thomas C. Johnson, who owned a 979-acre farm near Lexington, Mississippi. He had cleared 315 acres for farming, had a 45-acre cotton allotment, and each year had in vain requested an increase from the ASCS committee. ASCS fieldworkers measured each farmer's planted acreage to ensure it complied with the farm's allotment. The compliance measurer found that one of Johnson's fields that had always measured 3 acres suddenly measured 4.3 acres, and a remeasurement reported 4.1 acres. Johnson not only had to pay $9.00 for the remeasurement but also had to plow up over an acre of cotton. Since no African Americans served on the county ASCS committee, it was impossible to learn if allotments were fair. When the NSF raised the question of why black farmers did not work in ASCS offices, Ray Fitzgerald, deputy administrator of state and county operations, blubbered, "The negroes [*sic*] failure to secure county office employment could stem in part from this failure of the negro to participate in these [county ASCS] elections." Charging that Fitzgerald was begging the question, the report argued that blacks did not vote "because they are positively and effectively discouraged from such participation by the local white community." Since the ASCS was a federal agency, it was legally obliged to be nondiscriminatory, but Fitzgerald's response ignored bias and blamed the victim. His reply revealed either his poor understanding of southern ASCS segregated offices and elections or an effort to placate the NSF with poorly worded platitudes. The NSF report suggested that the ASCS educate black farmers about programs, hire African Americans in its offices, and appoint only those dedicated to civil rights enforcement to supervisory positions.[32] These suggestions resonated with those of other civil rights groups, but over the next several years, the ASCS stonewalled and passively nullified initiatives that might have opened ASCS elections and programs to black farmers.

Across the South, observers chronicled the tight grip that whites held on county USDA funds. In 1961, South Carolina attorney Ira Kaye, who worked for the South Carolina Council on Human Relations, attended a conference on agricultural problems. He realized that without pressure from Washington powerful county agricultural interests would not enforce civil rights laws. State and county officials ignored African Americans in the administration of federal programs, Kaye learned, and this "county elite" decided what was "best for the community," and "that is what is going to be done." He succinctly explained that African Americans were "on the lowest rung of the economic ladder in these areas and the power elite control all of their sources of credit and of making a livelihood."[33]

As Kaye understood, the *Brown* decision sent tremors of anxiety through USDA county offices that encouraged even harsher treatment of black farmers, for if public schools could be integrated, so too could the powerful ASCS committees and other USDA county offices. The 1964 Civil Rights Act and the Voting Rights Act a year later, plus growing enthusiasm for civil rights throughout the country, increased apprehension among rural elites, who feared that, despite their hold over federal programs and funds, African Americans would gain a voice in federal policy. They hoped to counter civil rights initiatives by posing as representatives of all farmers while subverting black efforts to participate in and benefit from USDA programs. Black farmers were invisible in part because whites wanted it that way and in part because black farmers wanted it that way. To be conspicuous was to become a target.

John R. Salter, a young Tougaloo College sociology professor, paid close attention to the human costs of rural reconfiguration in Mississippi and understood that since the 1954 *Brown* decision, "this matter of automation has been pushed with great zeal by the landowners as a means of mass retaliation against Mississippi Negro efforts to secure basic American rights and to achieve human equality." Salter and his wife Eldri moved from Arizona in 1961, and in late 1962, he helped organize a boycott of Jackson businesses that discriminated against blacks. In January 1963, Salter and Reverend Willy Goodloe of Clarksdale toured the back roads of Tunica County, an area now noted for its gambling casinos. Salter interviewed fifteen families who lived in the vicinity of what he described as a "never-ending sequence of sharecropper and farm worker shanties" and "an occasional plantation home of luxurious appearance." Bent blues notes ran through the lives of farmers caught between the old and new ways of farming. Forty-two-year-old sharecropper Jeff Betts lived outside Dundee with his wife, two children, and mother-in-law. He owned a 1950 pickup but was $400 in debt to "various persons." Herbicides had ended the demand for spring chopping, and, Salter and Goodloe reported, Betts "has no money, no current job of any kind, and no prospects of any job." Thirty-five-year-old Mary Conrad lived alone near Lula but could no longer depend on chopping and picking cotton as she slipped further into debt. Cleo Cotton had eight children and had lived in the same house for fifteen years, depending on seasonal chopping and picking, but she was $300 in debt with no job prospects. Even better-off farmers could not escape the economic forces raging through the Delta. Widow Mary Edwards and her two sons owned an eighty-acre farm near Dundee, and her family drove a 1952 Chevrolet, but they were $6,000 in debt for farming equipment. Other families offered similar stories. Some paid their poll taxes, some had their electricity cut off, and most were in debt with no job prospects.[34]

What Salter and Goodloe found in rural Tunica County personified the human costs of the rural transformation sweeping the South, for without work, farmers' rural cycle of plowing, planting, cultivating, and picking lost its meaning. Former tenants and sharecroppers who had barely survived doing labor-intensive seasonal farm work not only were idled by machines and chemicals but also were punished if they joined in civil rights activities that might address their problems. The USDA championed its modernist agenda but flavored it with discrimination against minorities. It was a wrenching tune being played in the Delta and across the South.

*The layers and layers and layers of racism that the*
*department has picked up through the years, I feel, is the big reason*
*why it's going to be so hard to undo it.*
*—Aaron Henry, Mississippi civil rights leader*

# 2

# EVIDENCE

In the winter and spring of 1964, even as the revolutionary transformation of rural life and culture tore through the South and the Student Nonviolent Coordinating Committee (SNCC) prepared for Freedom Summer, the U.S. Commission on Civil Rights expanded its spectrum of discrimination investigations and focused on the USDA. Created in 1957, the commission investigated discrimination, collected information, and published reports, but it had no enforcement power. John A. Hannah, president of Michigan State University, headed the six-member nonpartisan commission and immediately launched an investigation of voter registration in Montgomery, Alabama. Eventually USDA discrimination came to its attention, and at a meeting on February 13, 1964, Marian Yankauer, a forceful commission attorney, and several staff members met with Ronald Watts of the NSF. The NSF had evolved from the annual National Sharecroppers Week, which enlisted support for the integrated Southern Tenant Farmers Union. By 1964, the NSF had staff members and contacts throughout the South and had filed nine discrimination cases with Secretary of Agriculture Orville Freeman. The discussion revolved around deep and enduring discrimination issues in the ASCS, FHA, and FES. Freeman had promised to invalidate any discriminatory county ASCS elections, a pledge that wilted that autumn when civil rights activists challenged county elections. The FHA, the commission was told, had good leadership, but too few loans went to poor farmers who most needed credit. The

FES, everyone agreed, was extremely complex and evasive and would require careful scrutiny.[1]

The commission embarked on a major research project that would canvass USDA offices in Washington and county and state offices throughout the South, interview USDA employees and African American farmers, and result in the publication of *Equal Opportunity in Farm Programs* the next year. On March 3, the commission's Dean W. Determan visited William M. Seabron, the USDA's point man on civil rights, and secured several complaints and memorandums from his files, among them Secretary Freeman's eleven-page response to Congressman Charles Diggs Jr. reporting that all Mississippi USDA agencies "denied discrimination," the preposterous tune that played throughout the department. There were complaints about discrimination in acreage allotments, surplus food distribution, and employment and numerous criticisms from civil rights groups, "all of which were answered with the same form letter." Determan found Seabron cooperative and agreeable, and a year later, when the commission released its report, Seabron, an African American, moved from the personnel office to become assistant to the secretary for civil rights. By focusing on civil rights violations in the rural South, the commission touched the white South's tender segregation nerve and challenged southern politicians and some Kennedy and Johnson administration personnel.[2]

The commission tour was unique in that interviewers interrogated southern USDA workers about discrimination, probed how programs were structured and administered, examined office facilities and furnishings, and discovered how information flowed through the bureaucracy. The interviews, then, explored obscure but powerful southern USDA county offices that operated without scrutiny and opened a window on how USDA discrimination affected the lives of African American farmers. Mississippi civil rights leader Aaron Henry's astute observation about the many layers of racism suggested the way prejudice regenerated and exercised power as policy moved through federal, state, and county offices. Henry served on Secretary Orville Freeman's Citizens Advisory Committee on Civil Rights.[3] State administrators, for example, whited out any problems with county offices and sent sanitized reports to Washington, and even if an inspector general's report showed a violation of civil rights, the Washington office would find no blame.

Ten years after the *Brown v. Board of Education* decision and a century after Emancipation, white male USDA personnel continued to make decisions and implement policy without input from blacks, other minorities, or women, except in the Negro Extension Service. White hands disbursed the millions of dollars that poured through all-white county agricultural committees in the South, and these committees decided who received loans, the size of acreage allotments, and which farming methods to promote. The default setting for USDA policy makers, state and county employees, and land-grant university staff was white and male. Only in the Negro Extension Service did African Americans possess any authority to shape policy, and whites closely monitored its programs. As machines and chemicals transformed agricultural work, access to government programs and credit became crucial to survival. Without credit, for example, farmers could not buy the fertilizer, seeds, and pesticides to start the growing season, and while better-off farmers dealt with banks, many small farmers scraped by with credit from merchants or with FHA loans. By the 1960s, however, the FHA program had been corrupted to serve more solvent farmers and to punish farmers active in civil rights, placing many small farmers, especially African Americans, at a serious disadvantage.

On March 2, 1964, U.S. Commission on Civil Rights field investigators Richard M. Shapiro and Donald S. Safford started their swing through the South in the rural area around Valdosta, Georgia, only miles from the Florida border. They began with successful African American farmers and spoke with Joseph Wiley Register, a veteran who had made a transformation from cotton, tobacco, and peanuts to cattle. The forty-one-year-old Register had discerned the importance of beef cattle while serving in the military. "He indicated that he had not received any assistance from the Department of Agriculture in coming to this decision," the interviewers stated. Although he was up to date on farming trends, Register complained that he had not been informed by the USDA of the Rural Area Development (RAD) program and only recently had learned about it. When he had asked about the program, the black county agent had refused to discuss it, and Register complained that the Extension Service failed to educate black farmers about new trends or to provide information. "He indicated that the Extension Service is supposed to help educate farmers who get into poultry, milk cows and their own produce garden," the interviewers reported, but said they were "not fulfilling this responsibility." Register had some contact with the FHA through his construction business, but he had never asked for an FHA loan for his farming operation or received information about FHA loan programs, and he suggested that one must know a committee member to get a loan. He judged that his tobacco allotment, although reduced over the years, was fair, but he had scant contact with the ASCS committee and never received notices of elections. Indeed, he thought committee members were appointed. Register insisted that his success depended on his own sagacity, not on county-agent advice or federal programs.[4]

Next, the team interviewed seventy-one-year-old James Register, Joseph Register's uncle, who owned a 125-acre farm near Valdosta. The interview team learned that he grew tobacco, cotton, corn, and produce and had 32 hogs and 100 chickens. While he voted in ASCS elections, his one appeal for an increase in his 2.24-acre tobacco allotment had been unsuccessful. He had no reliable information on the FHA and borrowed from local banks. Mrs. Register had belonged to a home-demonstration club, but she complained that "the members have rusted out." Twenty-four-year-old Jerry Register was their only child who farmed. Two of his brothers had moved to Ohio and another to Philadelphia, and they were all doing "quite well."[5]

Shapiro and Safford were impressed with Andrew Blakeney's "clean and well-kept" brick home. Blakeney had worked as a brick mason before an injury ended that career, and his income from farming did not equal his wife's salary. Mrs. Blakeney taught at a county high school, and her Georgia State College School of Education diploma hung prominently on the wall. Their three children had gone to college in Michigan. Blakeney owned 47 acres, rented land from his mother's estate, had small cotton and tobacco allotments, and had participated in the Soil Bank program. He bought farm supplies from the Farmers Mutual Exchange, a cooperative that allowed black membership. Blakeney was "extremely critical of the Negro county agent," the interviewers learned, echoing Joseph Register's remarks. The white agent came to Blakeney's farm the very day Blakeney called to request technical information, but the black agent had never visited his farm. Neither agent had told him about FHA loan programs. Blakeney attended Extension Service educational programs at the Valdosta armory some half a dozen times a year, and blacks composed about a quarter of the 100 or so farmers in the audience. Seating was segregated, and he observed that when he arrived early and sat in the front row, no whites would join him there. When supper was served, white home-demonstration workers served whites and black workers served blacks.[6]

On March 3, Shapiro and Safford talked with sixty-five-year-old George Miller, who owned a 160-acre farm and rented an additional 70 acres that had a 4.46-acre tobacco allotment and a 3.9-acre cotton allotment. Miller split the cash crops 50–50 with his landlord and was allowed to plant corn and other crops for his own use. On his own farm, Miller raised 100 hogs and had allotments of 10.2 acres of cotton, 9.6 of peanuts, and 3.64 of tobacco. He also grew corn, sweet potatoes, and watermelons. He estimated that he grossed between $3,000 and $5,000 a year on these enterprises. In 1943, he had started out as a sharecropper, and in 1948, he bought a farm and later refinanced it for $7,000. He did not participate in USDA programs such as the Soil Bank or RAD, nor did he vote in ASCS elections, although he insisted that he could. He hedged on how much help he got from the black extension agent but admitted that black and white extension agents helped him with certain problems. His wife knew the black home-demonstration agent but saw little of her. Miller borrowed from a local bank with no difficulty. He had also received an increase in his peanut and cotton allotments after informal conversations with an ASCS committeeman, but he suspected that blacks were not given fair tobacco allotments. Although Miller was a successful farm owner, he had scant contact with the county agricultural hierarchy.[7]

These successful black farmers had carved out a niche in the segregated South and prospered. Their children left the farm and received educations outside the South, where they remained. The farmers obtained credit from sources other than the FHA, made only informal requests of ASCS committeemen, and relied primarily on themselves rather than extension agents or federal programs. Going it alone was risky, but so was being drawn into a web of debt through FHA loans, antagonizing ASCS committees with complaints, or depending on extension agents for advice. Discriminatory USDA policies put African American farmers in a bind because to survive in the machine and chemical age they needed USDA subsidies, but accepting them could lead to ruin, a dilemma that put them at a disadvantage to white farmers. Complaining about discrimination could cause even more trouble, for a message to USDA headquarters in Washington often made the round trip back to the local agency, sometimes leading to punitive action against the complainant. Even though these successful farmers chose to farm with minimal USDA contact, many black farmers had no idea they even had access to USDA offices. Commission investigators focused on farm owners, prominent officials, and bureaucrats, not on sharecroppers.

While in the Valdosta area, Shapiro and Safford interviewed Reverend W. H. Hall, pastor of the Ashley Street AME Church, president of the town's NAACP branch, and a member of the Southern Christian Leadership Conference. Hall recounted the black community's protests over segregation but expressed a cautious approach to integration. He also commented on the Negro Extension Service's annual Ham and Egg Show held at the city auditorium, which gave black farmers the opportunity to display and sell their products. Hall observed that the event was segregated, reflecting local custom. Although Shapiro and Safford judged that Hall was quite conservative, he caught them off-balance when he commented on "the lack of Negro members" in their field team.[8]

On March 4, Shapiro and Safford focused on the Lowndes County, Georgia, ASCS office in Valdosta and talked extensively with Freeling Scarborough, who supervised the county committee and eight community committees, each with three members. He provided a wealth of information about the operation of his office and the numerous ASCS programs. Scarborough explained that the members of the community committees elected the county committee. Referendums to approve acreage-allotment programs for cotton, tobacco, and peanuts required farmers to vote in person in their communities. Scarborough's office employed four full-time clerks, two additional clerks at peak season, eighteen compliance reporters for summer measuring (no blacks had ever applied), and a black janitor. When the interviewers asked Scarborough about nondiscrimination memorandums from the USDA, he searched his files "sweating profusely" but could not locate the material.

Freeling Scarborough had run the Lowndes County ASCS office for twelve years, and he managed some half a dozen programs. Chairman Leland Herring had served on the ASCS committee for twenty years, Vice Chairman Louis Wisenbaker had served for twelve years, and J. T. Stalvery had served for about seven years. Committee members awarded crucial acreage allotments and distributed released acreage accumulated from farmers who had decided not to use their full allotments. Unsuccessful applicants for released acreage were not notified, so Scarborough and the committee exerted substantial and confidential discretionary power. The ASCS office had also run the recently completed Soil Bank program, which paid farmers to take land out of production, using its discretionary power to value the land and thus deciding how much a farmer would receive. The Feed Grain program, similar to the Soil Bank, paid farmers to retire at least 20 percent of their crop acreage and receive payments based on the yield for 1959 and 1960. In addition, farmers enjoyed price support. A farmer whose tobacco was auctioned off below the support price, for example, could reject the offer and sell the crop at the support price to the Commodity Credit Corporation (CCC). If the CCC later sold the crop at a price higher than the support price, the farmer got a prorated share of the profit. Farmers could also store crops on their farms in CCC-approved structures, and the ASCS committee made loans to farmers to construct such storage facilities.[9] Storage of surplus crops evolved into a lucrative and sometimes corrupt program, one that begs closer historical scrutiny.

The Lowndes County ASCS committee's tight hold on communications ensured that only favored farmers heard of programs supposedly available to all, and many committees throughout the South carefully controlled the flow of information. Released acreage became crucial for many tobacco farmers because the almost annual acreage cuts left them with idle barns and packhouses. While black farmers seldom complained to Washington, white farmers often wrote to the USDA about discriminatory policies. Mary B. Gibbs complained to Secretary Freeman in September 1964, for example, that in order to care for her eight children, she needed more than her .95-acre tobacco allotment. Her farm near Enigma, Georgia, had a barn and packhouse, but primings from her tiny allotment would not fill her barn, forcing her to pay a neighbor to cure them in his barn. For four years, she had asked the ASCS committee for 2.5 acres, but the committee never granted her request. "I do know the ones that need it don't get it," she wrote. Her fate, she prophesied, was to sell the farm and go on welfare. Snow Hill, North Carolina, farmer John A. Thomas suggested that the released-acreage program should end. "That acreage goes to the Committeemen and big farmers—or to the ones with the 'pull,' as we say around here," he observed. "We are not idiots," Thomas wrote, "for every new committeeman puts himself up a new tobacco barn just as soon as he gets on."[10]

A complaint from a white Texas physician/farmer provided a glaring case of withheld information. In April 1960, Dr. E. A. Weinheimer of El Campo charged that the Wharton County ASCS office had not publicized the availability of released cotton acreage and that only 200 out of 2,300 county farmers had applied. Some farmers, Weinheimer complained, got "as much as 60 to 70 acres per farm or person." He caustically asked if more farmers did not apply because they were "so stupid or lacking in intelligence." The ASCS newsletter had not carried an announcement, nor had the newspaper, TV, or radio. As a result, only a select group of farmers knew of the released acreage. The reply from USDA headquarters in Washington stated the obvious—that farmers had been given the right to apply, that 200 had applied, and that farmers interested in extra acreage would have found out about it. The distribution had been handled "in accordance with instructions and regulations," the USDA contended.

Weinheimer's claim that only 200 out of 2,300 farmers applied when farmers were hounding ASCS offices for extra acreage strongly supported his charge of favoritism.[11]

Among the thousands of letters complaining of bias in allotting acreage, few were clearly identifiable as from African American farmers, no doubt because they had no expectation of redress. Freeling Scarborough apparently assumed that he should favor more prosperous farmers and went about his business with little thought of African Americans or poor whites. He explained to Shapiro and Safford that his ASCS office also handled the Agricultural Conservation Program (ACP), which subsidized irrigation, drainage, dams, and other projects. It provided part of the cost, and farmers made up the difference. All applicants, at least all white applicants, shared some of the county pie. State committee fieldman Leo Roberts worked as the liaison between the secretary of agriculture–appointed state committee and the county committees, and he worked closely with Scarborough. At a meeting attended by Shapiro and Safford on March 5, Roberts explained to the county committee the intricacies of a new program that allowed tobacco farmers to lease allotments from those not interested in growing tobacco.[12] Scarborough described a complex web of programs to Shapiro and Safford, but he had not shared that clear explanation with most of Lowndes County's farmers. Harboring complex information on USDA programs gave men such as Scarborough immense power to award favors and shape rural life.

The Lowndes County ASCS office had substantial discretionary power to reward and punish, and it used complex bureaucratic jargon and secrecy to cloak favoritism. Elite white farmers assumed that they would benefit from the programs, and in many ways, poor whites and African Americans assumed they would not. Shapiro and Safford did not detect overt discrimination either among the committee members or in the programs because white administrators were practiced in abiding by the rules and, when needed, using their discretionary powers. Few farmers voted for community committees, and incumbents remained in office year after year. As the civil rights movement gained momentum in the early 1960s, such white leaders assumed, as they always had, that they knew what was best for both African Americans and poor whites.

The commission decided that ASCS acreage-allotment policy deserved an intense analysis, especially the released acreage in cotton- and tobacco-growing areas. Commission attorney Marian Yankauer consulted with a very responsive James Hunt at the Washington ASCS office concerning which counties to study. She also learned that Native American tobacco farmers in Halifax and Robeson Counties in North Carolina had complained of discrimination in allotments, triggering investigations. USDA statistician Calvin Beale recommended that Yankauer include Robeson County in the study because it had "three different groups of people, the Lumbee Indians, the Negroes and the whites, each about equal in number." Discrimination, the commission was learning, was not restricted to African Americans. Yankauer had grown up in New York City, attended New York University, and earned a law degree at Brooklyn Law School and later a master's degree in public administration from NYU. Before joining the U.S. Commission on Civil Rights, she worked for the NAACP Legal Defense Fund.[13]

The FES also directed numerous programs, and its mission was to educate farmers and their children about better farming methods and encourage participation in community organizations such as 4-H and home-demonstration clubs. The Extension Service grew out of efforts to combat boll weevil destruction of cotton as the insect crawled across the South in the last years of the nineteenth and first two decades of the twentieth century. Led by Seaman A. Knapp, an enthusiastic advocate of scientific agriculture, agents taught farmers to maintain cotton production by disrupting the insect's life cycle. Organizationally, the FES relied on federal, state, and county funds, and complex memorandums of understanding set the parameters of responsibilities.

In normal times, the arrangement worked satisfactorily, but after the *Brown* decision and more intensely after the Civil Rights Act of 1964, the Extension Service used its trifurcated federal-state-county identity to foil civil rights initiatives. Although it was a federal agency, its programs operated out of white land-grant universities, and because counties contributed to funding, there was also a substantial local dimension. USDA civil rights initiatives passed through these strata much like laundered money. Washington administrators announced support of civil rights, land-grant universities gave lip service but recast directives to suit their own purposes, and by the time regulations reached counties, they were unrecognizable as federal civil rights currency. County agents and other USDA office managers controlled hiring, assignments, and salaries. Time after time, investigators discovered that the FES, the FHA, the ASCS, the Soil Conservation Service (SCS), and other agencies flagrantly violated civil rights laws, yet except for an occasional slap on the wrist, the USDA ignored or denied infractions. Bureaucrats in USDA headquarters not only abetted discrimination in the hinterland but also generated it in Washington.

Until the Civil Rights Act of 1964, the Negro Extension Service had been a segregated arm of the FES. African American county agents were enmeshed in confusing and often contradictory relationships with white agents and administrators. Shapiro and Safford encountered the system as it came under civil rights stress. They interviewed Cornelius Wallace McIver in Thomasville, Georgia, on Friday, March 6. McIver explained that he reported to the African American state leader in Savannah, but in Thomas County, he worked under the white county agent, William L. Whittle. McIver had slight contact with the RAD program and spent most of his time working with 4-H groups and adult clubs. The county encouraged blacks to vote in ASCS elections, McIver said, although he was unfamiliar with the nominating procedure. He did not attend county committee meetings and was unsure of how released acres were distributed. Whittle handled McIver's business with the county committee and obviously kept him out of the decision-making loop. While the white agent with six years' work experience received $375 per month, McIver with fourteen years' experience earned $212. Both black and white extension workers had offices in the federal building, and the county paid the phone bills and supplied good furniture for the offices. McIver's tentative and guarded replies suggested that the white extension agent intimidated him and denied him information on USDA programs.[14]

L. Monroe Jackson, assistant county agent for Brooks County, worked in a separate building from white extension workers and had no secretary; he did most of his own typing. The interviewers gathered that he spent the bulk of his time doing 4-H work but also responded to requests from farmers. Jackson met weekly with the white extension workers but reported to administrators at Savannah State College. He had never attended an ASCS committee meeting, nor did he understand its electoral procedure. The white county agent stated erroneously that Jackson was on the RAD committee and had attended an organizational meeting in Quitman, no doubt hoping to demonstrate to the commission team an instance of integration. Apparently no blacks were involved in RAD in Brooks County. White agents moved in an extensive sphere that included committee meetings and gatherings in Athens and Tifton, and blacks circulated in a much smaller sphere. In Thomas County, home-demonstration agent Sarah Martin Clark spent an estimated 65 percent of her time with the county's eighteen 4-H clubs, which were composed of 313 girls and 272 boys. She had also organized five home-demonstration clubs serving about 100 members. Some of the club members attended the black 4-H state convention, but none went to the national convention.

Her $40 per month expense money never seemed to stretch far enough, so she paid for the balance out of her pocket. Her clubs vied for prizes furnished by local organizations and were forced to compete with white 4-H programs for funds.[15]

In most southern counties, white agricultural officials never addressed blacks by courtesy titles but simply used their first names, an enduring and demeaning custom. In Brooks County, former extension agent J. B. Stevens, who served from 1930 to 1943 and from 1951 to 1961, never recalled being notified of ASCS elections or being asked to attend ASCS or other agricultural-policy meetings. White agents had not shared information with him, nor did he receive equal in-service training. Instead, he spent his time on benign 4-H projects to the neglect of black farmers. Blacks were given inferior offices, scant staff, few demonstration materials or tools, no vehicles, and feeble local business support, and they were forbidden to attend national conventions. The complexity of the Extension Service's nebulous federal-state-county persona encouraged white personnel to ignore civil rights rules. Stevens explained that all extension agents received two checks, one from the state and one from the county.[16]

As the weekend approached, Shapiro and Safford traveled to Fort Valley State College south of Macon to attend a conference on March 7 and 8 sponsored by the NSF. The conference brought together farmers, Fort Valley State College staff, and USDA officials from Washington, state, and local offices. The NSF booked Shapiro and Safford into a local motel, but when they discovered that it was segregated, they stayed instead in a dormitory at Camp John Hope, named for the first African American president of Morehouse College, a lifetime civil rights advocate. The next morning, they learned that the camp was a segregated Georgia state park that excluded whites. Unwittingly, they had participated in a "breakthrough." They later reported that several SNCC workers gathered the names of conference participants, and they were especially interested in the developments at Camp Hope. This was the sole mention of SNCC in commission interviews, even though its wrenching campaign in Albany, a city to the south, and its activism throughout the South had gained national attention. Shapiro and Safford were unhappy that the NSF's Jac Wasserman had booked them into a segregated motel, and in their report, they suggested a "reappraisal of our relationship with the National Sharecroppers Fund."[17]

At the conference, the commission team interviewed James F. Hughes, a special assistant at the SCS who recruited African Americans for USDA positions. He stressed that it was important to enlist the very best job candidates, not just those who could pass the civil service examination but also those who were willing to prove themselves. He warned, "You just have about to be a Jackie Robinson to succeed." A black extension agent told conferees that there was a major "bottle neck" in securing information to pass along to black farmers. The investigators sensed that black extension workers were "fearful" as nervous laughter greeted some requests for better information, but an undercurrent of insubordination swelled as blacks challenged white speakers. White USDA personnel warned African Americans to stay away from civil rights groups, misinforming them that attending civil rights meetings violated the Hatch Act. Civil rights workers, Shapiro and Safford discovered, "are one of the major channels to provide new information for farmers in the State," so if black extension agents did not receive information from white agents and avoided civil rights meetings, "the situation is obviously an extreme one."[18]

Dr. L. A. Potts, a special assistant to the secretary of agriculture who had for years served as a dean at Tuskegee Institute, seemed baffled when attempting to explain current USDA programs at the conference. "If he is a major communications link concerning RAD and other new departmental programs to the Negro committees throughout the Nation," the commission team judged, "this is a very weak link indeed." Black farmers questioned why there were no black extension agents in their counties and why white agents paid them no attention. White extension officials reacted to these questions with "immediate hostility and defensiveness." Dewitt Harrell spoke for the state extension service and, the interviewers reported, cited "voluminous statistics, all very much beside the point." Indeed, most talks by USDA staff and white extension officials were uninformative and some were "in part insulting." The white speakers adopted the demeaning USDA attitude when dealing with African Americans. At one point during the Saturday evening session, discussions got "hot and heavy" as blacks articulated their civil rights agenda, and Harrell referred one question to a ranking black extension agent; the agent then referred it to Potts, who had no answer. After the session ended, Shapiro overheard Harrell boast to a federal official, "We have more niggers in our service than any other federal agency."[19] The boring talks, indigestible statistics, and defensiveness implied a more significant issue, namely, that the Extension Service mission to spread information foundered on ineptness, meaningless jargon, and bias. Much of the material in Extension Service files chronicles social gatherings, bland meetings, self-congratulation, and strategies to evade integration orders.

The interviewers used the occasion of the Fort Valley meeting to explore the workings of technical-action panels, and D. D. Slappey, head of the FHA real estate loan division in Atlanta, explained that each county FHA supervisor acted as chairman of a large panel consisting of "the representatives of the various agriculture agencies in a given county." Each agency submitted the name of a member to serve on the panel, and since the request went to the University of Georgia, the FHA was not held accountable for the fact that only white agents were notified. The segregated panels offered technical assistance to RAD committees.[20]

Shapiro and Safford remained in Fort Valley through the Monday following the conference and talked with a variety of farmers and bureaucrats. Cozy L. Ellison, director of agriculture at Fort Valley State College, explained that at the American Association of Land-Grant Colleges convention each year, an all-white Extension Service subcommittee set policy. The chain of command ran from the state extension service in Athens to an area supervisor to a white county agent to the black county agent. He explained that the African American state leader at Savannah State College had minimal responsibilities, all outside the decision loop, but failed to mention that he and Fort Valley's president were plotting to steal the extension program from Savannah State (see chapter 6). Ellison argued that the 1890 Morrell Act creating segregated African American land-grant schools should be held unconstitutional, and he wanted these schools to receive equal funding, have autonomy over black extension agents, and receive information from the USDA and experiment stations in a timely manner. Ellison had his own problems with the local ASCS committee, which had declared that grain he had stored in a warehouse was infested with weevils and would be sold to the highest bidder. He removed the grain, which was not infested, and sold it himself, suspecting collusion among the inspector, the ASCS committee, and white farmers. When he began growing peaches in 1957, whites were hostile and packers would not deal with him. After he contracted with a broker who had been recommended by whites, the broker absconded with most of his profits. He suspected that members of "the white power structure" had conspired to get him out of peach production, which they considered a white man's crop.[21]

The interviewers talked with James Mays about his experiences with county USDA personnel in Leesburg, the county seat of Lee County, near Albany. Mays had taught school and farmed 250 acres of his family's 700-acre farm, and in 1959 and 1960, he had secured operating loans from the FHA office in Leesburg. In February 1961, another loan was approved, but in March, a new county FHA supervisor reversed the decision and informed Mays that his application had been made too late. Mays believed that the loan was denied because of his civil rights activity and his role as a teacher and PTA leader in defending a student essay criticizing the school. The school principal reported the incident to the county school board and suspended the student. At a PTA meeting called to protest the suspension, Mays, acting in his role as president, encouraged the PTA to continue its protest. He was asked to resign his teaching position in June 1962. That fall, he was denied an operating loan when the FHA questioned his ability to repay it and cited a negative character check. Mays did not appeal. He applied in October 1963 and was again denied. Unable to get credit from the FHA or from private sources in the county, he went outside the county for his loans. Two of Mays's brothers were also denied loans, presumably because of their civil rights activity. All three Mays men had canvassed with SNCC workers during the summer of 1962. When Mays took purchase orders drawn up by the ASCS office to an approved community business, the dealer refused to honor them, again because of his civil rights activity. The treatment of the Mays family for its civil rights work epitomized the pressure that both the USDA and the business community brought on African Americans who transgressed their assigned place in the community.[22]

White USDA officials sometimes lied to African Americans to prevent them from benefiting from USDA programs. Daniel W. Young asked a Walton County, Georgia, SCS staffer if he could place some of his 136 acres in a recreational program. The staffer consulted a more senior administrator visiting from Decatur, who stated that the program was not available in Walton County. Not only was the program available in the county, but the county actually returned unused funds for the program in 1963, suggesting that rather than allowing a black farmer to participate, the SCS had returned the appropriated funds. Young also complained to the ASCS about his yield figures, the basis of calculating payments, and the committee raised the yield a small amount. Even when black farmers asked the right questions and followed the proper procedures, USDA personnel could simply provide incorrect information.[23]

Thomas L. Delton, who worked for the USDA's RAD office in Georgia, denounced the FHA "character test" that had led to the denial of James Mays's application, for it was unlikely that a county committee would go against a white man's evaluation. As Mays's case made clear, being labeled a civil rights activist put credit at risk. Oliver W. Robinson, the only African American FHA staffer in Georgia, told his friend Thomas Delton that when he attended a meeting for professional workers in Perry, Georgia, he was met at the front of the hotel and asked to go to the back door. Delton revealed that Robinson had not participated in county committee meetings, although he told Shapiro and Safford that he had. Discrimination also froze black workers at middling ranks. While white FHA agents could easily move from a GS-7 to a GS-9 rating, Robinson, who had an M.A. degree, was stuck at the lower grade with no hope of promotion, which, Delton argued, "kills the initiative of Negro employees."

Delton insisted that FHA loans went not to poor farmers for whom the program was intended but rather to better-off farmers. "Apparently," Shapiro and Safford learned, "the administration of this program has been subverted so that this group of farmers does not actually benefit to the extent originally intended." County FHA committees had three members who served for three years on a rotating schedule, and each year, the county FHA supervisor submitted the names of three white men to the state office for it to make a selection. The state office also appointed county supervisors. Delton was a GS-7, and he had watched whites pass him by with regularity when he worked for the SCS. He covered Georgia, Florida, and South Carolina in his RAD job but only worked with blacks. The RAD program was segregated, and Delton was excluded from mixing with decision-making white committees. "It should be noted and emphasized," Shapiro and Safford stressed, "that Delton, a very competent individual, has not received a promotion since 1950." He had been chosen for his RAD job because a Washington supervisor learned of his excellent work with the SCS. Still, he was excluded from technical-action panels and important RAD meetings and, the interviewers concluded, was "out of the chain of command." Delton had mastered the RAD literature and depended on his reserve military friends to mimeograph material for distribution to black county agents, vocational agriculture teachers, and farmers. He attempted to work with black state extension personnel at Savannah State College but reported a "breakdown" there in getting material distributed. Delton had witnessed retaliation against black farmers who engaged in civil rights. "He feels that a Negro cannot be a successful farmer and also participate in civil rights activities," the interviewers learned. Delton's hard work and competence brought him to the attention of SCS administrators in Washington who tried to tempt him away from RAD with offers that to his chagrin involved working only with blacks.[24]

The commission's Marian Yankauer learned about the SCS's few black employees, their low grade, and their lack of promotion possibilities. She was one of the commission's most gifted employees. F. Peter Libassi, director of the commission's federal programs division, described her as "a most unique and unusual person to say the least. She combines sensitivity and commitment with a perceptive and quick mind." Libassi in 1966 would move to the Department of Health, Education and Welfare's Office of Civil Rights, where he had a distinguished career. Yankauer learned that of the 64 blacks with GS-5 grade or above, 33 worked in the South, 11 in North Carolina. Thomas Delton's GS-7 rating was stagnant, she observed, because at a higher grade he would supervise whites. He had been recommended for a state position, where, of course, he would work under whites, but he decided to remain in Fort Valley. Another African American from Louisiana was promoted to the Tennessee state office, moving into a higher position but not supervising whites. Yankauer suggested that the commission compile statistics on how long talented African Americans remained in the same grade, how many whites were promoted above them, and the normal promotion channel.[25]

Shapiro and Safford learned that as pressure increased to appoint or elect African Americans to agricultural committees, whites resorted to at best tokenism and at worst duplicity. Vocational agriculture teacher John H. Rollins taught at Peabody High School in Eastman, Georgia. The city manager contacted him and asked that he serve on a committee but told him neither the committee name nor the duties. He added "that it was mandatory to have Negroes on the committee." Rollins was never invited to a meeting or given information on the committee's function. The interviewers observed that even when they were appointed to committees, blacks were "kept very much in the dark as to the purpose of the committee and the operation," and they judged that such appointments were "perfunctory and merely surface compliance rather than a meaningful compliance." Still, white members could assure USDA compliance personnel that the committee was integrated, and black appointees no doubt were reluctant to push for inclusion.[26]

While Freeling Scarborough worked closely with the state fieldman and committee members in Lowndes County, Earl Anderson, ASCS office manager for Brooks County, boasted that he could function quite well without a committee and assured investigators that he hired all personnel in his office, including a black woman who did custodial duties. Although the chair of the ASCS committee had held his position for a decade, Anderson ran the Quitman office with little committee input and complained that useless regulations limited committee power. He had never met with the African American extension agent, and other than a few write-ins, no blacks had run for community ASCS committees. "The complexity of the procedures," the interviewers concluded with understatement, "suggest that a small farmer with a lack of education might well be discriminated against in the process and might be discouraged by the office manager from even applying for allotments."[27]

Shapiro and Safford were usually received politely, but commission interviewer Thomas T. Williams's visit to the office of Alabama state ASCS executive director Bernard L. Collins on May 11, 1964, exposed a glaring residue of prejudice beneath an awkward attempt to maintain decorum. Collins's staff had been told a "Dr. Williams" would arrive representing the U.S. Commission on Civil Rights but not that he was an African American, and when he approached the secretary, she "asked in a superior tone, 'What do you want?' Her facial expression," Williams reported, "was quite unbecoming to a secretary to the Director." Williams identified himself, and the secretary retreated to Collins's office, returning transformed "from arrogance to warmth and friendliness." Collins, Williams learned, had not grown up on a farm but held a degree in agriculture from Mississippi State University. Williams was well into his distinguished career. Born in Roanoke Rapids, North Carolina, he graduated from North Carolina A&T College in 1948 and, after earning a master's degree in 1949 from the University of Illinois, received a Ph.D. from Ohio State University in 1955. After teaching at Tuskegee Institute from 1950 to 1952, he became chairman of agricultural economics at Southern University, where he focused on development issues and small farmers. He was an adviser in the Lyndon B. Johnson White House and in 1966 won a Fulbright fellowship to the University of Pertanin in Malaysia.[28]

Williams had been asked to focus on the ACP and to determine whether black farmers had signed up. Admitting discrimination, Collins insisted that the ASCS was working to enroll black farmers in the ACP but that few black farmers had money to supplement the government funds. "However, you should know that I am a Catholic," he blurted out, "and you already know that the Catholic schools are to integrate this year," as if that would atone for his office's discrimination. When Collins could not come up with statistics showing black participation in the ACP, he told Williams, "Many Nigger farmers who profit from the ACP program never show up in the records because the Nigger farmer is a sharecropper and the landlord's participation in the program increases the productivity of the land he is cultivating." He did not explain that improvements accrued to the landlord, not the sharecropper. As for ASCS elections, he stated that blacks voted but did not run, or in his words, "Never had any Niggers running but they do vote." Collins alluded to the conventional fear that the election of an African American to a county committee would ruin the program when he observed that if there had been a "concentrated effort" to elect blacks, "we might lose our program." As for the ACP, "The small farmers and particularly the Nigger farmers believe the program can't do anything for them." Larger farmers, he stressed, "are good business people" and are eager to have the government fund 50 percent of a project.

After admitting that blacks often did not even apply for programs they assumed were for whites only, Collins expanded on the advantages of large farmers. He observed correctly that government programs "are like a closed corporation. If you are in, it's good; but if you are out, you can't get in for it would upset the apple cart." This was especially true for the allotment program. "The established farmers have the money. They know when to apply, where to apply and the details. Thus, they are in a position to take advantage of these programs." Collins personified the USDA administrator who saw black and poor white farmers as expendable and wealthy white farmers as privileged, and his constant use of the word "nigger" in the presence of Williams showed his contempt for African Americans. At the end of the interview, Collins "appeared quite uncomfortable," and Williams speculated that he might be second-guessing some of his answers.[29]

Conversations with ASCS office personnel across the South revealed a major effort in 1962 to enroll more farmers in the ACP. In Hale County, Alabama, a development group that included the county agent, the FHA supervisor, a soil conservationist, a forestry service representative, the ASCS office manager, and county committeemen met to discuss the ACP. No blacks attended the meeting, nor was the black extension agent tasked with providing ACP information to black farmers. A black agent employed in Hale County a few years earlier reported that unless a white person supported a black applicant, the committee would not approve the request. In the past, he recalled, "larger farms were most likely to participate in the program and small farmers did not know about the program." ASCS offices failed to keep records revealing the number of black farmers getting ACP funds, but some office managers recalled that a substantial number of blacks participated in 1962 and used the funds for pasture improvement. In Kemper County, Mississippi, ASCS office manager Jimmie Rodgers reported that in 1962 there was a major push to sign farmers up for the ACP. "The Community Committeemen in each county who obtained the most participants won a trip to a convention on the Gulf Coast," he recalled. Rodgers employed eleven compliance reporters during the summer, but he said he would hesitate to hire an African American for fear that white farmers would not allow him on their land. Office manager Obie Turner of Jefferson Davis County, Mississippi, observed that tenants were loath to participate in the ACP "since they hesitate to contribute to the improvement of someone else's farm." In all of the counties they visited, commission personnel found black extension agents isolated from decision-making committees and rarely interacting with whites in agricultural offices.[30]

Thomas Williams learned that the FHA had hired a few African Americans to work with black farmers. In the mid-1960s, the Alabama FHA made Tuskegee Institute graduate George Parris a state program specialist, a title change that failed to include a promotion or salary increase. He spent two days a week in his Montgomery office segregated from white employees by a bank of file cabinets and three days traveling across the state assisting black farmers with their loan applications. Parris had suggested that blacks be placed on FHA committees in 1950, he told Williams, "but the State Director merely laughed it off." He also understood that African American FHA employees moved on a different track from white personnel. The accepted promotion route, he revealed, began with an Auburn University degree, an appointment to the state office, a county assistant supervisor position, experience supervising farm loans, and a promotion, and then "the sky is the limit to where he can go." A degree from Tuskegee consigned African Americans to segregated and secondary positions. Parris insisted that only pressure from Washington could force an end to discrimination.[31]

A white farmer and FHA administrators ordered Parris to pressure NAACP member and voting rights activist Aaron Sellers of Midway, Alabama, to abandon his civil rights work. The white farmer, who had an $18,000 mortgage against Sellers's farm, told Parris that if he persuaded Sellers to give up his NAACP activities, "he would see that his loan was approved." Parris unsuccessfully attempted to meet with the county committeemen to discuss the case. State FHA director J. T. Lunsford actually threatened Parris with losing his FHA job if he did not convince Sellers to stop his NAACP work. Ultimately, the Masonic Lodge refinanced Sellers's loan. "Mr. Parris' answers were practically in a whispering tone and at all times he appeared rather afraid," Williams observed. No African American, he told Williams, had ever addressed his supervisor "in the same manner in which I had, although I merely talked to him like I would talk to another person." George Parris survived at the pleasure of whites and walked a narrow path surrounded by white treachery.[32] Despite their prejudice, Alabama USDA administrators offered Williams diffident respect, for his commission credentials intimidated them and his doctorate outranked them.

In Eutaw, Alabama, Williams interviewed black county agent Frank Jackson. Even though he knew nothing about the RAD program, African American state extension leader Bailey Hill had insisted that Williams mention it in his annual report, evidently hoping that participation would look good to federal officials evaluating inclusion of blacks in federal programs. White extension specialists who in the past did not meet with blacks had recently inaugurated annual meetings. "The reason for greater help from specialists in recent years," Jackson pointed out, "has been pressure from Washington, D.C." Jackson worked out of a new air-conditioned building but had no typewriter or mimeograph machine and sent all such work to Tuskegee. He and the home-demonstration agent served 283 county farmers, while the white agent, home-demonstration agent, assistant agent, and secretary served 320 white families and worked out of a fully equipped office. "A Negro won't go into dairying because he just won't work on Sunday and Saturday," Jackson complained. He would put "every Negro farmer in the Army for two years" to teach discipline.[33] It did not seem to occur to Jackson that it was his job to teach black farmers better husbandry.

In its analysis of interviews in Tuscaloosa, Greene, Sumter, Hale, and Dallas Counties in Alabama, the commission found that white cattlemen could take their bulls to Auburn for fertility testing but blacks could not. Nor did blacks share equally in the cotton acreage released each year and redistributed by the ASCS committee. White county agents worked on Cattlemen's Association and Farm Bureau projects, but African Americans could not belong to the Cattlemen's Association and rarely attended presentations by specialists in field crops, dairying, or livestock. One white agent in Hale County generously offered, "We don't run them off if they come."[34]

Charles Scott worked as the African American county agent in Dallas County, Alabama. The black extension staff formerly worked out of a downtown Selma office building that had been built in the late 1930s by African American contributions and then deeded to the city, Thomas Williams learned. City officials took over the building for "recreation activities for Negroes," and the extension staff migrated to a smaller office on Lapsky Street provided by Selma University. Scott suspected that white leaders moved them out because the extension agent at that time was involved in voter registration. That agent's dismissal and the move were a punitive lesson to discourage civil rights work. The Lapsky Street office was for-merly the Selma University president's home, an inadequate space that lacked parking and air-conditioning. "We almost burn up in here dur-ing the summer," Scott complained. It was a bare-bones office with one restroom, one telephone, no secretarial help, one typewriter ("We peck out our own typing"), and a mimeograph machine that Scott paid for out of his pocket. All 4-H, home-demonstration, and committee gatherings were segregated. Scott's assistant and the home-demonstration agent spent nearly all of their time with the 892 youths in 4-H clubs, while Scott handled all adult business. Sears Roebuck contributed $40 annually (raised to $80 in 1964) to fund poultry prizes for black youths. Black 4-H members could attend a summer session at Tuskegee, but they were not invited to the Chicago or Washington conventions. Scott was proud of his work with young people. "I feel that my kids can stack up with the whites because of the training of the Negro County Agent's staff," he boasted. Black farmers, he explained, received no guidance from Auburn experts on eggs, broilers, or dairying. Like many other black administrators, Scott faulted black farmers for poor work habits. "The Negro farmer lacks inge-nuity," he said, but he added that many were "old and they are set in their ways of doing things." He hoped that the 4-H program would bring fresh ideas into farming.[35] The goal of training black youths to farm with inade-quate resources and information in a system that was rapidly changing faced heavy odds.

In mid-May, Williams moved on to Mississippi and in Jackson inter-viewed William Ammons, the Mississippi state leader for the Negro Ex-tension Service. He climbed the steep steps to Ammons's second-floor office, which was "sandwiched in between a drug store and a gasoline station" near Jackson State College. The office was cluttered, desks were awash in files, and Williams was not allowed to see the "junky" mimeo-graph room. Ammons had graduated from Alcorn College in 1933 and had taken courses at Tuskegee, Prairie View, and Hampton Institute. He nervously explained to Williams how the state program for black agents operated and stated that there were black extension agents in 57 of Missis-sippi's 82 counties. He painted a depressing picture of African American farmers, claiming that many were "too old to operate machinery. Their coordination is slow." Few went into dairying, he argued, because it was a seven-day operation and "Negro farmers won't work seven days." During the interview, Ammons's phone rang often, and from the conversations, Williams discerned that "he was checking with someone else about my presence in his office" and that "he had been instructed in terms of what to say." Although he allowed Williams to see a list of salaries, Ammons explained that he could not provide a copy because his secretary was sick. When Williams returned to the office after lunch, one of his former students introduced him to Ammons's secretary, who was clearly not ill. Two years later, Ammons became special assistant to the director of the Mississippi extension service and worked at Mississippi State University. His timorous deportment had not changed, and he signed an affidavit exonerating Mississippi's extension service for failing to provide equal opportunity as it integrated.[36]

On May 14, Williams interviewed Jasper Davis, the black extension agent in Jefferson Davis County, in his segregated office some distance away from the new white offices in the county building. The black home-demonstration agent shared Davis's office but also had an office at nearby Prentiss Institute. Davis's office consisted of two small rooms and a rest-room but no mimeograph machine or air-conditioning. Williams was struck by the smallness of the space, the lack of bulletins, and Davis's eva-siveness. When Williams asked to see a memo, Davis would find a reason to step outside to confer with someone. Williams commented that Davis "was the most fearful person I've seen." He was exceedingly protective of his calendar of events tacked to a wall. "He permitted me to read it from a distance and then stood between me and the wall where the Calendar hung." He disparaged black farmers' lack of initiative in taking up dairy farming and faulted them for wanting "to display junk at the Laurel Fair in September." Like other black extension workers, he shied away from civil rights activists. "I work with people and their farm operation," he made clear, "and any activities such as civil rights or desegregation are none of my business."[37]

Williams then called on Al Johnson, dean of Prentiss Institute, which had been founded in 1907 by Jonas Edward Johnson and his wife Bertha, who was serving as president in 1964. Johnson had a low estimation of black farmers and insisted that they preferred to sit on the front porch rather than work in the fields. Farming, he declared, "is becoming big business and Negroes aren't trained in the operation of big business." White farmers used agriculture as a way to evade income taxes, he said, not as a way to be productive. Echoing Booker T. Washington, he warned that blacks were better off in the South away from urban temptations such as drugs. "We developed Prentiss Institute through funds by tell-ing the northern white man that he helps us to keep the Negro in the South when he gives to the Institute." A more bleak evaluation of fund-raising or of African American farmers and their prospects would be hard

to imagine.[38] The steady defamation of black farmers in a way exonerated administrators for their failure to provide information and advice
on better farming methods, but it also showed the frustration of dealing
with a transformation that the USDA had decided did not include African
Americans.

As field investigators moved through the South, the Washington staff
began discussions with top administrators of the ASCS, FHA, and FES. The
USDA functioned in isolation from scrutiny, and the white male–dominated leadership had never been asked to answer for their discriminatory
policies. It seemed natural enough to ignore women and minorities and
never solicit their opinions. Commission staffers asked penetrating questions based on field interviews, and USDA leaders recoiled with anger
and denial. When commission attorney Marian Yankauer interviewed
FHA administrator Howard Bertsch on May 15, he glanced at her request
for loan data on black farmers and sighed that his overworked office staff
could not possibly comply with her request. Bertsch grew up in Corvallis,
Oregon, took an agriculture degree at Oregon State University, and did
postgraduate work at Kansas State University. He had worked from 1934
to 1954 in the Farm Security Administration and the FHA before taking
a position with the Ford Foundation consulting on rural credit in Teh-
ran, Iran. In 1964, he received the Distinguished Service Award from the
USDA.

In his meeting with Yankauer, Bertsch was disinterested in cooperation, a strange reaction from one who later claimed to have rehabilitated
the FHA after years of inaction. After more discussion and hesitancy, Yankauer reported, "I indicated that if pressure elsewhere was necessary to
get his staff to do it, we would apply such pressure elsewhere." Bertsch
was nervous, "very close to breaking down" and "almost in tears," Yankauer noticed. He insisted that commission members spend two weeks
in FHA county offices before writing their report and then quickly shifted
to boasting about his role in drafting the poverty bill. Bertsch defended
a much-criticized FHA initiative appointing African Americans as non-
voting alternate county committeemen and explained that they were
alternates because the appointments came after the deadline for regular
appointments. "I pointed out," the well-prepared Yankauer countered,
"that on my list most of them were appointed in July and August 1963.
He then said, that some went back to 1962, but I pointed out only a few
were in 1962 on my list." After Yankauer caught him out, Bertsch claimed
that the FHA was far out front on civil rights and digressed into the origins of the alternate committeemen plan. Still, he presented no plans to

appoint blacks as full committeemen, and without votes, the alternates were clearly tokens. Bertsch defensively claimed that white southerners would not accept blacks as full committee members. In an interview two months later, Bertsch portrayed the Kennedy administration as bold and infallible, but neither in that interview nor in his conversation with Yankauer did he articulate his lack of initiative on civil rights.[39]

A week later, Yankauer met with two of Bertsch's staffers, Donald Oberle and Lawrence Washington. Oberle became "very defensive, rather antagonistic" when Yankauer announced that she had discovered some southern counties that would accept African Americans as full committeemen. "He clearly resented our inquiring into the question of why Negroes were only alternates," she observed. Bertsch had the power to make such appointments, and his reluctance reflected not only USDA apprehension of southern racists but also the convenient excuse that white objection ended the argument. Yankauer suggested that if her staff could find counties that would accept black committeemen, Bertsch and his staff could certainly find more.[40] Her challenge did not lead to a policy change. Bertsch and his office staff digested their embarrassment and pique and settled into complacency.

After Oberle left the May 22 meeting, Yankauer talked with Lawrence Washington, an African American. She brought up a Franklin Parish, Louisiana, case involving an FHA clerical worker who, against the state director's wishes, purchased 200 acres for $8,000 from a black couple earlier denied an FHA loan. When a black FHA staff member interviewed the farm couple before the sale, the white FHA clerk presumptuously sat in on the discussion. The clerk had earlier assured the couple that she would allow them to stay on the land and farm the cotton allotment. The clerk's actions raised ethical questions, and Washington had recommended that Bertsch fire her but understood that at least an inspector general investigation would result. "I tried to discuss with him the fact that this was a chance to really look at the entire situation in FHA to determine what kind of service was given to poor Negroes," Yankauer reported, and "that firing one woman didn't really get at the rest of the situation." Washington said that he would notify Bertsch that the commission "was onto" the case. Writing about what seems to be the same case, NSF fieldworker Ocie Smith reported that just as the state FHA was ready to move on the case, Washington arrived and bungled things. Firing staff for discrimination was evidently not in the USDA office manual. Several months later, Bertsch visited Clarendon County, and he talked with South Carolina Council on Human Relations executive director Alice Spearman on the phone. When she told him that there were no black staffers in South Carolina FHA offices, he meekly offered to appoint a secretary but did not follow through.[41]

On May 12, Yankauer met with Joseph L. Matthews, who directed the FES's research and training program; J. Neil Raudabaugh; and Albert Bacon. She discussed the commission's project to acquire information on services provided to black and white farmers and found Matthews and Raudabaugh "extremely resistant" to the study. When she raised the issue of separate training, "they said that Negro agents might not have requested to be trained, that Negro agents might not want certain training, that Negro agents might not in fact, be ready for certain training." After that string of weak excuses, Bacon joined the chorus that belittled black farmers, claiming that they were "not innovators," they were "very often afraid to change and to take any risk," and they fell behind the times. Born in Brooks County, Georgia, Bacon was well acquainted with rural life, and in 1931, as a Smith-Hughes agricultural student at the Brooks County Training School in Dixie, he set two records in corn and cotton production. After graduating from Savannah State College in 1938, he taught vocational agriculture in several counties and in 1952 earned an M.S. degree at the University of Minnesota. Before he moved to Washington as an assistant to the FES administrator, he held high positions in the SCS and the Extension Service. By a strange coincidence and contradictory to Bacon's characterization of black farmers, Shapiro and Safford had interviewed several quite competent African American farmers in the Valdosta area near where Bacon grew up. After some heated discussion, Matthews stated that extension agents could not "violate the mores of the community" or "be responsible for changing local customs." All three FES staff members evaded the implications of the segregated 4-H clubs, which offered opportunities to white youths and denied them to blacks. When Yankauer raised the issue of segregated offices, Raudabaugh offered that "after all it would not be very democratic to impose our outside ideas upon the local community and that the FES was there to work with the people and there was not really much to be gained by that kind of a change." Segregated offices, Yankauer retorted, meant that many black agents did not receive FES literature, which provoked Matthews to laugh and say, "Most of it is junk anyway." Neither Matthews nor Raudabaugh took the request for information on discrimination "very seriously." They assumed, Yankauer concluded, "that Negroes and whites in the same community, growing the same crops, would have different and unrelated needs and desires for and interest in services from a County Agent."[42] In office after office, the commission staff met resistance, obstinacy, and an attitude that civil rights were a passing fancy that could be avoided by delaying any programs that would include African Americans.

John W. Banning, assistant director of 4-H and youth development at the FES, admitted to Yankauer in May 1964 that "Negro youths were not getting the kind of service that they needed" and that he was "particularly concerned at exclusion resulting from a bias in favor of the economically favored operating in the program." There were two national meetings, he explained. Fourteen hundred youths attended the National 4-H Club Congress, which met annually in Chicago. Sponsored by agricultural supply companies, the meeting hosted 4-H prizewinners, who, according to Banning, "were the wealthy youth"; all were white. Also, each state sent four delegates to Washington every summer to attend the National 4-H Club Conference. In 1964, one black member from North Carolina attended, evidently the first African American to do so. The National Negro 4-H Club Convention had been eliminated, so black members were effectively excluded from conventions and from competing for national prizes. Yankauer described Banning as "hopeful, concerned and intelligent and not at all frightened," unlike her assessment of other USDA officials. He proudly showed her a pamphlet prepared for distribution at the National 4-H Club Congress in Chicago, which on the first page "contained a picture of a white girl, a Negro boy, a white boy, and an oriental girl holding the 4-H flag." When southern 4-H executives received the pamphlet, Banning got calls threatening his job, yet there had been no retribution. Nor had there been any commendations. Interestingly, the controversial photo presented a public relations fantasy of interracial cooperation that was lacking in 4-H programs. No doubt it made good public relations for the Extension Service, at least in some parts of the country, but on the ground, such scenes never happened.[43]

The FHA, like the Extension Service, failed to distribute relevant information to African Americans. Given the numerous loan categories and the sheaves of forms to be filled out, uneducated farmers needed assistance from FHA offices. First, of course, they needed to know about the loans, but since FHA offices ignored black farmers and poor whites, loans increasingly went to literate white farmers who were within the circle of USDA information distribution. The class structure favored elite farmers with representation on agriculture committees and with benefits; family ties and friendships, so the conventional wisdom went, also contributed to favoritism.

Marian Yankauer immediately analyzed the interviews and information on USDA programs flowing into the commission office. It was true that small farmers were losing out everywhere, but African American farmers, she argued, were segregated and "consistently outside the decision-making process," and any assistance "has been phrased in terms of a separate economy." She located serious discriminatory actions in the FES, ASCS, and FHA. In Washington, the FHA employed one black professional, and in southern states, the few African American FHA employees worked in segregated offices and dealt only with black farmers. With the ASCS, Yankauer wrote, "we move from large scale segregation into absolute exclusion." Only one GS-9 African American worked in the Washington ASCS office, and otherwise African Americans worked only as custodians. Information suggested that ASCS committees shorted black farmers on allotments, but since they had no representation on county committees, there was no conclusive evidence. Allotments were "bitterly contested," she observed, and, importantly, "the total value of ASCS expenditures dwarfs other expenditures in the county both in dollar volume and in impact upon the people." Despite its funding and oversight of the FES, Yankauer learned, the USDA "completely disowns responsibility for the way in which the Extension Service is operated." Support for African American extension "is distinctly inferior and provided entirely on a segregated basis." She cited examples from the interviews. The issue of civil rights had not penetrated the walls of the USDA, and personnel, many of whom belonged to segregated organizations, blissfully avoided discussing racial discrimination.[44]

As Shapiro prepared to visit Louisiana in June, he and Yankauer met with FES director Lloyd H. Davis and two of his deputies, John Cox and Albert Bacon. Davis had moved from a top extension position at the University of Massachusetts and had extensive experience. Born in Dyersburg, Tennessee, he grew up on a farm in northern Pennsylvania near LeRaysville. He took his Ph.D. at Cornell University and was an army major in World War II. Shapiro asked Davis for the names of likely people to interview in Evangeline, St. Landry, Tangipahoa, and St. John the Baptist Parishes. The commission was curious why there were so few black extension agents. John Cox "emphasized considerably the great difficulty in working with people as poor as these Negroes were" but agreed with Shapiro that the FES needed more staff for this work, suggesting that some of the "poverty money" be used to help African American farmers. The three FES administrators avoided talk of integration but "were looking for ways to improve service within the existing structure" and eliminate the "inequities." When Shapiro pointed out that black Alabama extension workers had to send their stencils to Tuskegee Institute for mimeographing while the white office had mimeograph machines, both Davis and Cox "shook their heads sadly." When Cox stressed that change had to come at the local level, Yankauer asked if it was not the role of the FES to raise the sights of local agents and make them see broader issues. Davis claimed that they were trying. At one point, Cox "said some idiomatic thing about not plunging into the water too fast, to which the reply was made that it was the only way to get into cold water, and he responded with surprise that of course that was so."[45] Such forced amicability, unconvincing cooperative attitude, and poorly disguised bias would be seen over and over, and it hid a grim and determined agenda to maintain segregation.

As the commission gathered and analyzed information on discrimination, Freedom Summer unfolded in Mississippi. By July, many activists looked to the Civil Rights Act to remedy discrimination. In late July, the NSF's Jac Wasserman met with the USDA's William M. Seabron and Assistant Secretary for Administration Joseph Robertson to discuss the USDA's civil rights efforts and discovered that Robertson, who had only worked for three years at the USDA, was unfamiliar with many of its programs. Robertson grew up in Glen Dean, Kentucky, fought in World War II, graduated from Western Kentucky State College, and earned an M.A. degree at the University of Alabama. He had worked in Orville Freeman's Minnesota office and moved to Washington to serve as assistant secretary for administration. He expressed surprise when told of the few black FHA workers in the South but countered that there were few qualified black applicants. Wasserman suggested that if the black land-grant schools were not putting out qualified students, "they should try to correct the situation by integrating the white and Negro colleges so that they have the same courses, as well as integrated faculty and student bodies." Robertson agreed that the USDA had issues but argued that until the Civil Rights Act he had no authority to correct them. Wasserman suggested to Fay Bennett, executive director of the NSF, that they meet with the administrators of the FHA, FES, and ASCS and explore how they intended to implement civil rights. Bennett grew up in Massachusetts and graduated from Simmons College in 1937. In 1951, she took a part-time job with the NSF and a year later became executive secretary. She was extraordinarily active in organizations that worked with farmers and migrant laborers.[46] Each office that commission and NSF staff visited seemed taken unawares and shocked that anything was required of them as far as civil rights were concerned.

*Fay Bennett, executive director of the National Sharecroppers Fund. Courtesy of Walter P. Reuther Library, Wayne State University.*

In May 1964, Yankauer had discussed the issue of segregated county-agent organizations with Albert Bacon, the African American assistant administrator of the FES. Bacon served as adviser to the National Negro County Agents Association, created fifteen years earlier, and had suggested efforts on the state level to meet with the white National Association of County Agricultural Agents (NACAA) to explore a merger. Membership in the white national group depended on state membership, and in most southern states, blacks were not allowed to join. Blacks especially feared that their leadership positions in the African American group would not be honored in the white organization and that they would have no voice in state and national NACAA bodies. African American home-demonstration workers, on the other hand, had been integrated into the white National Home Demonstrators Association and had attended an integrated meeting in Hot Spring, Arkansas.[47]

Sensing, accurately as it turned out, that blacks might pay a terrible price when black organizations disappeared and members attended hostile white meetings, the commission encouraged full integration of white organizations that had frozen out black USDA workers. As the annual National Association of County Agricultural Agents meeting in New Orleans approached in October 1964, the commission sought assurances that blacks would be given equal treatment. Six years earlier, Major Elmer McCoy, a black county agent in Pine Bluff, Arkansas, who was secretary of the National Negro County Agricultural Agents Association, asked to attend an NACAA meeting in Seattle, where he was visiting relatives. The NACAA president invited him as an observer, but a white county agent from Pine Bluff accused him of attempting to integrate the meeting, and after receiving a phone call from his Arkansas supervisor, McCoy left the meeting. The commission warned that unless "assurances should be obtained that they [African Americans] are eligible to attend all business and social functions listed in the program," USDA officials should not attend the convention. In 1965, the NACAA convention went to Pittsburgh, and Fletcher Lassiter, a North Carolina extension agent who was head of the state's African American county-agents organization, questioned the Commission on Civil Rights about the segregated meeting and the commission passed his queries to the USDA.[48]

Despite statements to the contrary, African American county agents were often at the beck and call of white agents. On June 11, Shapiro and Safford interviewed Arthur Britton Sr., who until a few weeks earlier had worked as a county agent in Richland Parish, Louisiana, since 1955. They learned that he had ended up there because all former black agents had been unacceptable to Doles, the white county agent. Doles demeaned Britton by persuading him to serve guests at his barbecues so he "would get to meet the white leadership in the community," but Britton became "an object of ridicule among the Negro community" and stopped helping at white gatherings. When Britton entered Doles's office, he could not hang his hat "where the white folks hung theirs" but had to either put it on the floor or hold it. Britton's office had no telephone, mimeograph machine, secretary, restroom, or running water. Doles accused Britton of "becoming involved with integration leaders," manipulating parish politics, and failing to submit forms, and despite Britton's efforts to get along, Boles fired him, effective June 30. When Britton requested a meeting with state extension leaders, Doles threatened to fire him immediately if he attended, and he was terminated on May 31. At the meeting, an extension official said, "One thing is, you haven't made Mr. Doles like you," and he supported Britton's termination because of his "unwillingness to accept supervision." Such humiliating treatment typified the experiences of many black extension workers. "The fellers from the Civil Rights Commission caused quite a row," the NSF's Ocie Smith wrote in June 1964. Britton worked closely with them and told them "all that the others were afraid to divulge."[49]

W. M. Bost, director of the Mississippi State Cooperative Extension Service, described to commission interviewers in August the complex arrangements and uneven service of county offices. Bost admitted that offices were segregated but denied that blacks had inferior space. The vestiges of segregation haunted the extension program, and according to Bost, African Americans insisted on preserving autonomy or rather the benign neglect of the segregated system. For example, he offered, the black supervisor usually approved raises, yet blacks did not regularly get raises "because the supervisor does not feel that they qualify for a raise." It was best that advisory committees remained segregated, Bost inexplicitly claimed, because "if they were integrated, we would lose many Negro members." Farmer meetings, demonstrations, and agent training were usually segregated, and Bost claimed that integration would take "authority away from the Negro State Leader which we do not want to do." Black extension leaders and agents, of course, only had second-class autonomy, but that was better for many agents than what integration would offer. Bost continually tied his policies to school integration and the ideology of the white community. "I can't move faster than the local community," he hedged, knowing that the white community did not want to move at all.[50] Bost's self-serving and exaggerated claim of black agents' autonomy was not informed by discussing the issue with black agents, nor were black agents privy to setting up the guidelines for integration a year later.

Bost's easy generalizations poorly masked extension's discrimination. According to a 1964 statistical study, Mississippi's rural population of 1,375,336 lived on 82,719 white and 55,423 African American farms. Every county in the state had white extension and home-demonstration agents, many having 2 or 3, while only Bolivar County had more than 2 African American agents, and only 11 counties employed 2. Blacks in 48 counties had no county agent, and 47 had no home agent. Many of these counties had a substantial number of black farmers. Black extension and home-demonstration agents had a much higher caseload than whites. In Sunflower County, for example, 3 white agents served 1,145 farmers while the sole black agent ministered to 2,009, and the 2 white home-demonstration agents served 3,151 families while the lone black agent had 5,333 clients. Counties that lacked black agents had no 4-H program for African American youths. White FES administrators constructed and defended this discriminatory system, and Bost and his white colleagues saw it as ideal.[51]

In the summer of 1964, the commission studied extension personnel and services in 25 Georgia counties with large black populations and discovered a similar imbalance. In these counties, 36 white agents served 15,715 white farm operators, while 12 African American county agents served 8,238 black farm operators, or, put another way, the ratio of white farmers to agents was 436 to 1 and African American farmers to agents 681 to 1. In these counties, there were 30 white home-demonstration agents and 10 black agents. The ratio among home-demonstration agents was 1,330 to 1 white and 2,894 to 1 black. Some counties with large black populations had no home-demonstration agents or extension agents and therefore no opportunities for the 8,081 boys and 10,284 girls to participate in 4-H programs.[52] Despite these glaring inequities, FES personnel from counties, land-grant schools, and Washington were unwilling to bring African Americans into the decision-making loop. Instead, the FES obstinately opposed efforts to correct its segregated system or implement integration.

With its rich trove of interviews, the commission's staff continued to analyze USDA policies and, not surprisingly, found grievous problems. Even as the civil rights movement flared across TV screens during the summer of 1964 with news of violence and the Mississippi Freedom Democratic Party's challenge at the Democratic National Convention, the USDA continued its discrimination against black farmers and even punished African American employees for participating in civil rights activity. The commission's 1965 report would reveal pervasive USDA discrimination and offer suggestions for reform, but the combined strength of all civil rights organizations could not erase USDA discrimination, for its bureaucrats would react to the report first with denial, then with vows to do better, and finally with passive nullification, that is, refusal to carry out civil rights initiatives even as they announced support. It was this large and unwieldy bureaucracy that SNCC workers faced in the fall of 1964 when they attempted to gain access to federal programs for African Americans. Although they started late, they focused on ASCS elections and made a significant bid to get African American farmers elected to county committees.

*Daily the Negro is coming more and more*
*to look upon law and justice, not as protecting safeguards,*
*but as sources of humiliation and oppression.*
—*W. E. B. Du Bois,* The Souls of Black Folk

# 3

# FREEDOM AUTUMN

From the moment of the Greensboro sit-in by four African American students from North Carolina Agricultural and Technical College (A&T) on February 1, 1960, civil rights protest burned brighter in the South. The next day, twenty-three students from A&T and four women from Bennett College, an African American women's college founded in 1873, sat at the counter, and within days, white women from what was then Women's College of the University of North Carolina joined in. It was as if Greensboro lit a fuse among young Americans, for sit-ins quickly exploded in some 100 southern cities, engaging roughly 70,000 people. About 3,600 demonstrators were arrested, and 141 students and 58 faculty members were dismissed or suspended. Impatient young college students, nearly all African American, challenged the painstakingly slow pace of civil rights court cases, the isolated nature of most demonstrations, and segregationists' iron fist. The NAACP, Southern Christian Leadership Conference (SCLC), and Congress of Racial Equality (CORE) were eager to harness this youthful enthusiasm for their own agendas.[1] The sit-in at the Greensboro lunch counter generated waves of protest that broke across the South for years.

Ella Baker, who at the time of the sit-ins was working with the SCLC in Atlanta, called a meeting for April 16–18 at Shaw University in Raleigh to allow students to meet, discuss strategy, and organize for further action. Shaw University had hosted A&T, founded in 1891 as a black land-grant school, until it moved to Greensboro in 1893. Baker, a Shaw graduate, had moved far beyond her North Carolina roots and settled in New York, brilliantly organizing for the National Association for the Advancement of Colored People (NAACP) and then working with Martin Luther King Jr. at the SCLC. Baker's diminutive stature, conservative wardrobe, and impeccable manners camouflaged her radical views—until she spoke in a commanding voice and articulated ideas that challenged the moderate pace and tactics of the civil rights movement. She embodied a unique style of leadership seldom witnessed in the 1950s and 1960s, and her long career established her as one of the most important civil rights leaders of the twentieth century. Heeding Baker's caution, the 120 black activists from 56 colleges and a dozen white students at the meeting decided to remain independent of the major civil rights groups and form the Student Nonviolent Coordinating Committee. In May, they opened a small office in the SCLC's Atlanta headquarters, put out a newsletter, and began attracting an exceptionally gifted and diverse group of people.[2] Not since the 1920s, when African American college students protested antiquated rules of conduct and white intrusion into school governance, had black students shown such enthusiasm for protest.

Most SNCC members were born during the World War II era; grew up with the Cold War, atomic bomb, and Korean War; and shared a post-war restlessness for change. The *Brown v. Board of Education* decision in 1954 striking down school segregation, Montgomery bus boycott and elevation of Martin Luther King Jr. as leader of the SCLC, and crisis at Little Rock's Central High School in 1957 helped shape their insistence on equal rights. In their minds, time had run out on segregation and discrimination, and they were not tempered by well-meaning but cautious liberals. Most SNCC workers came from the working class, a background they shared with blues and rock 'n' roll artists and stock car drivers, who challenged the weight of segregation and lethargy and left their mark on American culture. Historians, novelists, reporters, and photographers too numerous to catalog but too significant to ignore joined them to create an unlikely canon of southern action, words, songs, and imagery.[3]

SNCC staff melted into the southern landscape and listened to people who were usually ignored by black leaders and white do-gooders. They wore bib overalls and work shirts, walked along town and country roads, sat on front porches, listened, occasionally worked in the fields, escorted people to courthouses for voter registration, and explored ways to end the system that deprived African Americans of equal opportunity and the vote. SNCC staff did not stir up trouble and leave local people to face hostile whites but instead set inspiring examples of bravery. Sam Block from

Cleveland, Mississippi, for example, made it clear that he was in Greenwood to stay. The sheriff confronted him and asked where he was from, and Block replied that he was from Mississippi. When the sheriff bragged that he knew all the "niggers" in town, Block asked him if he knew "any colored people." After spitting at Block, the angry sheriff told him to pack his clothes and leave town. "Well sheriff," Block calmly replied, "if you don't want to see me here, I think the best thing for you to do is pack your clothes and leave, get out of town, cause I'm here to stay."[4] Needless to say, white sheriffs were not accustomed to back talk from African Americans, and such behavior combined with SNCC's voter-registration work provoked anger and retribution.

In the twilight of Freedom Summer in 1964, civil rights activists inaugurated their Federal Programs Project, which sought to take advantage of the numerous federal programs that often excluded African Americans. As they talked with farmers about crops and discrimination, they uncovered the importance of powerful USDA county committees, the centers of rural white economic power. Many black farmers, they learned, had scant knowledge of ASCS committees, FES programs, or FHA loans. Because members of county ASCS committees were elected, civil rights workers hoped to invade the clubbish white meetings with the only tool they had: a majority of votes. Every contact with USDA agencies reminded African American farmers that Washington's promises, memorandums, and laws made little difference in their daily lives. Although it never was as successful as voter registration, the effort to elect black members to powerful county ASCS committees attacked the white financial nerve center. This significant assault on a key USDA program attracted little national attention, but it vividly illustrated SNCC's savvy understanding of local politics and its tactical skill.[5]

During the mid-1960s, rural transformation, discrimination in the administration of federal programs, and civil rights activism, especially voter registration, created enormous tension in the South. As machines and chemicals decimated the rural labor force and the civil rights movement accelerated protest, whites attempted to drive blacks out of the South by denying them jobs and relief. Whites cynically turned federal agriculture programs, the distribution of surplus food, and access to welfare payments into weapons to punish African Americans, but as civil rights laws and community-action programs proliferated, blacks gained experience in management and ultimately challenged white control. The migration of rural blacks to northern cities pleased southern whites but came to alarm their northern cousins, especially after urban riots began in the mid-1960s. During these years, unrest in northern cities challenged the notion that urban African Americans inhabited a promised land filled with good jobs and adequate housing. The conventional wisdom held that riots were the direct result of poor southern blacks' inability to adjust to city life, although most rural blacks adjusted quite well. Whites targeted former sharecroppers rather than analyzing the discrimination that drove the riots.[6]

The effort to win seats for African Americans on county ASCS committees offered a democratic avenue to substantial federal economic resources and power, and it triggered bitter white opposition. Obviously, black farmers had not participated widely in ASCS elections in the past, nor had many grasped the substantial economic power vested in the committees. They grumbled but fatalistically accepted the small allotments parceled out by the committees. On the county level, the white men who ran federal programs showered contempt on black farmers, as did their office staffs, who put out what amounted to a white-only welcome mat at the office door. Those challenging the ASCS in Mississippi, Georgia, Louisiana, Arkansas, and Alabama confronted committee members who were comfortable with the operation of the ASCS and, alarmed by black opposition, eager to use intimidation, fraud, and purposeful ineptness to divert black voters. Cracking white control of local agriculture programs was the best chance that black farmers had of sharing power and fairly apportioning federal ASCS allotments and funds.

Since its inception as the Agricultural Adjustment Administration, the ASCS exerted enormous control over acreage allotments and the distribution of federal funds. Nearly every county in the nation had an ASCS office, and from the beginning, county elites dominated those offices. According to a 1947 study by Robert Earl Martin, as AAA programs were first established in counties across the country, FES county agents appointed committee members, but quickly the AAA set up its own system of elections. At first farmers met and voted for candidates by acclamation or a standing vote. Even so, Martin argued, county agents maintained enormous control over the committees. Over time, the elections became more formal.[7] The county committee, an ASCS administrator explained, "serves as a board of directors, sets policy within the national regulations, makes certain hardship adjustments on cotton and rice acreage allotments, reapportions cotton acreage from acreage released by farmers who do not desire to plant cotton, establishes yields and indexes under the cotton, feed grain and wheat program."[8]

As New Deal agricultural agencies proliferated, county offices provided well-paying and prestigious jobs not only for office managers and assistants but also for secretaries and clerks. Many offices gained reputations for officiousness, although most were accommodating, especially to wealthier white farmers. Attending local churches, putting their children through school, and usually living in rural communities, ASCS personnel understood the customs attached to segregation and hierarchy. These full-time federal workers, in cooperation with wealthy farmers, basically directed county farming operations, controlling the acreage-reduction programs, parceling out loans, running 4-H programs, distributing information, and signing up farmers for programs. There were no African Americans in decision-making positions, and the few black employees swept the floors and tidied up the offices.

When the AAA set up the machinery to control the amount of acreage farmers could plant, it stressed that farmers would elect the county committees that made such decisions and it boasted of grassroots democracy. In the rural South, some farmers were black, Native American, or another minority, some white; some literate, others unlettered; some rich, others poor; some greedy, others generous—the list of variables goes on. Farmers preferred to elect literate and generous representatives, and given the nature of southern discrimination, that meant white. Over the years, ASCS elections were based on just that equation, but considering the power vested in these committees to set acreage allotments, funnel federal funds, and make crucial decisions affecting county agriculture, the committees were not free of greed and chicanery. Indeed, the archival files of the secretary of agriculture bulge with complaints about unfairness and favoritism. In most instances, however, the Washington office stayed out of local disputes, and county committees and office staffs interpreted and executed federal rules without interference.[9]

From the beginning, farmers elected AAA committeemen who were landowners with large farming operations. In his 1947 study, Robert Earl Martin focused on Wilson County, North Carolina, and Darlington County, South Carolina, two flue-cured-tobacco-and-cotton-growing areas. In neither county did sharecroppers serve on committees, and only a few tenants gained seats. Since serving on a committee required calculating acreage allotments, filing reports, and understanding complex and ever-changing farm programs, younger educated farmers were conspicuous on committees. Once the pattern of landownership and education was set, the same men, or class of men, were elected year after year. Sharecroppers, tenants, and small farmers soon lost interest in elections and deferred to larger farmers. African Americans, who were eligible to vote and run for committee seats, did not challenge the color line by standing for election, although they insisted that they should have representatives. Martin asked whites why they failed to vote and got answers such as "I just left it to the big fellows" and "They run it the way they want to." To the same question, blacks replied, "I felt it was for white only," "Didn't know they wanted colored," and "'Cause they are all white and don't give us a chance." African Americans understood that whites made and enforced the rules without their input. By the mid-1940s, interest in AAA elections declined to the point that USDA officials sometimes had to drum up farmers to get a quorum, and Martin reported that this indifference caused concern in Washington. What had begun as an enthusiastically heralded democratic program had evolved by the end of World War II into a program plagued by lassitude and dominated by county elites.[10]

Neither the erosion of voters in ASCS elections nor unprincipled efforts to prevent black farmers from participating stopped USDA leaders from hailing farmer-elected committees as the embodiment of democracy. President Franklin D. Roosevelt's secretary of agriculture, Henry A. Wallace, termed it "economic democracy." The method of electing ASCS committees varied over time, but by the 1960s, each community in a county elected three farmers who convened with other community representatives to elect three members to the powerful county committee. The secretary of agriculture appointed from three to five farmers to a state ASCS committee, with the state director of extension an ex officio member. Farm organizations, land-grant university deans, extension directors, state commissioners of agriculture, and other state leaders vetted state ASCS appointments. Given the complexity of the committee system and the constantly amended USDA programs, the ASCS strayed far from its intended grassroots charter and provoked testy challenges. In 1955, for example, there were 14,000 review proceedings. The ASCS records in the National Archives are full of complaints and documentation of violations, and a body of case law adds further testimony to the contentious nature of allotments. The contradictory policies of reducing acreage while intensifying research to increase yields made sense for larger farmers who could manipulate USDA programs to fit their operations. Better-educated and landed farmers also profited from increasingly complex and lucrative government programs and were not averse to seizing federal money and acreage allotments intended for sharecroppers and tenants.[11]

Most disputes traveled up the bureaucratic chain to Washington and then back down. The ASCS chain of command in 1964 ran from adminis-

*Horace Godfrey,
Agricultural Stabilization
and Conservation Service
administrator. Courtesy
of American Sugar Cane
League.*

trator Horace D. Godfrey to Ray Fitzgerald, deputy administrator of state and county operations (DASCO), who supervised six area directors and the fifty state ASCS offices that oversaw county ASCS committees. Godfrey grew up in Waxhaw, North Carolina, attended North Carolina State College, began work with the AAA in 1934, and in 1961 became head of the ASCS. Godfrey was a southern New Deal liberal, and while his views on integration were ahead of those of most white southerners, he was constrained by both hard-line segregationists in the USDA and recalcitrant southern congressmen. Ray Fitzgerald, a South Dakota native, earned a B.A. in business from Xavier University in Cincinnati. Before moving to Washington in 1961, he was South Dakota's secretary of agriculture.[12]

Godfrey and Fitzgerald seldom intervened in state and county affairs, even when events spun out of control. In 1963, the Arkansas state ASCS committee descended into such chaos that it tested Washington's usual hands-off policy. Arkansas ASCS executive director A. C. Mowery Jr.'s interests became "dominated almost entirely by his association with State political leaders," DASCO Ray Fitzgerald learned, and the laxness of other members of the state committee had caused "a number of serious deficiencies." Lacking state oversight, some county committees ignored fraudulent acreage measurements, and farmers failed to destroy overplantings. Investigations in Woodruff County showed that rice and cotton farmers had "planted and harvested acreages of rice and cotton far in excess of the acreage allotted." Ultimately allotments in thirty counties were checked, and serious irregularities were found in eight. In Craighead County, the farm of the Craighead Rice Milling Company had overplanted by 209.5 acres, and one cotton farm had 133.8 excess acres. In Monroe County, farmers did not destroy overplantings. In Jefferson County, the Swan Lake Plantation had overplanted its 1,028-acre cotton allotment by 411 acres. In Lincoln County, the county ASCS chairman overplanted his cotton allotment by 23.8 acres.[13] Obviously, the ASCS program in Arkansas was riddled with favoritism and fraud, but there is no documentation of punishment of offenders.

Also in 1963, warring political factions disrupted Sharp County's Community C election. When Arkansas ASCS fieldman Bill D. Fowler arrived on January 7 to monitor the election, Wayne Orr, chairman of the community committee, threatened him, and, Fowler reported, the chairman of the county committee, J. T. Orr, "was very belligerent toward me and the State Committee." The election was held in a trailer outside a country store eight miles from centrally located Ash Flat, and voters approached either by fording a river from one direction or taking a narrow road from another. That afternoon, Jim Shelby Ferguson entered the trailer and clumsily stuffed ballots into the box while pretending to study the voting register. Observant monitor Bill Fowler immediately pulled out some dozen ballots still protruding from the box. J. T. Orr, standing in the doorway, told Wayne Orr to grab the ballot box and shouted, "This whole god dam election is illegal." The Orrs took the ballot box to the nearby store, burned the ballots, and declared the election over.[14]

The Arkansas state committee reviewed the election, deemed a new election necessary, and recommended that the Orrs be suspended, and DASCO Fitzgerald ruled that the inept state committee should handle the election. Community C held a new election in Ash Flat, a more central location, and on March 19, all of the community representatives met to elect the county committee. Irregularities and arguments nullified this meeting, and joined by state administrators and vocal rival factions, community representatives met again on March 27. That meeting ended when a committee member, noticing the sheriff's presence, adjourned the balloting, declaring that it could not be continued under armed guard. When state officials failed to intervene, Fitzgerald at last removed authority from Sharp County ASCS officials and moved the election to Little Rock. That fall, a Sharp County farmer warned Secretary Orville Freeman that he distrusted the local committee. "In the past, they have just mailed ballots to the ones they feel will support them rather than to all qualified voters."[15] Official correspondence did not elaborate what provoked Sharp County factions or spell out what the Orrs had at stake, but Washington's reluctance to intervene in an obviously fraudulent county ASCS election suggested the autonomy of even out-of-control community and county ASCS committees. The control of acreage allotments, federal funding for projects, and no doubt patronage loomed large in rural areas, and politically powerful farmers sometimes sought unfair advantages. Such autonomy and ineptness would later hinder SNCC's attempt to gain seats on community and county committees.

The deterioration of the ASCS in Arkansas demonstrated serious flaws in the state administration and awkward hesitation in the Washington DASCO office. There were also formidable structural faults in the committee system. In the early 1960s, Morton Grodzins, a prominent political science professor at the University of Chicago, served on a committee that evaluated the ASCS committee system, and he submitted a critical minority report. Author of a major critique of Japanese internment during World War II, Grodzins was a keen student of federalism. The majority report claimed that ASCS committees "exemplify grass roots democracy at its best: elected neighbors serving neighbors, local control avoiding the evils of a national bureaucracy." The ASCS program was vast: 3,000 counties, 26,000 communities, and 90,000 farmers participated in the program. Committees not only supervised programs but also convinced farmers of their value. "*Democratic forms*," Grodzins cautioned, "*may camouflage central control.*" It was in the interest of the USDA to preserve committees "in order to mask central control, or make it more palatable," but in actuality, the USDA twisted "democratic forms in an *authoritarian manner.*" Grodzins discovered that in 1961 less than 23 percent of eligible voters participated in elections, and he found committees undemocratic and inefficient. More disturbing, committees often sat in judgment of "quasi-judicial matters," and he wisely questioned whether such complex matters, settled in courts of law before the establishment of the AAA in 1933, should be determined by farmer committees. Apparently the USDA had never critically examined the judicial powers of committees, despite the fact that moving judicial issues from courts to committees raised serious constitutional questions. Grodzins suggested that the secretary of agriculture should evaluate the committee system and either revise it or abolish it in favor of a more effective program. He made a strong case for abolishing ASCS committees and for encouraging USDA personnel to work more closely with county executives.[16]

The ASCS farmer committees and the complex bureaucratized appeal process that replaced court jurisdiction resulted in a morass of tangled and uncodified decisions. Prior to the New Deal, landlords and tenants settled disagreements over crop-lien laws, control of crops, and other disputes in courts of law that relied on precedent. Allowing committees judicial power removed the legal ground from underneath farmers, for decisions no longer rested on precedent but were shaped to fit each case. Under the bureaucratic committee system, farmers often received confusing information from county offices that could put them in conflict with the Code of Federal Regulations and result in appeals to county and state committees and finally to DASCO. Neither county committees, appeal committees, nor the DASCO office kept systematic records, and thus no body of case law existed in published form. To get a sense of precedent, lawyers later filed freedom-of-information requests and analyzed DASCO's letters to farmers. As Karen Sorlie Russo reported in 1991, the 1,200 DASCO letters sent in 1989 were inconsistent in their rulings and contained "curiosities . . . too numerous to list." Placing judicial power in the hands of ASCS committees and bureaucrats who did not rely on precedent or the principle of *stare decisis*—that is, the application of one case decision to a similar case—resulted in inconsistency and favoritism. The ASCS thus wrote and executed the codes and then sat in judgment of disputes. The funds and benefits at stake were substantial. The Commission on Civil Rights' May 1967 report on twenty-six Alabama counties showed $30.5 million spent on ASCS programs, enough to generate numerous disputes.[17]

In 1967, at the peak of African American challenges to county committees, Grodzins's critique was supported by a *Columbia Law Review* article that analyzed ASCS election procedures and concluded that the USDA "permits to exist, and by its neglect of its duties encourages, a system which gives white southern farmers vast power over the economic well-being of their Negro neighbors." The article explained in detail the election process, white control, committee discretion, irregularities, and the appeal process. "To the illiterate, cautious, poor farmer the prospect of facing successive committees, filling out many forms, and then undergoing the expense and time of a court proceeding is so formidable that it is unlikely he will even begin the process," the essay argued. Paradoxically, whether ASCS discrimination was challenged under the Constitution, the Civil Rights Act of 1964, executive order, or USDA regulations, "the people truly victimized by the system will not be in court," for they lacked the resources to challenge elite control. It was quite a conundrum: "A plaintiff could not succeed unless he alleged, in effect, that he was incapable of filing a suit." Clearly the ASCS committee system was flawed, but it proved impossible to unwind the tentacles of white elite control.[18] As both Grodzins and the authors of the *Columbia Law Review* article argued, the ASCS election system only dressed and spoke as democratic. When the Council of Federated Organizations (COFO) in Mississippi and SNCC in several other states challenged white control of county committees in 1964, ASCS offices resorted to draconian measures to defeat them, exposing the democratic fault line.

SNCC activists, of course, saw discrimination from a rural perspective. As they moved among farmers, they developed a keen understanding of agricultural policy and its biases. In August 1962, SNCC began voter registration in Sunflower County, Mississippi, the home of Senator James O. Eastland. In November, SNCC's Bob Moses, who had been organizing in Mississippi since 1960, wrote Fay Bennett, executive director of the NSF, that three families had been evicted for attempting to register. He said that one woman, Fannie Lou Hamer, "left the Marlowe plantation after she was threatened if she did not remove her name from the applications at Indianola." Moses reported that Reverend Jeff Surney and Boss Foster were also evicted. Foster had grown up on the Livingston plantation and had spent thirty years there and "was in charge of the tractor drivers." Moses had called off voter registration in the vain hope of convincing the Justice Department to sue plantation owners and force them to take back the evicted workers. Quietly intelligent, Moses had attended an excellent high school in New York and then graduated from Hamilton College. In 1960, he moved south to work with the SCLC but quickly discovered the tiny SNCC office and, urged by Ella Baker, went to Mississippi, Alabama, and Louisiana to recruit people to attend SNCC's fall meeting in Atlanta. Moses had been abroad, both to Europe and to Japan, and he had earned an M.A. degree in philosophy from Harvard University. He shared with Ella Baker a conviction that local people should be trained and encouraged to work for civil rights and that SNCC should avoid a central order-giving structure. No one leader could speak for the organization, and thus no one person could start or stop demonstrations. Tactics and strategy emerged from intense discussions at meetings that were notoriously lengthy and boisterous. A legend grew around Moses, who combined patience, intelligence, bravery, and unsought leadership. In 1962, even the vatic Bob Moses had no idea that Fannie Lou Hamer would emerge as one of his most powerful colleagues.[19]

SNCC workers Charles Cobb and Charles McLaurin, also working in Sunflower County, witnessed the complexities of survival in the cotton

*Fannie Lou Hamer addressing the Mississippi Freedom Democratic Party Convention, 1971. Photograph by Pete Daniel. Smithsonian Institution, 87-3858-17.*

culture. Landlords and credit merchants guarded accounts of sharecropper purchases and debts, figures often questioned by farmers. Sharecropper Willie Mae Robbinson farmed near Ruleville, and in 1962, she picked twenty bales of cotton that Cobb and McLaurin calculated should bring $3,740, her share being $1,870, but Robbinson's landlord, after deducting expenses, figured her share at $3.00. A neighbor told the SNCC workers, "I know that she hasn't eaten what would have come out of ten bales." The SNCC team discovered that sharecroppers earned $300 to $400 per year and day laborers $150 to $160.[20]

In the spring and summer of 1963, nearly a thousand protests in southern towns and cities led to arrests and dramatized civil rights momentum. The March on Washington culminating in Martin Luther King Jr.'s "I Have a Dream" speech at the Lincoln Memorial again demonstrated SNCC's outsider status as old-line civil rights leaders urged John Lewis, SNCC chairman, to tone down his criticism of the Kennedy administration for ignoring unlawful arrests and for its weak and languishing civil rights bill. SNCC insisted that the fight for equal rights required more than rhetoric. Lewis, who ultimately would serve in the House of Representatives at the other end of the Mall, understood that the old order's litany of faith in the existing political order largely ignored southern African Americans. The press portrayed SNCC as irresponsible and infiltrated by communists, but the organization welcomed anyone to join in the fight for civil rights. Charging that these young African Americans and their white friends were under communist influence when their policy was hammered out late at night in fierce debates was absurd. By 1964, SNCC's unique approach to challenging the South's white power structure left it nearly isolated from many civil rights organizations and from the Johnson administration.[21]

In the fall of 1963, SNCC leaders in Mississippi took up the controversial idea of bringing in hundreds of northern white students whose visibility would ensure publicity and prod federal officials to protect them. At this time, the staff was 80 percent African American, young, college- or high school–educated, and from working-class families. SNCC only reluctantly decided to bring white students into Mississippi, realizing that while the Federal Bureau of Investigation (FBI) undermined African American activists and the press ignored or pilloried them, whites would attract a different kind of attention. As SNCC's Lawrence Guyot observed in 1963, "We could not move unless we bumped into an FBI agent as long as there were white people involved." When they came, enthusiastic and well-educated white students created tension, not only because some were naive and had little idea of the potential violence surrounding movement activists but also because, as predicted, they offered opinions and advice that sometimes challenged more experienced but less formally educated SNCC leaders. Dividing work among blacks and whites, men and women, northerners and southerners added to the strain.[22]

In the summer of 1964, SNCC and its allies challenged segregationists with voter registration, demonstrations, freedom schools, school integration, and glaring publicity. If SNCC's unique program to enlist black community leaders to challenge segregation's grim brutality was not audacious enough, the Mississippi Freedom Democratic Party (MFDP) was founded in April 1964 to challenge white Democrats for seats at the August national convention in Atlantic City, New Jersey. The events that summer, especially the participation of northern white students in civil rights efforts in the South and the MFDP challenge to party regulars, captured headlines and dominated TV screens across the nation. It was the same summer that Nelson Mandela was sentenced to life imprisonment during apartheid in South Africa.

Several civil rights groups coordinated their forces in Mississippi and formed the Council of Federated Organizations, in SNCC leader John Lewis's words, "an indigenous, state-wide movement in which all national groups active in Mississippi, all local political and action groups, and some fraternal and social organizations participate." CORE, SNCC, and the MFDP furnished most of the muscle, with important support from the NSF and the NAACP. Aaron Henry, an NAACP stalwart, served as president, SNCC's Bob Moses as program director, and CORE's Dave Dennis as assistant program director.[23]

Before traveling to Mississippi, SNCC volunteers attended orientation sessions in Miami, Ohio, that acquainted them with SNCC's program and prepared them for Mississippi's unique culture. Muriel Tillinghast, a Howard University graduate who had been active in civil rights demonstrations on Maryland's Eastern Shore, prepared for canvassing in Washington, Issaquena, and Sharkey Counties. She learned about the roads, town layouts, officials, and resources. "This you needed to know like the back of your hand," she wrote. "Your life depended on it." Once there, she observed rural life, which she labeled "fairly humdrum," broken up on weekends with "dancing at the juke joint, church on Sundays." The farmers' cycle of work and leisure—juke joint, church, plowing, planting, chopping, and picking—held no charm for Tillinghast or for most nonrural people. She also learned that small-town and country residents were sensitive to anything different, for even a visiting automobile produced a discernible alien sound. Police entered African American homes without knocking and searched without warrants. Housing SNCC workers put hosts at grave risk, she discovered. "When they allowed us to sleep on their floors or in their beds, whatever the accommodations were, black Mississippians were risking economic reprisals, the loss of their jobs, or worse." One hardworking family of seven near Hollandale, she judged, was "as close to slavery as I hope I ever see in life." When farmers registered to vote, she observed, they were not only denied surplus commodities but also "dismissed immediately, losing both their means of employment and their homes." She learned that prosperous white farm owners received USDA subsidies and larger acreage allotments, so that fall, she helped organize black farmers to stand for seats on ASCS committees.[24]

Mary E. King had watched TV coverage of the civil rights movement in 1962 during her senior year at Wesleyan College, and at the urging of Ella Baker and the historian Howard Zinn, she joined SNCC. Although seldom visible, Baker continued as SNCC's nurturer, always listening, asking questions, and offering encouragement. Although King helped direct publicity operations in Jackson, her assignments, she wrote, "took me into rural homes insulated with pasted-up newspapers and the wearily weatherboarded houses of worship of those who had been among the most physically and psychologically brutalized of all the diverse peoples who have made their homes in the United States." Rural people greatly impressed her. "The individuals with whom I worked possessed dignity, magnanimity, courtesy, respect for others and were often contemplating cosmic questions: the meaning of life, whether forgiveness is necessary for reconciliation, and whether there is a force working for justice in the world." To survive, many farmers took on second jobs, and King admired their fortitude in the face of immense odds. "Such lessons," she concluded, "do not fade."[25] It was in such remarkable local people that Ella Baker and Bob Moses placed their faith.

White women such as Mary King who violated the South's color line lived under tremendous stress and chilly exclusion from the southern white community. Into her fourth year with SNCC, Jane Stembridge, who was white, reflected on the psychological implications of working in the Delta, "out there," as she put it in a May 1964 letter. In SNCC, black and white men and women united in common cause, often stirring southern white men's deepest sexual anxieties by violating segregation codes meant to keep black men and white women apart. When Stembridge was jailed for trumped-up traffic violations in October 1963, her Greenwood jailor woke her at 5:00 A.M. "to ask me did I sleep with Sam Block. It was buggin [sic] him." Local police were viciously attentive to SNCC.[26]

At night, Stembridge said, she walked "up and down the rutted dirt roads" where "there was nothing . . . between me and the powers of evil," echoing Robert Johnson's chilling blues image of the devil materializing at the crossroads. On June 23, as the SNCC staff continued to orient volunteers in Ohio, Stembridge learned that three SNCC workers were missing, and the next day, she wrote of the "burned-out car on the edge of a swamp." A week later, the SNCC staff and white volunteers spread across Mississippi. "These are the battlefields now," Stembridge wrote.[27] SNCC workers resembled guerrilla fighters with no visible weapons or protection, and it took breathless courage to face hostile, leering lawmen caressing their pistols and nightsticks. With the temperature in the nineties, Freedom Summer descended on Mississippi.

Even as voter registration continued, the Mississippi Freedom Democratic Party challenged the established party by opening membership to all people and insisted that it, not the segregated party, represented Mississippi. It elected delegates to the Democratic National Convention in

Atlantic City, carefully following procedural rules. On August 22, less than two weeks after the bodies of the three civil rights workers were exhumed, the convention's credentials committee heard Fannie Lou Hamer's haunting account of being evicted, shot at, and beaten. She left the panel members and a national TV audience breathless. "All of this on account of we want to register, to become first-class citizens," she insisted. "And if the Freedom Democratic Party is not seated now, I question America." They were not seated but were offered two symbolic seats, which they rejected.[28]

Among MFDP activists and their COFO allies who had risked their lives to challenge white segregationists, the failure of the Democratic Party to seat their delegates was demoralizing. In Atlantic City, some black leaders and Johnson administration personnel condescended to Hamer and other SNCC staff, deigning them politically naive and unlettered. Less than a month after the convention, National Council of Churches, NAACP, SCLC, and CORE leaders met in New York with several representatives from SNCC, COFO, and the MFDP. The tone of condescension from Atlantic City continued, and establishment leaders pontificated about what should be done in Mississippi.[29] They intended to contain COFO's fiery ideas or put out the fire altogether.

Ella Baker and Robert Moses had found forceful leaders among local people, and SNCC had ventured where the old guard feared to tread and forced the country to confront vicious Mississippi racism. The task remained unfinished, but in popular memory and mythology, civil rights activists fought successful battles and marched to victory, with school integration, voter registration, and election triumphs marking the way. Yet as numerous historians have pointed out, the road was uneven, and the zeal born of the sit-ins cooled by the mid-1960s, in part because four years of living on the edge of violence exhausted the staff and in part because of internal stress and external vilification. Fittingly, in October Martin Luther King Jr. received the Nobel Peace Prize for his equal rights achievements.

While some COFO and SNCC staff attended the New York meeting in September, others, including John Lewis, Bob Moses, Julian Bond, and Fannie Lou Hamer, traveled to Africa. SNCC also planned a retreat for mid-November. The staff had recently led one of the most significant protests of the twentieth century, and many had risked their lives to bring civil rights to Mississippi. When staff attended the retreat near Waveland, Mississippi, pent-up emotions and frustrations erupted, weakening the organization and driving a bitter wedge between both blacks and whites and men and women. The meeting took place at Gulfside, an African American Methodist resort founded by African American Methodist bishop Robert E. Jones in 1923 as a haven where African Americans could worship, study, learn trades, and enjoy the Gulf waters. Gulfside had endured financial crises, fires, and storms and was one of a number of segregated beaches that provided recreation for blacks in the age of Jim Crow. SNCC's Elaine DeLott Baker recalled secluded well-constructed buildings, a chapel, and a path to the beach. Recently back from Atlantic City, New York, and Africa and unsure of what they had accomplished, lacking an agenda, low on funds, and smarting from critics, the SNCC staff did not see a bright future. The press had undermined SNCC's accomplishments by claiming communist influence, and liberals looked askance at its radical agenda and determination to remain disconnected from mainstream civil rights groups. At Gulfside, Chairman James Forman's insistence on centralizing SNCC leadership, which challenged the very soul of the organization, did not sit well with leaders such as Bob Moses, who maintained that SNCC should continue to work with local people who would emerge as community leaders. Forman, then, was rejecting the loose and inexact caucuses that had fueled SNCC and was advocating a centralized organization, ironically much like the structure SNCC had opposed at its founding in Raleigh in 1960.[30]

An anonymous statement concerning the treatment of women in the organization that Casey Hayden, Mary King, and Elaine DeLott Baker wrote and circulated at Gulfside grew out of intense discussions in the energized climate of Freedom Summer. Women, black and white, tended to SNCC's daily business and kept the train on track, while black men held the dangerous leadership roles and strong women such as Ella Baker and Fannie Lou Hamer were esteemed by the entire staff. During Freedom Summer, the volatile combination of eager and sometimes naive white volunteers, hardened veterans, transgressive (in Mississippi) sexual relations, strains on black and white friendships, and constant danger created enormous friction. Black women were angered by black men sleeping with white women, which they saw as rejection. As the summer wore on, black women pressured black men to abandon their relationships with white women, and even when evening liaisons continued, black men no longer acknowledged them. Some white women resented that their intimacy with black men was hidden. They also felt that they were underutilized and often targeted for clerical and support jobs. The statement began with a list of eleven incidents illustrating male dominance and insensitivity. "Assumptions of male superiority are as widespread and deep rooted and every much as crippling to the women as the assumptions of white supremacy are to the Negro," they pointed out. The authors did not expect rapid change and, indeed, suspected that their statement would provoke laughter. In the future, though, they hoped that women in the movement would demand an end to discrimination "and start the slow process of changing values and ideas so that all of us gradually come to understand that this is no more a man's world than a white world." Like other causes that grew from the civil rights movement and reshaped culture, the emerging women's movement was powered by ideas embodied in the women's statement.[31]

The unsettled emotions roiling at Gulfside included a gnawing unease among many of the African American staff that working with whites was counterproductive and that whites should be canvassing white racists to defuse their prejudice. College-educated whites, it was alleged, were unable to relate to poor and uneducated African Americans, and white women working side by side with black men triggered white hostility. Black SNCC workers also yearned to break free of the historical echo of whites making decisions about every aspect of their lives and, despite their friendship with whites, sought to consider future actions solely from a black perspective. Forman's push to centralize leadership, the deteriorating relations between blacks and whites, the white women's statement, and impatience with nonviolence dominated retreat discussions. In many lengthy meetings, SNCC staff had ultimately reached consensus, but the disagreements at Gulfside were never resolved and in time contributed to SNCC's demise. An autumn 1964 report from the Jackson, Mississippi, SNCC office indicated that "discipline has broken down" and that "white workers are often subject to severe racial abuse." COFO weakened and in the spring of 1965 merged with the MFDP.[32]

The idea of challenging ASCS committees originated a year earlier when, in November 1963, SNCC's Charles Cobb, a Howard University student who was organizing in Greenville, Mississippi, learned that in Issaquena County, "a sizable number of independent Negro farmers . . . for the most part have no awareness of what kinds of federal aid programs are available them." He determined that, first, he needed to understand federal programs and present them to farmers in language they could comprehend, and then he would canvass black farmers to go to the ASCS, FHA, or extension office and demand participation.[33]

Despite the chaos within SNCC, in the tumultuous autumn of 1964, COFO hastily organized a challenge to ASCS elections in a handful of Mississippi counties. In Clay County, where blacks did not win any seats, COFO workers had only two weeks to organize the vote. In other counties, there was more lead time. When white Freedom Summer worker Mike Kenney arrived in Holmes County, Mississippi, in the summer of 1964, he commented that it was "not at all what I pictured Mississippi to be like. We are in the fringe of the Delta here." Plantations did not dominate the county. Holmes County black farmers, he boasted, "are the backbone of the Movement," for they owned their land and "cower before no man."[34]

Indeed, 800 Holmes County African American farmers owned their land, many taking advantage of the Farm Security Administration's (FSA) Mileston Project, which divided over 9,000 acres into 106 farms available to African American farmers. Ralthus Hayes had worked on plantations until 1941, when he secured both a long-term FSA loan to purchase fifty-four acres and a low-interest loan for housing material, a mule, and implements. Hayes was fiercely independent, along with other black farmers who bought land and vowed to defend it by force if necessary. SNCC's Sam Block had been canvassing in Mileston, and Hayes immediately supported the movement, as did Hartman Turnbow. Turnbow traced his landownership back to Emancipation, when his grandmother's former master sold land to the son he had with Turnbow's grandmother. In April 1963, Hayes, Turnbow, and twelve other farmers arrived in Lexington, the county seat, to register to vote. The fourteen black farmers faced an army of white law enforcement officers and town officials. Attempting to intimidate the farmers, the deputy sheriff asked who would be first. Turnbow stepped forward, and officials allowed only him and another farmer to fill out paperwork. The other twelve returned the next day, but none was registered. A month later, whites threw Molotov cocktails into Turnbow's house, and he then engaged his attackers in a firefight. In what was a familiar scenario, Turnbow; Bob Moses, who arrived the next morning to take photographs; and three other SNCC workers were arrested for setting the fire. Without federal intervention to prohibit such distorted law enforcement, civil rights workers understood their vulnerability, but they also knew that all of the black farm owners were well armed.[35]

Mike Kenney may have had these "First 14" and Howard Taft Bailey in mind when he praised black farmers' bravery. Bailey was a landowner and community leader. His grandmother, a nineteen-year-old slave, arrived in Mississippi from Virginia in a covered wagon. She lived to be 117 and passed in 1924. Once she told an overseer who threatened to whip her unless she worked in bad weather, "I ain't going out. Just kill me." When the weather was good, Bailey recalled, "she could get out there and beat 'em all working." He referred to his father as "a fifth grade genius" who did wage work and eventually saved enough to buy 160 acres. Bailey completed high school, joined the NAACP in 1929, and then farmed the home place. "I worked like the devil in the Civil Rights Movement," he boasted, and he invited SNCC workers to stay at his house. The bank cut off his credit for hauling people to the courthouse to register to vote. "And then you take the Ku Klux Klan," he told historian Lu Ann Jones in 1987. "When I was in the Civil Rights Movement, they wanted to get to my house but they knew I was going to kill 'em if they come up there." Bailey refused to take off his hat for the Holmes County sheriff and warned him that if he tried to hit him, "I'm gonna tote a gun and kill you quick as I would a snake." Like many rural blacks, Bailey had guns for hunting and guns for self-defense.[36]

With farmers like these for allies, Mike Kenney took heart. He stayed in a two-bedroom house with a husband, wife, five young girls, a baby, three teenage girls, an eleven-year-old boy, and a grandmother. Kenney and another volunteer shared one of the bedrooms. Unlike in the Mileston community nearby, in Tchula black people lived in fear that any activism would cost their jobs. Kenney's work in Tchula introduced him to the violent history of the town, the senseless killings, the discrimination, and the climate of fear. As ASCS elections approached in Holmes County, Kenney held a meeting at the Pilgrim's Rest Church in Durant. The twenty-five people at the meeting enthusiastically nominated two men to run in the community committee election, and the group met every week and reported on canvassing.[37]

Even as COFO prepared for ASCS elections, a delegation of Mississippi African American farmers, civil rights workers, and college students gathered at the National Theater in Washington, D.C., on November 14, 1964, to air their complaints. Several farmers cited discrimination by USDA agencies. Holmes County farmer Cris Dixon grew eight acres of cotton and twelve of other crops and usually grossed about $2,500. His twelve-member family lived in a one-bedroom house. Dixon broke his shoulder in January, but the doctor in Tchula refused to set it because he could not pay, and then another doctor set it improperly. It was October before it was set correctly, and he missed the entire crop year before finding wage work on a plantation. Fannie Lou Hamer pointed out how whites in Ruleville controlled the distribution of food and that in the past, planters signed for workers to get food, but with mechanization, former plantation workers had no one to sign for them. The delegation went to the Department of Agriculture and talked with several administrators before having an audience with Secretary Orville Freeman.[38] COFO workers had impressive access to officials in Washington, but back in Mississippi, they lacked federal support.

Shortly before the election, a SNCC worker complained that the Holmes County ASCS committee had combined communities in order to confuse black farmers and place them at a disadvantage. On election day, 109 African American voters from the South Lexington community checked in at COFO headquarters before voting, yet black candidate Earl Travis received only 92 votes and Thomas C. Johnson, also black, 77. In the Durant community, ASCS officials would not allow a blind black man assistance in marking his ballot.[39] Kenney tallied up only one black community chairman and two black alternates in Holmes County, but he was proud of the hundreds of black farmers who had, despite intimidation, voted in a significant federal election for the first time. The COFO workers who monitored the vote count sat at a table with white men and were handed contested ballots for their judgment. This, Kenney suggested, was the first integrated meeting in the county. A few weeks later, he sat on Christmas Day in front of a butane heater unable to defeat the seeping cold air and wrote a letter under a naked lightbulb, furnishing depressing context for life in rural Holmes County. This house, he wrote, was "near luxurious" compared to some he had visited while canvassing. "One was heated by an old wood burning heater/stove; another with one oil lamp for illumination was heated only by the fireplace with ragged, almost sullen children huddling around it." As he took his holiday sponge bath, he watched TV. "Barbara Walters on the Today show was sitting in a $100,000 completely chinchilla covered room," stressing the advantages of "a $75 chinchilla covered coat hanger." Rural Mississippi was another world.[40]

Ultimately, the USDA investigated charges from Holmes County that a large percentage of petitions for African Americans to run for community committees were denied. In one instance, whites claimed that even if all eligible blacks had voted, it would not have changed the result. "Win, lose or draw," the USDA's Kenneth Birkhead replied from Washington, "the Negroes should have a chance to vote." Birkhead, who was formerly the finance director of the Democratic National Committee, was looking over his shoulder at other concerns. Since the USDA was not asking for another election in Holmes County, he strategized, "Whitten could not use this document as a means to try to undercut anything we are planning." He doubted that Mississippi congressman Jamie Whitten "is in any way trying to be vindictive about activities in Mississippi" or that Holmes County whites would visit "retribution" on people who complained of the ASCS election, for "this word would quickly get back to us and would cause them considerable harm." Birkhead's naive analysis ignored Whitten's ruthless exertion of political power and revealed his lack of recognition that southern whites were more adept at retribution than the USDA was at enforcing civil rights. The Johnson administration, of course, relied on southern Democrats to support its legislative program and obviously calculated what actions might jeopardize that support. Birkhead's unwarranted trust of local USDA officials proved fatal to the hopes of African American farmers, for without black representation on ASCS committees, federal benefits would accrue only to the white elite. The U.S. Commission on Civil Rights' Marian Yankauer, never one to mince words, suggested that anyone guilty of "trying to interfere with the right of Negroes to run for office can be denied participation in the program in the following year." The USDA never implemented such enforcement.[41]

Throughout twenty-one Mississippi counties, COFO workers moved among farmers with information and encouragement to run for ASCS seats. At times, they picked cotton to earn money for gas and expenses. Local ASCS offices countered the threat of democratic elections with bureaucratic ineptness, evasion, deceit, and obfuscation, while Washington administrators offered mock concern but no enforcement. African Americans challenged county ASCS hierarchies and threatened white control over acreage allotments, several lucrative support programs, and the pace of change. Although black farmers had not participated in crafting the New Deal's grand rural design for regulating production and disbursing subsidies, gaining a voice in ASCS committees could alter the course of federal funds and ensure the fair distribution of acreage allotments. As in other civil rights challenges, whites fought back fiercely, not only using bureaucratic cunning and inertia but also allowing, and in some cases encouraging, intimidation.

In October 1964, an MFDP newsletter, the Benton County *Freedom Train*, announced the formation of the Citizens Club of Benton County, which would focus on helping farmers vote in the coming ASCS elections. "We all know that Negro farmers in the county have been cheated in their cotton allotments, and we intend to correct this discrimination, by putting Negroes on the A.S.C. committees." The group would also investigate discrimination in granting FHA loans. COFO got the attention of ASCS bureaucrats in Mississippi several weeks before the December 3 election when Aviva Futorian wrote to the state office from Holly Springs about Benton County's ASCS office manager, Gordon Stone. Although he was polite to COFO workers, he had failed to notify "between one-third and one-fourth of the eligible Negro voters about the coming elections," a violation that Futorian passed on to the U.S. Commission on Civil Rights' Marian Yankauer. Stone announced that the farmers he "forgot" to put on the mailing list could come by his office, an option Futorian ridiculed as "clearly unrealistic, given the fact that until very recently many farmers in Benton County were kept ignorant of the elections altogether." She suggested that Stone was either discriminating against African Americans or inept at running his office.[42] Both charges fit many of the USDA county offices that ignored black farmers and lost records. Ineptness could serve many masters.

C. W. Sullivan, the head of Mississippi's state ASCS office, predictably denied any discrimination and insisted that all eligible voters had been mailed a special notice on November 3. Stone, he replied, denied ever saying he "forgot" to put some voters on the list, and he vehemently disagreed that black farmers until recently did not know about ASCS elections. "We do not classify farmers according to race," Sullivan huffed, "and after diligent inquiry we have found no indication of any discriminatory action against negro farmers or other farmers in administration of ASCS programs in Benton County."[43] Denials such as Sullivan's became the standard response throughout the South as the transgressions of county ASCS staff members were always forgiven and sanitized at higher state and federal levels.

Most COFO workers in Mississippi were in their late teens or early twenties and were remarkably resilient, and some had substantial experience with SNCC. Eighteen-year-old white college student Penny Patch had already been arrested several times before she arrived in Albany, Georgia, in May 1962. "The force of the movement was electrifying," she recalled, and she did not return to college that autumn. In early 1964, she went to Jackson, Mississippi, to work with COFO preparing for the arrival of white volunteers and then moved to Panola County to canvass for ASCS elections that fall. She stayed with Robert and Mona Miles, who owned a 167-acre farm with 24 acres planted to cotton. In the late 1950s, Robert Miles had started a voter-registration drive in the county, and he was a founding member of the Panola County Voters League. He housed COFO volunteers and was outspoken on civil rights, and in retaliation, nightriders bombed, teargassed, and fired into his home. Patch heard stories of violence against black farmers from the Miles family and from C. J. Williams and his wife, who farmed 40 acres in the hill section of the county. In March 1965, after the trial of several demonstrators in a

*Penny Patch (left) canvassing in Panola County, Mississippi, 1965.*
*Photograph by Elaine DeLott Baker.*

very tense Batesville, nightriders shot into the Miles home as Patch and Chris Williams, another white COFO volunteer, sat watching TV with the family.[44]

Williams and Patch canvassed black farmers and told them about the ASCS election. When Felix Webb decided to run in the ASCS election, three of the white candidates called on him. "They came right out into the field where I was working," Webb recalled. "Asked me, did I know what I was doing." They told him to withdraw from the election. Five other nominees were told they would lose their jobs unless they withdrew. Three dropped out immediately, and two asked for time to think it over. COFO notified the USDA and the FBI, and shortly afterward the chairman of the county ASCS committee announced that the men could run. Williams canvassed the forty sharecropper families on the 3,000-acre Carlin Hays plantation; Hays was chairman of the county ASCS committee. Despite a good turnout of black farmers, whites carried all communities in the county.[45]

On cold and wet election day, Patch and a young black female SNCC worker observed the vote in a small country store located on a dirt road. Two officials presided from behind a table near a woodstove, and Patch and her friend sat nearby. "It was terrifying to be so totally isolated in that little store surrounded by people who hated us," she recalled. White men strolled in, looked them over, voted, and said nothing, while black farmers quietly voted and left. Some white teenagers "threw a live snake at our feet," provoking the woman who owned the store to yell at them. The snake, Patch observed, "slithered under the wood stove." Patch admitted that she was scared then and that the longer she worked in the movement, the more it affected her. She and her friend displayed enormous courage to sit all day in the hostile store. "We were young, we were living in wartime conditions," she wrote later. "We were always afraid; we never knew whether we would see one another again." The Sardis community vote took place in the courthouse, and as black farmers marked their ballots, planters sitting with checklists in the back of the chambers sent runners to corral white voters to ensure victory.[46]

Patch also observed the fragile tenure of many black sharecroppers. She had focused on the forty families on the Carlin Hays plantation not only for the ASCS vote but also for voter registration and demands for better housing, higher pay, and accurate settlements. Although Hays had not evicted anyone, Patch reported, he was "slowly squeezing people out, not giving work to those people on his plantation who are active in the Movement." Sharecroppers had no bargaining chips, for prolonged protest would simply prompt Hays to "buy himself a mechanical cotton-picker." By 1964, sharecroppers and wageworkers realized that it was only a matter of time before machines arrived on every plantation. Local people had always protested against white control, Patch learned. "The Movement was there before we came," she recalled, "and it went on after we left." Already, though, Patch felt the tension between black and white SNCC workers, and she resigned from SNCC in April 1965, worked a few more months in Panola County, and then left for California with Chris Williams.[47]

On November 20, Ray Fitzgerald addressed the possibility that there would be a large number of write-in votes and suggested procedures to ensure that they were counted correctly. He also handled a complaint from COFO's JoAnn Ooiman about the ASCS office's delay in furnishing a list of eligible voters in Madison County. Ooiman called Fitzgerald collect at his Washington office and sent him several copies of her complaint. Attempting to head off charges of unfair ballot counts, Fitzgerald assigned state ASCS personnel to watch the voting in fourteen counties, in essence assigning foxes to watch the chickens.[48]

Despite confusing voter lists, the election in Madison County proceeded. County, state, and federal ASCS officials guaranteed COFO workers the right to observe the election, yet in Madison County, at some balloting sites, poll watchers and black voters faced loitering and sometimes hostile whites, intimidating law enforcement officers, and belligerent store owners. A Mississippi State Sovereignty Commission investigator monitored the situation and reported that unless whites kept control of the ASCS committee, future elections "will be dominated by COFO and local Negroes, and this same danger exists in other counties where Negroes out-number the whites and are presently qualified to vote in the ASC elections." Created in 1956, the Sovereignty Commission attempted through public relations to polish the state's image and through its investigative arm to spy on and undermine civil rights activists. At Hawkins Store, Community A's polling place, the owner told COFO poll watcher William H. Forsyth Jr. to leave at 9:30. Euvester Simpson arrived at 2:30 to replace him, and the owner asked her to leave. She did but later returned. On her third return, she was arrested for trespassing. Outside the Community B Health Center polling place, ruffians broke poll watcher Marvin Rich's nose. At Ballards Gin, the Community C polling place, COFO's Eric Orr was assaulted. Some African American farmers decided not to vote because of the chaos around the polling places. At McDonalds's store, the Community D polling place, all three observers were arrested, in-

*Rosie Nelson with sons Carl and Marvin, Panola County, Mississippi, 1965.*
*Photograph by Elaine DeLott Baker.*

cluding George Raymond. Things went better at the Communities F and G polls, and except for hostile whites, the intimidating law enforcement presence, and complaints that black voters were instructed to vote for five candidates instead of as many as five, COFO raised no objections. Instead of maintaining law and order, Sheriff Couthen and his deputy arrested COFO poll watchers while allowing the moiling white mobs to have their way with black farmers and COFO workers.[49]

Twenty-two-year-old Elaine DeLott Baker went to Mississippi from Radcliffe College in the summer of 1964 to teach summer school at Tougaloo College but was quickly recruited by SNCC to analyze federal programs available to African Americans. During the ASCS election in Madison County, she was a COFO poll watcher at the R&N Grocery, the Community E polling place. Baker explained that when illiterate black farmers arrived to vote, she and one of the ASCS community committeemen accompanied him to the booth, read out the candidates' names, and showed him where to make his mark. She felt no hostility but sensed nervousness from the ASCS officials and suspected that they found it awkward to deal with a young white woman and obvious outsider empowered to observe the polls. As for the farmers, "They really were scared, but very brave." The election went well until Sheriff Couthen arrived and told Baker to observe from outside the store. She left, and a SNCC staffer waiting outside used his walkie-talkie to notify leaders in Canton, who passed the word to Washington. Baker mentioned to one of the ASCS committeemen that kicking her out could invalidate the election, and he invited her back inside. Concerned ASCS committeemen asked Baker to confirm with Washington that she was back inside. Then Sheriff Couthen returned and told the ASCS staff that while they might take orders from Washington, he did not, and he gave Baker the choice of observing outside or going to jail. When she asked the sheriff why she had to go outside, he told her that some white boys might rough her up. She asked why she would be safer outside and was told that outside she could run. COFO leaders advised her to go back inside, and when Sheriff Couthen returned, he arrested her. In the jail, Baker made acute observations of the other prisoners, the nasty conditions, the talk, and the music. George Raymond and a female COFO worker were also in the jail, and Raymond was taken out of his cell and beaten. Baker was released about noon the next day.[50] The threats, beatings, and arrests failed to stop COFO's fearless workers, and the sophisticated communication network and access to Washington gave whites pause.

*Elaine DeLott Baker, 1971. Photograph by Roberta Price, © 2004, 2010. Courtesy of Elaine DeLott Baker.*

After all the canvassing, dealing with stubborn ASCS staff, and poll watching, only fourteen African Americans were elected to community committees in six counties. None made it to the powerful county committees. Charges of intimidation and fraud in the Mississippi ASCS elections generated complaints, denials, and investigations that lasted into the spring. What could have been settled quickly by decisive ASCS action to hold and closely supervise new elections instead dragged on with intentional inefficiency and confusion. In mid-December, F. Wainwright Blease, director of the south-central ASCS area, dismissed complaints from Issaquena County and steeled himself to a flurry of criticism. A native of Saluda, South Carolina, Blease had worked for thirty years with the USDA. On January 7, 1965, George Raymond, COFO project director in Madison County, who had participated in freedom rides and worked with Fannie Lou Hamer in Sunflower County, made a statement to two investigators from the USDA inspector general's office about his experience in Madison County. Poll officials had instructed black farmers to vote for five candidates, not up to five, he wrote, and because there was only one black man on the ballot, whites were guaranteed to get four votes. In one instance, a boss accompanied his black worker and cast the worker's vote without consulting him. Some black farmers could not read the ballot, and poll officials, Raymond recalled, hatefully prohibited COFO staff from reading it to them. Both black and white voters turned up at the wrong polling place because current maps had not been posted, an example of ASCS duplicitous inefficiency. In one community, COFO workers were run off before the count, in another they were told there was not enough room for them to observe, and in yet another their sight line to ballot counting was blocked. On the day after the vote, COFO workers learned that the woman who ruled on voter eligibility "appeared to be biased in her attitude regarding the question of eligibility of Negro challenged voters."[51]

Two days later, Raymond amplified his formal statement by observing in a letter to Secretary Orville Freeman that although county ASCS committees were elected, they had been run "solely by whites for whites since [their] inception." Whites got the large allotments that they requested, he continued, while black farmers not only got smaller allotments but also were often told they had overplanted and then forced to plow up part of the growing crop. As for the Madison County election, he explained, black farmers were enthusiastic about their candidates and turned out heavily on voting day. When it became obvious that they might carry the vote, ASCS officials scoured the area for white voters while intimidating black voters and poll watchers. A store owner described as "the local dirty old man" threw out a woman poll watcher, he informed Freeman, "warning that he would 'kick her in the ass' if she came back in." When she returned, a highway patrolman arrested her for trespassing. When one black voter entered the polling place, an official snidely asked if he wanted to go to jail. The bosses, in his words, were "hardly standing around polling places to encourage Negro voters." Poll watcher Marvin Rich complained to Freeman on February 17 that his "nose was broken by a small mob of white youths while I was observing the ASCS election," and he had reported this to the FBI. On March 1, ASCS administrator Walter L. Bieberly assured Rich that the inspector general's office was investigating.[52] The level of fraud, duplicity, and violence revealed the value that whites placed on controlling the ASCS.

As complaints piled up, ASCS administrator Horace Godfrey in January 1965 requested that Inspector General Lester P. Condon make a thorough investigation. The ASCS's goal, Godfrey stressed, was to make sure that all black and white voters had the opportunity to vote. He promised to institute new procedures to enroll voters, evaluate challenged ballots, train county committeemen in correct procedures, and assign a state or even federal ASCS representative to each county if needed. His sudden interest in reform acknowledged that the old system was riddled with flaws, but he shrewdly navigated the class dimensions of ASCS committees. Although he boasted that fourteen black farmers were elected to community committees or as alternates, he surely realized that this would do nothing to diminish the power of the white elite.[53]

Despite Godfrey's reform talk from Washington, Mississippi ASCS executive director C. W. Sullivan claimed inaccurately on February 25 that testimony exonerated the Madison County committee, which had "complied with the regulations in the conduct of committee elections last fall." Still disgruntled at a December 18 USDA press release that suggested irregularities, Sullivan demanded a corrected release and exoneration. He also claimed, again inaccurately, that the USDA inspector general had found that the Madison County election was conducted in conformity with ASCS regulations. Sullivan and other ASCS officials were adept at misreading reports to their advantage.[54]

Long after the election, southern ASCS officials searched the inspector general's report not for violations by ASCS personnel but for any hint of wrongdoing by COFO workers. Wainwright Blease alleged in March 1965 that Elaine DeLott Baker had violated the ground rules by offering aid to illiterate farmers and by watching them mark their ballots, omitting that ASCS committeemen had cooperated with her. He also implied that she was arrested for these actions rather than for refusing the sheriff's order to leave the R&N Grocery. Walter L. Bieberly, acting director of ASCS's south-central area, targeted SNCC's Aviva Futorian and suggested that she had violated election procedures. A native of Kansas and a World War II veteran, Walter Leo Bieberly started his USDA career in 1948. Although Futorian's work in Benton County involved sorting out community maps, ballot delivery, voting by deceased or imprisoned farmers, and other problems caused by ASCS ineptitude, Bieberly implied that a more thorough ASCS investigation might reveal "more flagrant irregularities" by COFO workers. Nothing came of this, but it showed his eagerness to discredit and intimidate COFO workers while ignoring the ASCS's inadequacies. Turning the inspector general's report upside down to suggest COFO irregularities was much like arresting Hartman Turnbow for firebombing his own home. Bieberly hoped to mute the larger implications of ASCS duplicity and intimidation by blaming COFO. As the commission's Marian Yankauer shrewdly observed, intimidation was aimed not so much at black farmers but "at the civil rights organizations that were assisting them in understanding how to vote for Negroes."[55]

Despite the inspector general's report documenting beatings, arrests, intimidation, and purposeful ineptness, Mississippi county and state ASCS executives distanced themselves from all disruptions. "ASCS officials in Madison County," Wainwright Blease reported in a four-page letter to Raymond Fitzgerald in April 1965, "denied that they had any part in the arrest of the COFO workers or that they had requested the presence of officials at the polling place." As if to make his explanation credible, he added, "This information was confirmed by the sheriff." Three pages of Blease's letter included details from the inspector general's findings that supported most of COFO's complaints of violations in all communities. This narrative of harassment and intimidation contradicted Blease's denials and hand washing. Fitzgerald set new elections in only Communities B and F, this time by mail ballot.[56]

William H. Forsyth Jr., forced to leave Hawkins Store at 9:30 on voting day, penned several biting complaints about the ASCS's decision to hold re-elections in only two communities. On May 4, he advised Mississippi ASCS head C. W. Sullivan that black farmers would boycott the election. He strenuously objected to the county committee arbitrarily placing black farmers on the ballot, raised questions about the eligibility of farmers' wives, and cited ASCS regulations with authority. Forsyth sent copies of his letter to, among others, Secretary Freeman, USDA Inspector General Lester Condon, the U.S. Commission on Civil Rights' Marian Yankauer, and local and federal ASCS officials. In a separate letter to Freeman on May 5, Forsyth argued that if the USDA was serious about ending discrimination, "some more is to be expected than the half-hearted acknowledgement of injustice."[57]

The Madison County ASCS committeemen took the liberty of placing African American farmers on the Communities B and F ballots without consulting them, hoping to confuse black voters and dilute their votes, a tactic that would become commonplace in the fall of 1965. As Forsyth predicted, many blacks boycotted the May election, protesting that there had been violations in all Madison County communities. Still, the COFO threat prompted more white farmers to vote. The white vote count in Community B increased from 96 in December to 153 in May, and the African American vote declined from 60 to 51; in Community F, the white vote increased from 111 to 147, while the black vote fell from 105 to 47. No African Americans were elected; indeed, the same whites who had won in

December won again in May.[58] Freeman's pledge to hold fair ASCS elections, it turned out, had no teeth.

It took ASCS officials several drafts to compose a reply to Forsyth. One note attached to the correspondence file stated, "Forsyth, I understand, is a 17 year old boy and is getting mail to & from almost everybody." That denigrating statement did nothing to diminish Forsyth's penetrating questions and his mastery of ASCS regulations; nor did it explain why only two communities were voting again. Another rejected ASCS draft failed to address Forsyth's crucial question about the eligibility of tenants' and sharecroppers' wives. Finally, on May 27, Ray Fitzgerald signed a tightly wound bureaucratic letter citing ASCS regulations and explaining wives' eligibility. It was unclear whether Fitzgerald had consulted or invented regulations or whether such rules had previously been applied to white wives. Given the apathy surrounding ASCS elections before 1964, the issue probably had never arisen until civil rights workers challenged black women's exclusion. Owners' wives, Fitzgerald judged, must have their names on the farm deed to be eligible. The eligibility of tenants' and sharecroppers' wives, he added, depended on having "a distinctly separate business interest in the crop or livestock production." Just working the crop would not suffice. "In order for such person to be eligible to vote as a tenant or sharecropper," Fitzgerald amplified, "she must perform in her own right and individually and separately from her husband as a tenant or sharecropper." It was very difficult in a culture that seldom relied on signed documents to establish such an interest. If these rules existed at the time of the December election, they were unclear to the county staffs. Likely, DASCO hastily concocted them to reply to Forsyth. It was obvious that COFO would get no relief from Ray Fitzgerald or the county and state Mississippi ASCS staff. Enthusiastic COFO organizing and the high expectations in December petered out by June, but COFO looked toward the next ASCS elections with experience and boldness.[59]

The ASCS voting and appeals came as the U.S. Commission on Civil Rights was digesting its research and interviews. When *Equal Opportunity in Farm Programs: An Appraisal of Services Rendered by Agencies of the United States Department of Agriculture* appeared in March 1965, the USDA was mired in discrimination complaints even as Secretary Freeman boldly issued his memorandum demanding equal treatment for African Americans. Freeman had good intentions when he backed his 1965 memorandum by creating the Citizens Advisory Committee on Civil Rights and appointing African American William M. Seabron, who had been in the USDA personnel office, as his assistant for civil rights. A native

*William M. Seabron,
assistant to the
secretary of
agriculture for civil
rights. Courtesy of
Walter P. Reuther
Library, Wayne State
University.*

of Chicago, Seabron graduated from the University of Iowa with a degree in chemistry and also attended DePaul University and the University of Michigan. He worked with the CIO during World War II, the Minneapolis Urban League from 1945 to 1950, and the Detroit Urban League the next five years. When he left for Washington in 1962, he was deputy director of the Michigan Fair Employment Practices Commission. In the aftermath of the Civil Rights Commission report, Seabron attempted to implement Secretary Freeman's edicts. His agenda was bold—integration of the Extension Service, the appointment of blacks to several state ASCS committees, the hiring of blacks in temporary jobs in ASCS offices, and the creation of biracial FHA committees in all states with a significant number of black farmers by July 15, 1965. On April 2, Secretary Freeman named three African American farmers—George W. Spears of Mound Bayou, Mississippi; Caldwell McMillan of Annapolis, Maryland; and John Gammon Jr. of Marion, Arkansas—to serve on state ASCS committees. He also announced the creation of his Citizens Advisory Committee on Civil Rights, which would confront USDA discrimination.[60]

Seabron reported to Joseph M. Robertson, the assistant secretary for administration. Civil rights liaison personnel in twenty USDA offices reported to Seabron, including Victor B. Phillips in the ASCS, Floyd Higbee in the FHA, and John B. Speidel in the FES. The Civil Rights Commission's Richard Shapiro warned that "with few exceptions" liaison personnel were "protective of their agencies and not committed to civil rights policy implementation." He pointed to Speidel at the FES as an example. After a meeting with FES administrators on April 22, 1965, the commission's William Payne observed that FES head Lloyd Davis appeared receptive to recommendations but added, "Perhaps he was just being politic." In a May 22 meeting with Secretary Freeman, the commission's F. Peter Libassi stressed that "it is important that the civil rights liaison officials in the agencies share the Secretary's deep commitment to the elimination of discrimination." Shapiro had earlier stressed that only Secretary Freeman could pressure the bureaucracy to implement civil rights. The pressure never materialized, and Seabron encountered persistent problems with the civil rights liaison staff and executives more intent on defeating civil rights edicts than on implementing them. Despite impressive civil rights machinery, Seabron never found the switch to set it in motion, which, given numerous instances of obstructionism, may have been the design.[61]

Freeman hoped that by appointing Seabron and bringing citizens into his circle of advisers he could address the embarrassing issues raised by the U.S. Commission on Civil Rights. The commission also cooperated with the NSF and the NAACP, sharing complaints and suggesting approaches to end discrimination. These initiatives and especially the commission's 1965 report challenged agrigovernment's white male hegemony and provoked white USDA employees at the county, state, and federal levels first to resist implementing civil rights and then, hiding behind cooperative smiles, to drive more black farmers from the land. The tracks of discrimination led from local committees and agriculture offices to state offices, to underfunded black land-grant schools, to flush white land-grant schools, to experiment stations, and on to Washington to disappear into the trackless bureaucratic wilderness where untamed prejudice flourished and staff alienated from the land punished the clientele they were hired to help.[62]

Despite staff opposition both in Washington and throughout the southern states, on March 22, ASCS head Horace Godfrey ordered county committees to hire nonwhites for summer crop measuring and office jobs at the same percentage of nonwhites in the county's farm population. Traditionally the ASCS hired thousands of summer workers, usually high school or college students, to measure land and keep records of acreage allotments. Godfrey recalled that he had intended to move faster on integrating ASCS office staffs in the South, and the commission's report provided welcome cover. He called a meeting of state committees in Atlanta and, after explaining his plan, advised anyone who could not carry out his orders to resign. No one resigned, although Godfrey heard grumblings from staff that before they would sit beside an African American in a meeting they would quit. This plan went far beyond tokenism for it opened summer work for young African Americans.[63]

State ASCS leaders began executing this extremely controversial order. In North Carolina, Marcus B. Braswell, chairman of the state ASCS committee, first briefed his office staff and then called three regional meetings for fieldmen, county committeemen, and county office managers. Despite describing the reaction as "a severe shock," Braswell said the reception among county committeemen had been "unusually good" and the press coverage reasonably objective. "There is no county in the State in which racial feeling has been any deeper or any more tense than in Halifax County," he judged, "a county with 71 percent of its farmers in the non-white category." Still, when he explained the policy to the county committee on April 5, it "was received with a degree of tolerance and understanding that seemed almost unbelievable." Indeed, several Halifax committeemen offered to find qualified blacks for the crop-measuring positions and said they would welcome them on their land. Braswell had more problems hiring black county and state office staffs, for he had no "trained reservoir of Negro persons to draw from," so the new hires would need training.[64]

Heavily black counties presented problems that reflected past discrimination. Halifax County in the eastern part of the state, for example, would have to add eighteen people to its office staff of thirteen and seventy-five to field positions. Since no African Americans had experience in ASCS offices, all would need training, and to avoid chaos, Braswell recommended that an experienced white employee work with inexperienced blacks even if it upset the ratio. He fretted over where he would find 860 qualified black workers and especially young black men with cars to travel to the farms. He pointed out how blind whites were not only to the implications of segregated and underfunded black schools but also to their own assumptions about black competence. Braswell showed good faith in attempting to hire African Americans for positions that had always been reserved for whites.[65] His bold implementation suggested that rural whites were willing to accept blacks in the ASCS workforce if prodded by the federal government and if hiring black workers was presented as inevitable. Braswell's aggressive implementation of Godfrey's directive contrasted sharply with resistance in other southern states.

Few ASCS administrators acted as diligently as Braswell, and instead of carrying out the order in good faith, many ASCS committees explored ways to weaken or nullify it. In Mississippi, the Leflore County ASCS committee complained that census records on county farmers were out of date because machines and chemicals had replaced many black farm laborers. In 1959, there were 1,712 farms in the county, 1,161 of them nonwhite, while 1965 ASCS records showed only 750 farms with a far lower percentage of nonwhite operators. The committee pledged full cooperation with Godfrey's plan but insisted on using the ASCS farm count, which would greatly reduce the number of black employees. In Georgia, twenty-nine county committees diligently counted the number of farmers and found the percentage of black farmers had declined drastically. In Henry County, for example, the percentage dropped from 26 percent to 12, and in Meriwether, from 49 percent to 29.[66] The 1964 Census of Agriculture reported that black farmers in Lowndes County were 65.5 percent of the farm population, and in 1965, black farmers represented 58.8 percent of the eligible ASCS voters. A year later, the Civil Rights Commission found that only 438 or 39.9 percent of the eligible voters were African American. No other county in the state had such a drastic decline, but the commission offered no clue as to how such statistics were generated or whether they represented reality.[67] Regardless of the accuracy of these counts, tens of thousands of black farmers were moving away from the land, and as they left, whites faced less opposition to their control.

The drastic decline in the number of farms dramatized the revolutionary transformation sweeping through the South, especially the flight of small farmers who could not afford machines and chemicals. The collision of the civil rights movement, mechanization, the end of sharecropping, and the demise of small farms shattered rural traditions in the South. Even as the civil rights movement offered aid to black farmers, at least in theory, the impact of science and technology and discrimination in federal programs pushed them off farms. Promises of ASCS jobs, pledges to clean up county committee elections, and civil rights rhetoric emanating from the USDA masked the human dilemma in the rural South. In the cotton culture, spring planting should offer chopping for day laborers, but herbicides had wilted the demand for such work. Mechanical pickers had nearly ended hand picking in the late summer. Despite the odds against them, on April 9, 1965, forty-five black day laborers, domestic workers, carpenters, and mechanics met in Shaw, Mississippi, and formed the Mississippi Freedom Labor Union, demanding $1.25 an hour, overtime compensation, and health and accident insurance. It was an inopportune moment to challenge landlords, but within two weeks, 1,000 people joined the union. COFO and the MFDP encouraged the new union, but Claude Ramsay, president of the Mississippi AFL-CIO, warned that farm laborers were difficult to organize in the best of times and faced ruthless planters.[68]

The people who joined the union had few options in the spring of 1965, for the rhythm of labor-intensive work had been fractured. Union member Willie Mae Martin had chopped cotton to earn money to feed her family of seven children. As that work dwindled, she applied for welfare in 1961 and attempted to live on $35 a month, relying on food from the surplus-commodities program until that program ended in April, to resume in November. Edna Mae Garner, the union secretary, expected eviction any day from the three-room shack that housed her family of seven children. "No matter how bad you're starving and your kids are doing without, they don't care," she said. Formerly she had done domestic work, and her employer gave her dinner and let her off early to chop cotton. After James Meredith enrolled at the University of Mississippi, however, she was dismissed. Despite her dire situation, she refused to chop cotton for less than minimum wage. According to William Brewer, who hauled choppers to the fields, chemicals controlled weeds so little chopping was needed until June. Most tractor drivers in the area did not join the union because it would mean giving up their $7.50 a day wage. After the union was formed, the Shaw police increased traffic arrests and harassment. Several farmers admitted that they could no longer borrow money to keep farming, and increasing debt could lead to foreclosure. "One colored man had 13 acres," Beatrice Miller related, "but when his wife died owing $200 they closed him out." Another woman lost sixty acres because of a $900 debt. With few resources, the rural population around Shaw attempted to face down the planters, but unfortunately their labor was no longer needed. Ultimately their enthusiasm cooled under harassment and eviction, and the union dissolved in the summer of 1965.[69]

The distribution of surplus commodities through the Commodity Distribution Program generated many complaints about whites getting preferred treatment, blacks being shunted aside as whites were given food, and staff hatefulness. SNCC's Lawrence Guyot complained to Secretary Freeman in the fall of 1965 that in Forrest County, Mississippi, a black woman was not given any meat while the white woman in line behind her was. There were no blacks in the program administration, making it easier to discriminate, Guyot pointed out. He enclosed three affidavits. William Seabron distilled the complaints about this program in December 1965, documenting numerous instances of separate lines and whites receiving food instead of African Americans. Fannie Lou Hamer, for example, charged correctly that there were separate lines in Montgomery County, Mississippi, in March 1965.[70]

Meanwhile, ASCS committees hired young black men to measure planted land for acreage-allotment compliance, a job much preferred over less-well-paid farm work. Whites were quick to claim discrimination when blacks moved into these jobs. From Drew, Mississippi, Bessie Mims wrote to Senator James O. Eastland in April 1965 complaining that her son, who customarily measured land to pay for college, had not been rehired. The ASCS office in Indianola explained that it was "being forced" to employ a percentage of African Americans, and she was indignant that black schoolteachers had been hired, "releasing the white boys of their summer jobs." She entreated Eastland to do his part to preserve states' rights.[71]

Writing from Elberton, Georgia, Mary Anne McGee complained to her congressman that although she had six years experience working for the ASCS, "now I can not even be considered for temporary or permanent employment with ASCS because I am white." Her experience, she judged, would have saved taxpayer money that would be needed to train blacks. "Today," she observed, "we hear so much about discrimination against the Negro, but discrimination can work both ways." The county ASCS office manager had agreed to hire her "until Mr. Godfrey's ruling was announced, now I do not meet the first requirement—be non-white."[72]

Sandra S. Greene had an unfortunate run-in at the Oglethorpe, Georgia, ASCS office. Greene, who was white, arrived when the office manager was out and asked the clerk if she could apply for an office job. "She smiled and replied you can if you like, but it wont [sic] do any good," Greene reported. When Greene asked if she could leave her name and phone number, she got the same response. According to witnesses, she raised a commotion. She asked, "What good will it do us to press leaving a job application anywhere now until they can start hiring white people again[?]"[73]

U.S. Senator John L. McClellan heard from a constituent that his son, a student at Arkansas State College in Jonesboro, was not given the ASCS job measuring acreage that he had worked the two previous summers. Although he was assured the job both the summer before and in December, when he showed up at the office, "he found he had no job and was told his place had been filled by a young Negro man—without warning or notice." McClellan seemed perplexed that the young white man failed to get the job and claimed it was "discrimination because of race." The world was turned upside down for McClellan as well as for Mims, McGee, and Greene, who could not empathize with young black men or women who had been closed out of ASCS employment almost totally until 1965. From Bluffton, South Carolina, Edith Dickey Moses bristled at Horace Godfrey's recent claim in Myrtle Beach that the Johnson administration was not discriminating in favor of African Americans. "That in my considered opinion is a calculated, black lie," Moses fumed. She was outraged that "any number of capable white college students who have done summer work . . . for several years have been fired and supplanted by negroes this year by vote-hungry clowns in Washington." Her family came from Indiana and Ohio, she boasted, and ten great-uncles and two grandfathers had fought for the Union. She judged that the South was going through another Reconstruction era.[74]

Godfrey explained to Moses and others that in the past African Americans had held very few positions in the ASCS and that the jobs they held in the summer of 1965 were temporary. Summer workers did not have tenure, he pointed out, a two-pronged answer implying that whites had no lock on the jobs, nor would blacks the next summer. The policy was not discrimination, he insisted; it carried out goals set by President Johnson and Secretary Freeman "to assure that qualified Negroes participate at all levels in our activities including employment. We are attempting to alleviate an imbalance which has existed for many years." William Seabron announced that the USDA would strictly enforce its hiring policies. "If a farmer doesn't want his acreage measured by a bi-racial ASCS team," he warned in one of his most decisive statements, "he can't participate in the program." The commission's William Payne heard that an ASCS staffer had resigned "apparently in light of Godfrey's policy on Negro hiring."[75] Godfrey's hiring initiative turned out to be the most positive civil rights action taken by any USDA administrator in the 1960s.

In the minds of southern whites, hiring African Americans for federal jobs formerly set aside for whites affronted white supremacy. In Louisiana, the state ASCS director reported that the county office manager in Winn Parish had received "several anonymous telephone calls" demanding that the ASCS get rid of black employees. "These calls were considered as threats," he reported, and he notified the sheriff, who agreed to cooperate.[76] Given the activity of the Citizens' Councils and other racist groups, the threat carried weight.

Godfrey's initiative, while barely increasing the number of blacks on permanent ASCS staffs, did crack whites' hold on summer jobs. He reported that the ASCS employed over 4,300 nonwhite workers in the summer of 1965. To recruit qualified applicants, Godfrey sent his staff to a number of African American colleges and universities. Since the ASCS could not manufacture permanent jobs in county and state offices, African Americans would be hired permanently only when openings occurred. In truth, the ASCS used the number of temporary workers to inflate its claims of integrating the agency, but at the same time, hiring black temporary workers forced whites to work with blacks as equals. According to the National Sharecropper Fund's Jac Wasserman, Harold Cooley, the powerful chairman of the House Agriculture Committee, held a secret meeting with southern white committee members to discuss Godfrey's integration order. At a hearing that Cooley chaired, Godfrey endured withering criticism of his integration order and was accused of turning his back on his own people, hiring unqualified workers, and displacing deserving whites to hire blacks.[77]

County offices complained that qualified black candidates for permanent jobs could not be found, an old ploy to avoid hiring blacks, so the ASCS established a training program for clerical positions. Given the number of black colleges and universities in the South, it seemed unlikely that qualified candidates were lacking. Still, Mississippi established training centers in Rankin and Oktibbeha Counties for six trainees, all either college graduates or certified in secretarial work. The training would last for at least six months, and then the trainees would be placed in a permanent position when a vacancy occurred. Other southern states initiated training programs that, at best, epitomized tokenism. In June 1967, Alabama had five trainees, Georgia six, North Carolina four, and South Carolina eight. Of the 14,644 full-time county ASCS office employees in November 1967, 306 were black. North Carolina led with 58, followed by Mississippi with 56 and South Carolina with 29.[78] Godfrey's positive initiative shifted attention from the 1964 ASCS election crisis, but despite its morale problems, SNCC geared up for a fresh challenge in the fall of 1965.

In retrospect, civil rights initiatives in Mississippi and throughout the South in 1964 dramatized not only discrimination but also the extent to which southern whites would go to preserve segregation and white supremacy, even violating the clear regulations that required all farmers an equal right to vote for ASCS representatives.[79] Memories of the horrifying news of the burned-out car, the testimony of Fannie Lou Hamer, the violence surrounding voter registration, and numbing nightly TV news of southern whites' poor behavior accumulated to shift public opinion in favor of civil rights. COFO's ASCS campaign never captured headlines or even seats on powerful county committees, but it did educate black farmers about federal farm programs and allowed some to exercise their vote for the first time. It was too late, of course, for many black farmers to profit from federal programs.

*There are places we fear, places we dream,*
*places whose exiles we became and never learned it until,*
*sometimes, too late.*
—*Thomas Pynchon*, Against the Day

# CHEATING DEMOCRACY

USDA leaders rarely had contact with civil rights workers, or African Americans for that matter, and they had no grasp of the commitment that blossomed from the 1960 Greensboro sit-ins and no insight into why SNCC and later COFO challenged more cautious civil rights organizations. It was distasteful enough to work with the NAACP, the NSF, or even the U.S. Commission on Civil Rights, but USDA executives detested the amorphous grassroots style of COFO. "COFO is composed of the younger, more radical elements of the Civil Rights movement," Thomas R. Hughes complained to Secretary Orville Freeman. A graduate of Macalester College, Hughes had been Freeman's executive assistant in the Minnesota governor's office, and he served in the same capacity in Freeman's USDA office. In May 1965, Hughes admitted that he had found COFO "difficult, if not impossible to work with." He preferred the "more responsible people who clearly understand what can and what cannot be done" and shrank from COFO's impatience and stridency. "We aren't able to stop their ridiculous statements," he confided to Freeman, "but I think we can isolate their statements so that the other active groups, particularly in Mississippi, will not fall for their line." Hughes did not specify COFO's line but implied that he opposed its canvassing for ASCS elections and thus disturbing the status quo. Liberals such as Hughes and Freeman had no stomach for confronting radical young African Americans or upsetting southern politicians by threatening federal subsidies. The NAACP also lost patience with COFO and ended its support in March 1965, and in

June, COFO's remnants joined the MFDP. Hughes would no longer have to deal with COFO as such, but enough activists remained in the South to disturb his rest.[1]

Even as civil rights coalitions were shifting in 1965, confrontations along the color line left an unfortunate trail of violence. Malcolm X, who had broken with the Black Muslims, was assassinated on February 21 at the Audubon Ballroom in New York. On March 7, TV cameras captured police savagely attacking civil rights marchers in Selma, Alabama, as they headed for Montgomery to protest the murder of a young black man. Two weeks later, the march resumed, and prominent whites joined for the triumphant arrival in Montgomery. Even as civil rights protests continued to gain national attention, the escalating Vietnam War generated increasing protests, including a national teach-in in May. Over the summer, President Johnson increased the number of U.S. troops in Vietnam, part of an escalation that accelerated the draft. Campus activists focused increasingly on the war, staging teach-ins and demonstrations. Press coverage shifted from civil rights to the war and, of course, its opponents, and civil rights moved off page 1.

Hughes's flawed analysis of COFO and other Mississippi civil rights activists ignored SNCC's unique and constructive projects throughout the South that educated black farmers about USDA programs, efforts that were far more successful than those of federal, state, or county ASCS, FHA, or Extension Service employees. SNCC and its allies embodied the best hope for integrating ASCS committees, for they bravely endured threats and violence to advocate democratic change. The young radicals, as Hughes deemed them, understood that cracks in the segregationist wall would come only through confrontation and struggle and not through delusions about what could or could not be done.

USDA bureaucrats were more comfortable working with organizations such as the NSF, even though it lent increasing aid to ASCS challenges. In August 1964, NSF executive secretary Fay Bennett advised field director Jac Wasserman to use caution when dealing with Mississippi's extension leaders. "It will be easier for the State Extension people and the County Agents to cooperate with a strictly economic organization," she suggested, "rather than with a meeting set up by COFO which also works on voter registration and civil rights." She stressed that "our effectiveness has resulted from our keeping the two separated."[2] Despite NSF efforts to aid poor farmers, sharecroppers were almost extinct by the mid-1960s, and black farmers of all tenures were dwindling away. The problem with NSF gradualism was that gradually there were no more sharecroppers.

As the commission's report, *Equal Opportunity in Farm Programs*, circulated through the USDA in early 1965, Walter B. Lewis, director of the U.S. Commission on Civil Rights' federal programs division, analyzed the ASCS's civil rights record. In late July, he commented favorably that Horace Godfrey had "established the quota system for nominations to community committeemen positions" but suspected it would not benefit black farmers unless they understood ASCS regulations. The commission's USDA report, he suggested, "must be seen against the background of a century of alienation between the races in the South, of almost total absence of communication—certainly no communication as 'equals'—and of 'self-fulfilling prophecies.'" Two months later, he amplified his remarks about USDA prejudice. "There is a widespread belief that Negroes cannot profit from the programs offered—be it education, job training or technical assistance in farming—and this self-defeating conviction acts as a bar to the success of the program with Negroes." Given that attitude, he warned, flooding the South with federal funds would not be successful, for whites would siphon off the money. Lewis understood that many government employees believed that black farmers were incompetent, inferior, and doomed.[3]

When Secretary Freeman missed the opportunity to appoint blacks to state ASCS committees, Godfrey, in a ploy that resembled FHA head Howard Bertsch's segregated alternate committeemen scheme, created all-black advisory committees to assist white state ASCS committees. In August 1965, commission staff director William L. Taylor protested that segregated advisory groups failed to implement requirements for equal participation in federal programs. Taylor had grown up in Brooklyn, and after graduating from law school, he worked for Thurgood Marshall at the NAACP Legal and Education Defense Fund, then with Americans for Democratic Action, before becoming involved in the John F. Kennedy presidential campaign. Taylor joined the commission in the early 1960s, and he regularly met with a subcabinet group on civil rights that included the USDA's Joseph Robertson and William Seabron. Regardless of Godfrey's intent, Taylor argued, "separation is continued and equality of participation as well as access to decision-making positions remain restricted." This, he concluded, "perpetuates the evil which the Federal Government has now committed itself to overcome." The commission's Marian Yankauer had earlier suggested advising Godfrey that "the day of Negro committees is over and that the appointment of such a committee to implement the Civil Rights Act constitutes a violation of the Civil Rights Act in itself." In mid-September, Thomas Hughes warned Godfrey

*William L. Taylor. Courtesy of
U.S. Commission on Civil Rights.*

that the commission found separate committees "offensive to the spirit and intent of the Civil Rights Act" and suggested that the committees be integrated and weighted with a majority of black members. Godfrey obstinately claimed that the advisory committees were already integrated since they occasionally met jointly with the state committees when civil rights issues were discussed.[4]

William Seabron had a dubious role in creating and staffing these committees. At a meeting in the secretary's office, he strongly opposed all-black committees, but ASCS staff insisted on forming separate advisory groups. Seabron was asked to recommend committee members, but by the time his contacts replied, ASCS staff had already made the selections, some unsatisfactory to Seabron. A year later, Ray Fitzgerald insisted that Seabron had cleared the appointments with civil rights groups. "No appointments were made without his formal approval," Fitzgerald proclaimed. There had been complaints about all-black committees, Fitzgerald admitted, but most concerned the appointment of members who were not farmers.[5] The ASCS staff's insistence on creating all-black committees, its stiffing of Seabron on vetting appointments, and its misrepresentation of Seabron's input epitomized deviousness that would ensure the appointment of compliant committee members. It hardly mattered to Fitzgerald which African Americans served on the committees so long as they did not cause trouble, and he did not need Seabron to help find conservative urban blacks who could be trusted. Seabron apparently did not challenge Fitzgerald's account of his role.

Predictably, whites circumscribed the activities of the black advisory committees. In Mississippi, the state ASCS committee set the agenda in advance and limited discussion to issues relating to nonwhites. Advisory committee members could visit county offices so long as a white staff member from the state office accompanied them, they could offer opinions on policy affecting nonwhites, and they could encourage blacks to participate in ASCS programs. The agenda for the August 27, 1965, meeting included discussion of the Agricultural Conservation Program, the Production Adjustment Program, equal employment opportunity, and committee elections. One question to be addressed was "to what extent . . . county committees [should] consult with leading Negro farmers in determining Negro nominees for positions on community committees." Another was whether black leaders should be contacted about their views on election participation. Such puerile questions, as intended, derailed serious inquiry into discriminatory policies. Georgia's advisory committee met with the state committee on July 16 and focused on orientation, finding qualified blacks for ASCS positions, and travel reimbursement.[6]

By the summer of 1965, many southern communities had witnessed significant shifts in the relations between blacks and whites triggered by the Civil Rights Act of 1964 and federal programs that not only helped African Americans but also increasingly were staffed by them. Elaine DeLott Baker went to Panola County, Mississippi, early in 1965 to help organize the West Batesville Cooperative, and she took photographs of a meeting and many of the farmers and canvassers in the county. Her photographs of the spring meeting showed well-dressed men and women listening to an address by Robert Miles. The West Batesville Cooperative proved an enduring initiative that allowed middle-class farmers to pool supply purchases and cooperatively sell crops. Karel M. Weissberg arrived in Panola County in June to work as a volunteer for the MFDP and used the experience as the basis for her honors thesis at Radcliffe College, disguising people's names and setting the action in pseudonymous Cotton County. The Head Start program initiated in 1965 provoked a struggle for control of federal funds that resulted in the creation of separate white and black programs in the county and was to some extent divisive in the black community. Weissberg was a keen observer and quickly absorbed the black community's class structure. Robert Miles (whom Weissberg named

*Robert Miles speaking at a West Batesville Cooperative meeting, 1965.*
*Photograph by Elaine DeLott Baker.*

James Jones), a wealthy farmer, was the leader of a group of middle-class landowners and businessmen who were active in voter registration, the MFDP, the West Batesville Cooperative, and Head Start. Sharecroppers played a lesser role in civil rights efforts, and women composed about half of those attending the various meetings. SNCC workers encouraged sharecroppers to register and take a more active role in civil rights issues, but with landlords threatening to replace them with machines, many feared losing their tenuous jobs. MFDP meetings started with a song, Weissberg observed, and the organization and parliamentary procedure resembled a church business meeting. Membership in African American churches mirrored the wealth-based class structure, with landowners, the middle class, and sharecroppers finding solace in their separate congregations.[7] Weissberg's class analysis, which showed landowning farmers taking the lead in civil rights, contradicted later ASCS claims that SNCC was ineffective because it focused on sharecroppers.

During the summer of 1965, the NSF, SNCC, CORE, and other organizations were compiling information on ASCS rules and election opportunities. SNCC made careful preparations for the ASCS elections in some Mississippi, Louisiana, Georgia, Arkansas, Tennessee, and Alabama counties, stressing that ASCS committees dispersed large amounts of federal funds, assigned acreage allotments, handled CCC loans, and managed other programs. Since these were federal programs, a SNCC memo noted, "we can hold the Agricultural Department directly responsible for the entire ASCS program and its elections." SNCC's initiative to contest county ASCS committees fit well into its larger program of empowering local people.[8]

SNCC workers Barbara Brandt, Kay Prickett, and others translated opaque election rules into language that farmers understood. The "A.S.C.S. Organizers Handbook" clearly explained the community and county committee election process by imaginatively plotting community divisions on a county map and using stick figures to represent committeemen. It outlined committee duties, suggested how black farmers could win seats, clarified voting eligibility, provided examples of ballots, and supplied a timetable. For years, the Extension Service as well as such agencies as the ASCS and FHA had neglected to explain programs to poor farmers, and it took civil rights workers to decipher obtuse bureaucratic language that had put uneducated farmers at a disadvantage. Brandt later learned that ASCS staff had, with no attribution, used similar graphics in some of its publications.[9] Sadly, many farmers, black and white, were poorly educated, and even sadder, the South was in the midst of a revolution that would shrink the demand for labor-intensive work that had defined southern agriculture for centuries. For many such farmers, the train had already left the station.

As the ASCS elections approached in 1965, civil rights workers pestered USDA officials for information and procedures. Stung by complaints about the 1964 elections and desiring to present a better public stance toward civil rights, ASCS officials from the national to the county level promised cooperation with African American farmers. A new rule instructing county committees to place on the ballot the "names of Negro farmers in relationship to the percentage of Negro farmers in the parish and communities" at first glance seemed advantageous to black farmers, but it led to acrimony as white committees picked a slate of compliant candidates while blacks nominated by petition endured a cumbersome approval process. Voicing what became a common complaint, Joel Horowitz of the West Tennessee Voters' Project wrote from Fayette County that the all-white ASCS committee selected black nominees who "have reputations in the Negro community for cooperating with the whites to the detriment of Negroes." Ballots would thus include black farmers handpicked by county ASCS committees plus successful petitioners, which, as intended, would spread votes among numerous candidates.[10]

In practice, the ASCS's promised reforms did not work smoothly. In Louisiana's Claiborne and DeSoto Parishes, the ASCS committees placed

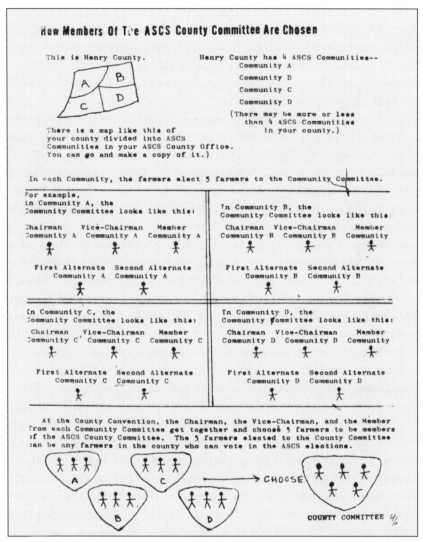

*Stick figures used by SNCC to explain ASCS elections, 1965. Courtesy of SNCC Papers.*

the names of deceased black farmers on the ballot, forcing new elections. CORE's Harold Ickes confronted Madison Parish's ASCS office manager, James B. Stewart, about why black farmers' wives had to obtain their ballots from the ASCS office instead of receiving them by mail like white women. Stewart's "courteous manner" soured, Ickes reported, and Stewart told him "to mind my own goddam business and he would run his office." Ickes complained to William Seabron that Stewart was ignoring ASCS regulations, and he strongly objected to Stewart's profane language. The complaint caused consternation in Washington, and Seabron dismissed a draft reply as unresponsive on several counts, including not addressing Stewart's profanity. In the final reply, Seabron assured Ickes that all ballots, for both men and women, should be mailed, and he reported Stewart's denial of his use of profanity and shamelessly boasted of the high caliber of ASCS employees.[11]

First reports indicated that the Louisiana ASCS election went smoothly and that mail ballots were counted with CORE observers present. Ickes helped tally votes in one parish office, and observers even took photographs in Caldwell Parish. ASCS personnel quickly undermined CORE's effectiveness. According to Wainwright Blease, "The pre-election activity by CORE brought little if any increased participation by Negro farmers. It did increase participation by white farmers, especially where CORE was most active." Increased participation of whites, of course, resulted from the CORE challenge. The Louisiana ASCS election yielded fifteen African American alternates and two community committeemen.[12] Blease's observation illuminated the plan to encourage more white participation while tactically reducing the black vote, always keeping a worried eye on civil rights workers.

Despite civil rights laws, black farmers often endured humiliating treatment. St. Landry Parish's African American farmers distrusted ASCS acreage allotments and measurements, but they had no recourse since they were forced to avoid the ASCS office run by a Mr. Wyble. Blacks had to take off their hats upon entering the courthouse, and Wyble often threw black farmers out of his office, refusing to inform them of ASCS programs. When they did apply for program benefits, Wyble told them there were no funds. A Mr. Foret replaced Wyble before the 1965 ASCS election and announced that the ASCS committee would place African Americans on the ballot. Many black farmers protested, reasoning that "those candidates who would be handpicked by the present committee could be controlled by those same people if they were elected." Foret sent ballots only to owners, not to eligible sharecroppers and tenants. When a disgruntled delegation went to his office, he announced that tenants and sharecroppers could come by the office and get ballots. After farmers tied up the office with demands for ballots and complained to the USDA about discriminatory policies, shortly before the election, Foret relented and mailed ballots to sharecroppers and tenants. Some of the black farmers placed on the ballot only learned of their nomination when they received ballots. The election lasted from August 1 through 10, and black farmers distrusted security since the ballots were not guarded after being returned to the office. As the election progressed, CORE workers circulated around the parish and learned that many sharecroppers, women, and tenants had not received ballots. All of the black candidates lost. CORE leaders reasoned that they had not started their campaign in time to deal with the issue of sharecropper, tenant, and wife ballots, nor had they focused efforts on problematic communities. Still, in a fair election, they would have done better.[13]

As Louisiana's 1965 ASCS elections moved ahead, whites continued to exact a price for black activism. Jack Allums and his wife got on the wrong side of whites in 1957 when he attempted to register to vote and she testified in New Orleans before the U.S. Commission on Civil Rights. Later that year, police stopped their son for speeding and beat him. In the summer of 1965, after a CORE worker was killed on the streets of Minden, Jack Allums transported CORE workers to a sit-in. The industrious Allums farmed, as well as working as a school patrolman, a drayman for the L & A Railroad, and a janitor for Montgomery-Ward. According to NSF field representative Ocie Lee Smith, Allums "was fired from each job, in succession. The intimidation was terrific." Someone had attempted to force Mrs. Allums's car off the road. Neither a local bank nor the FHA would lend Jack Allums money. The Allums family had a nice three-bedroom house with indoor plumbing. The couple had nine children, and one daughter was set to attend Southern University in the fall. Allums owed money on his tractor, but he owned a truck, a heifer, a mule, and a horse. "The Allums family," Smith reported on October 22, "has been existing on donations of all kinds, by community people." On December 13, William Seabron passed this narrative to the USDA inspector general and asked for an investigation.[14]

Many ASCS offices seemed inefficient by nature, but when faced with civil rights challenges, they became adept at purposeful ineptness. Will McWilliams, the office manager in Holmes County, Mississippi, finally furnished a list of eligible voters the day ballots were mailed, and only then did black farmers discover hundreds of missing names. In addition, black farm owners' wives did not receive ballots and were compelled to bring proof of their eligibility to the office. Farm owner Ralthus Hayes was chairman of the county MFDP at the time and traveled to Washington to complain. The county committee, instructed to send challenge ballots to those left off the list, did not comply with the order, and the MFDP advised farmers to obtain challenge ballots at the ASCS office. "It takes a courageous and protesting man," Hayes pointed out, "to stand up and demand his ballot when it is obvious that the white man, either the ASCS manager or his 'owner,' has kept him from getting his ballot in the normal manner." The ASCS office grudgingly gave out ballots but required contracts, receipts, and other records that many African American farmers who relied on verbal agreements lacked. ASCS vote counters used petty excuses to toss out black ballots.[15]

First-term congressman Joseph Y. Resnick, forty-one years old and hailing from a rural New York district, accepted the MFDP's invitation to visit Mississippi and observe the ASCS elections. Reporter Jack Anderson, who delighted in covering political junkets, praised Resnick for traveling to Mississippi to investigate discrimination. Most of Anderson's column, though, garishly reported increasing use of marijuana on college campuses, the spread of "beatnik garb," and references to sex. Anderson and many middle-class citizens viewed beatnik culture with distaste. Southern whites attached similar descriptions of beatnik attire, drugs, and sex to civil rights workers, and the Mississippi State Sovereignty Commission obsessed over possible sexual relations between black men and white women. Opportunistic whites, of course, seized on any demeaning image to discredit civil rights activists. Congressman Resnick, though, ignored negative portrayals of civil rights workers. He gave a talk at Tougaloo College on November 29, and the next day, he toured Issequena, Washington, Holmes, and Madison Counties. In Issequena County, he learned that black farmers who were denied FHA loans and were active in ASCS elections had their insurance policies cancelled. In Washington County, he visited a tent city of workers evicted from the Andrews plantation.[16]

Much to the delight of Holmes County blacks, Resnick grilled Will McWilliams, the Holmes County ASCS office manager, and caught him out when he claimed to have mailed ballots to black farmers' wives. McWilliams had refused to furnish challenge ballots unless farmers presented deeds, his voting rolls were not up-to-date, he did not accept gin receipts as proof of sharecroppers' eligibility to vote, and he was derelict in other duties. When Resnick learned that black women had to furnish proof of farm ownership to vote whereas white women simply got ballots by mail with no questions, Unita Blackwell and Annie Devine reported, "he called up Mr. McWilliams, the office manager, right then and there, and said he had a lot of people who should get ballots and had not gotten them." He put them on the phone to furnish their names and addresses. "The Freedom Democratic Party people were really impressed with the Congressman, and his concern about the elections," Blackwell and Devine wrote in their report. Unita Blackwell joined the movement in Issaquena County and had gone to Atlantic City with the MFDP. Annie Devine was a powerful leader in Canton, a former schoolteacher, a staunch church member, and an agent for a life insurance company, and she was contesting a U.S. House of Representatives seat. Resnick explained in a letter to Secretary Orville Freeman that he corrected McWilliams when "he gave a deliberately misleading explanation of which wives were entitled to a ballot," and he complained that McWilliams "refused to address the Negroes properly, insisting on calling them by their first names."[17]

That night, Resnick spoke in Canton about the irregularities he had discovered in ASCS voting, and on December 1, he held a press conference in Jackson. "Resnick announced that he had found so much discrimination that he was going to recommend that the elections be voided and that new elections be held under federal supervision," Blackwell and Devine noted, "unless ASCS officials did something to correct the irregularities." Resnick mentioned the evicted farmers he visited and reported estimates that "10,000 to 12,000 people will be thrown off the land this spring because the owners will no longer have any use for them. They will be replaced by mechanical cotton pickers." When asked why a New York congressman was visiting Mississippi, he replied that many of the folks said "he was the first Congressman they had ever seen."[18] He also seemed to be the only federal official to look closely at local ASCS elections and demand accountability. Mississippi's white politicians paid little attention to black farmers, and until Resnick's visit, no federal official had delved into the fraud surrounding ASCS elections.

Predictably, Horace Godfrey claimed that Resnick "listened only to one side without checking with State and county ASCS representatives to determine the truth of the allegations which were made," ignoring the fact that Resnick witnessed the abuses firsthand. In December, Ray Fitzgerald had ruled that several of the people for whom Congressman Resnick demanded ballots were ineligible, but he disregarded the broader charges of discrimination.[19] The press coverage of Resnick's tour and his letter to Freeman had little effect on Mississippi's ASCS offices. From Washington to rural counties, ASCS officials blithely reported that their programs worked harmoniously and provided equal opportunity, and they ignored evidence to the contrary.

Resnick's intervention did not occur in time to prevent Will McWilliams's sloth, lies, evasions, and duplicity. He mailed contested ballots to Holmes County farmers two days before the deadline to return the ballots, so most did not even receive ballots in time to vote. Although ASCS officials stated that the vote-counting process was open to all, they only allowed four observers in the counting room. COFO observers reported having to pass by ASCS secretaries who "kept a grudging eye to make sure only four were in at the time." Meanwhile, white farmers walked in and out at will. Ballot counters ignored tardy challenge ballots, strictly proctored the spelling of names, and found fault with the marks of illiterate farmers. That a federal office tasked with parceling out acreage, settling disputes, making loans, and holding annual elections lacked an eligible voters list, stonewalled black farmers, fudged community boundaries, and discriminated against black farmers suggested that over time ASCS committees had, in addition to exhibiting widespread incompetence, become cynical and ruthless in maintaining power. They in effect dared state and national offices to interfere.[20]

Ralthus Hayes's complaint to the USDA in December 1965 about irregularities in the Holmes County election led to an Office of the Inspector General (OIG) investigation, but the copy of the report that was promised to Hayes had not arrived eight months later. The Holmes County ASCS office, he charged, "used an incomplete and inaccurate list of eligible ASC voters." Since the next ASCS election was approaching, he stressed in August 1966, black farmers had already been to the ASCS office looking for an accurate list of voters. "We were not allowed to see anything," he complained. "We were told that the list had not been compiled *yet*," and the staff added that it would be at least a month before it was ready. William Seabron looked "with disfavor" on a September 9 ASCS draft to Hayes that admitted to no deviation from ASCS policy in the 1965 election. Yet in his final draft, Seabron explained away eligibility issues, wrestled unsuccessfully with the complex issue of challenged ballots, and reported that the OIG had found no irregularities or discrimination in the election. Seabron's final reply included a labored conclusion that "investigation reports do not substantiate the allegations submitted in your letter of complaint." Given the unforgivably tardy replies, misleading drafts, and failure to share the OIG report, the ASCS obviously had no intention of admitting violations in Holmes County. Seabron offered no explanation for not forwarding the OIG report, although it surely would have substantiated Hayes's complaint.[21]

Washington bureaucrats understood how county committees manipulated elections, but they refused to take corrective action or, in the case of Seabron's letter to Hayes, even to admit bias when faced with overwhelming evidence. Kenneth M. Birkhead, assistant to the secretary, succinctly summed up how county ASCS committees in South Carolina swayed elections. The committees placed the minimum number of whites on the ballot and as many blacks as possible, thus diluting black votes and concentrating white votes. This strategy, he suggested, put the better class of white farmers on the ballot and the poorer class of blacks, but, he insisted, "there is no evidence of discrimination." The same questionable tactics were being used across the South with the blessing of the ASCS.[22]

In Clay County, Mississippi, where no black candidates won community ASCS committee seats in 1964, the MFDP held precinct meetings in the summer of 1965, began organizing in October, and by November 6 had twenty-one candidates. James Mays of the NSF spoke at one meeting, and there were follow-up workshops. After the ballots were sent out on November 22, MFDP workers visited farmers to make sure they received ballots.[23]

MFDP workers in Clay County left a telling record of their work on the 1965 ASCS elections as well as a dismaying account of threats and violence. On August 8, a volley of rifle and shotgun fire left fifty-one holes in the Freedom House in Pheba. Sheriff Joe Ed Strickland responded by studying the tire tracks. The next Friday evening, a white man barged into Freedom House and held a young black worker at gunpoint until he broke away and fled to a car with a friend. Whites chased them and broke the car's windshield. FBI agents arrived Saturday morning and "interrogated rudely and would not let the youths speak." On Saturday night, whites again fired at Freedom House, but this time blacks returned fire. They knew who owned the white shooters' car, and Sheriff Strickland escorted two of the three suspects to Freedom House, where they imaginatively claimed that they had shot not at Freedom House "but at a rabbit that had crossed the road in front of their car as they were passing the Freedom House." When an FBI agent arrived, he insisted that "the FBI could only investigate." Civil rights workers had no one to turn to for support. "To visit SNCC field headquarters in these rural outposts of the Deep South," the historian and activist Howard Zinn observed, "is like visiting a combat station in wartime."[24]

After the election, MFDP worker Ada Mae Holliday sent a reasoned and poignant letter to the USDA. She insisted that the ASCS committee was "extremely important to our livelihood inasmuch as our entire existence is dependant upon our farms," and she provided insight into the workings of the Clay County ASCS office. "The ASCS committee (we are being generous in presuming that the committee has some voice in what the Office Manager, Mr. Emmett Hood, does) represented by Mr. Hood and the ladies in the office, have done everything in their power and out side of it (which is part of our objection) to keep us from running these candidates." This critique resonated with accounts of other African Americans about encounters with office managers and white women office staff.[25]

African Americans complied with Hood's unreasonable demands and deadlines and had everything in by November 17, the "last possible date to complete the slate of nominees." When Hood saw that twenty-five black candidates had met the requirements, he evidently panicked and, violating his own strict deadline, canvassed communities on November 18 searching for more black candidates to divide the vote. Ada Holliday took pride in her work with the MFDP. "We do not intend to let this work and effort go in vain," she warned. "We have risked the burnings of our homes and churches and the loss of our lives. We have been harassed and intimidated by our white neighbors and people working with the current ASCS committee." She pointed out that Sheriff Joe Ed Strickland, who planted 233 acres to cotton, and ASCS administrator Emmett Hood, also a planter, as well as a substantial list of large growers, had an interest in maintaining the white elite's control of the committee. These people would encourage violence, she suggested. "They will not be the ones to shoot, the ones to stomp through car windshields, the ones to burn crosses, but as in all Mississippi rural counties, they will be the ones to encourage and condone these acts." Holliday insisted that the black candidates be seated because they would have won if Emmett Hood had not stooped to illegality.[26]

As ASCS administrators sorted through complaints, Wainwright Blease suspected that although black farmers who signed with their mark caused problems with county authorities, he had learned that "a majority of the ballots declared ineligible in the Delta area were for those of partnerships and corporations" where the signer did not specify his office. This juxtaposition suggested the gap between illiterate black farmers disqualified for not affixing their X properly and inattentive, perhaps unconcerned, executives at Delta planting corporations. Basing his decision on the word of county ASCS staff, Ray Fitzgerald in mid-December dismissed all Clay County complaints. It took Inspector General Lester Condon, however, eight months and four pages on July 25, 1966, to dismiss Holliday's accusations, concluding that his investigation "failed to substantiate the charges you brought against the ASC Clay County Committee."[27]

William Seabron's reply to Ada Mae Holliday on July 25 was a disingenuous masterpiece that cited obvious discriminatory incidents while concluding that there was no bias. Clay County officials unsurprisingly denied that they had obtained additional black candidates after the

November 17 deadline, yet the OIG investigation supported Holliday's charges that on November 18 ASCS officials asked thirty African American farmers to run. The investigation revealed that the county office manager sent out letters without courtesy titles, but Seabron lamely stated that courtesy titles were used thereafter. Clay County ASCS officials denied other allegations of intimidation, financial threats, and discrimination in awarding acreage allotments. "We are determined," Seabron explained in his four-page letter, "that ASCS community and county elections be administered equitably and without regard to race, creed, color, or national origin."[28]

Constant pressure from civil rights workers resulted in SNCC, CORE, and NSF staff attending a November 9, 1965, meeting of the Mississippi State ASCS Civil Rights Advisory Committee that discussed elections, employment, and programs for poor farmers. While some blacks had been employed in temporary summer jobs, state ASCS director C. W. Sullivan stated that permanent office jobs, primarily typing, would open to African Americans only when whites retired. "The problem is," he added, "that there aren't enough of them qualified to fill those jobs," although a simple typing test was the only requirement and it was unlikely that black college graduates lacked typing skills. Barbara Brandt from SNCC observed that the advisory committee was composed of "rich Negroes" who believed in uplift and nodded affirmatively at white suggestions. After the meeting, she invited African American Victor Phillips, who had worked at the USDA for a dozen years and was a civil rights assistant for the ASCS, to meet with other SNCC workers. Phillips grew up in Dover, Oklahoma, and he graduated from Langston University and completed Ph.D. course work at Oklahoma State University. They asked him for election observers from the USDA since they did not trust county ASCS executives. The USDA could not send observers, Phillips pleaded, "because they were understaffed due to lack of funds." He recommended asking Horace Godfrey for federal observers.[29]

Jan Maedke's complaint from Carroll County, Mississippi, suggested both subterfuge and incompetence. "We feel that the entire election was mishandled by the Carrolton ASCS office," she noted. In addition to the handpicking of black candidates and the failure to give public notice of the election date, ASCS records were in chaos, as they had been the year before. Walter McGloithan and his wife received return envelopes for different communities, so he went to the ASCS office to straighten it out. The secretary explained that Mrs. McGloithan had received the wrong envelope, so her vote would not have counted. Some envelopes, the secretary admitted, were mailed out by mistake. "The counting gave me the impression that the entire election had been run in a careless manner and that the ASCS office would not admit that it was at fault," Maedke concluded. Carroll County's ASCS office, along with others, had mastered malign inefficiency.[30]

In Alabama, SNCC's Elmo Holder began preparations in May for the fall 1965 ASCS election, asking the NSF for pamphlets and for a volunteer to meet with farmers and explain esoteric USDA policies. The previous year, the Alabama SNCC staff had focused on five counties shortly before the election, which was conducted by mail ballot. Reports of ASCS staff failings were similar to those in Mississippi, as was violence; the home of a black candidate was burned in Lowndes County. SNCC worker John Liutkus met with top ASCS officials from Washington and Alabama and asked that candidates and those observing the count receive protection from violent whites. The request went to Ray Fitzgerald in early November, and he at first rejected it but then "made one meaningless gesture: he said that a state official would be in a county office at the time that the ballots were counted." The SNCC staff might have done more, but there was no money for gas, no paper for posters, not enough cars, and few telephones.[31]

After reading the NSF pamphlets, Holder requested information on the election from Alabama ASCS director B. L. Collins, and he was not satisfied with the reply. "We found your answers," Holder replied on July 19, "most unsatisfactory and uninformative." Collins had failed to clarify eligibility or residence requirements, to explain the term "insufficient participation," to guarantee ballot security, to set a date for the election, to set community boundaries, or to explain why white ASCS committees placed black farmers on the ballot. "In areas where Negroes put up their own candidates," Holder argued, "there is absolutely no legitimate reason why the committees should put up additional Negro candidates." Having a large number of African Americans and few whites on the ballot "will only be a device by which the Negro vote could be split and the whites will retain total control of the committee." Setting a date was important, Holder reasoned, because black farmers who petitioned to run "would not want to have publicity until the last possible moment—because publicity will mean reprisals from the white community." Holder asked that SNCC representatives be allowed to attend state ASCS meetings "so that we might keep abreast of public business."[32]

A consternated Collins telephoned E. W. Bayol, director of the southeast area, about the SNCC problem. Bayol grew up in Hale County, Alabama, and had worked at the USDA since 1933. Collins pleaded that he was in "no position to change regulations" or accommodate Holder's requests, dolefully suggesting that they "go beyond my authority." If Collins was unable to explain eligibility, community boundaries, ballot security, the term "insufficient participation," the election date, and whites putting blacks on the ballot, it suggested alarming incompetence. Bayol sent Collins relevant memorandums. On August 24, Holder and his wife Susan called on Collins to clarify ASCS rules. After discussing the issues, Collins explained to the Holders that SNCC tactics generated irritation and suggested how SNCC might soften its approach. "Our discussion was dispassionate throughout," a relieved Collins wrote to Bayol.[33] Collins had evidently envisioned Susan and Elmo Holder as dangerous radicals, and he was dreadfully anxious about meeting with them.

Gaining input into federal agricultural programs was part of a larger SNCC agenda to make federal programs available to African Americans. Since few black farmers understood the complex election regulations, much less voted in ASCS elections, starting in July 1965, SNCC held workshops even though guidelines furnished by the ASCS were confusing and contradictory. As the election approached, SNCC workers spoke in churches and at other gatherings about why participation in the ASCS was important. SNCC leader Stokely Carmichael recalled that he worked with eligible voters in Lowndes County and they nominated four African Americans from one community and five from several others. When voters received their ASCS ballots, however, the five community ballots contained 36, 17, 68, 29, and 9 African American nominees for the community seats, hopelessly confusing some black voters. Some black farmers received ballots for the wrong community. While Carmichael said he had "always been treated courteously" at the ASCS office, he reported that one black farmer who attempted to get a proper ballot "was chased out of the Lowndes ASCS County Office."[34] Carmichael's experience in Lowndes County revealed how intense canvassing, placing black farmers on the ballot, and teaching farmers how to fill in their ballots could be derailed by an ASCS office that, while friendly, sought to sabotage the black vote.

In Greene County, the ASCS office disqualified eleven candidates too late for them to appeal and gave black farmers incorrect boundaries for their communities. Four tenant families were evicted because they signed petitions for candidates. SNCC's John Liutkus appealed to state ASCS director B. L. Collins to have someone on hand to monitor the vote counting in Greene County. When Liutkus arrived for ballot counting, the Greene

County ASCS chairman told him, "The next time you call Collins, tell him he has no goddamn say in my county, and this is my city." The chairman's attitude revealed both the power and the hubris of ASCS committee members and suggested the tight, jealous control that committeemen and office managers exerted in rural communities. It was June 1966 before Seabron replied to Carmichael about complaints in the 1965 Greene County ASCS election, and predictably, he said the investigation revealed that "there were no irregularities committed by the Greene ASC County Committee in their conduct of community committee elections."[35]

Some ASCS staff were less than subtle in expressing their hostility to black voters. "The representative in Wilcox County told the county committeemen, in the presence of a SNCC worker, that he had been sent down to give the committee moral support," a SNCC report noted, "making it clear that complaints of irregularities in the elections would not be acted upon by him." African Americans made up 75 percent of the Wilcox County population, but blacks won only ten positions on community committees. In three communities, 120 voters cast ballots for the same three white men, suggesting astounding organization. In Hale County, the movement put up seven candidates, all disqualified, and several families active in ASCS organizing were thrown off plantations.[36]

County ASCS committees and office staffs frustrated African Americans by their failure to supply information, their duplicity, and their discriminatory treatment. SNCC's newsletter referred to Baker County, Georgia, as "Bad Baker," in part because it was the site of the infamous Screws Case. Sheriff Claude Screws and other law enforcement officers were prosecuted for beating to death Bobby Hall, a young black mechanic, while he was handcuffed. Screws was acquitted, and the memory festered in the African American community. According to a telephone complaint from SNCC Baker County fieldworker Isaac Simkins and a letter from SNCC's Fred Anderson, the ASCS office failed to publicize the fall 1965 election, and only black landowners, not eligible tenants or sharecroppers, received ballots. SNCC sponsored a write-in campaign for three candidates, but ASCS committeemen arbitrarily placed several other African Americans on the ballot to confuse voters and dilute voting strength. On vote-counting day, September 27, Anderson found the county agent's office deserted at 9:00 A.M., and later the office staff said they "knew nothing of any vote-counting." SNCC workers had helped forty-four black farmers prepare write-in ballots for three candidates, but ASCS vote counters tallied only twenty votes for the SNCC slate, "all in one way or another invalid." The Georgia ASCS office denied wrongdoing and shamelessly claimed that ballots had been counted at the ASCS office at 9:00 A.M. on September 27. As for the black farmers placed on the ballot, "these committeemen from their intimate knowledge of farmers in their respective communities determined that the Negro nominees were the type of individuals who were well aware of ASCS programs, the committee system and, if elected, would be willing to serve." As in so many instances of white privilege, the committeemen saw no reason to ask black candidates if they would stand for office. Whites knew "the type of individuals" they wanted to run. William Seabron found the ASCS report "inadequate" for it failed to address Anderson's questions, and he requested an OIG investigation.[37]

After the secretary's Citizens Advisory Committee on Civil Rights held its fourth meeting of the year in early December 1965, Seabron summed up its activities and digested progress reports from the ASCS, FHA, and FES. Stung by charges that African American farmers did not receive information, USDA offices were contemplating using TV, slide shows, and a pamphlet, "You Too Can Get Help." The ASCS boasted of African American state committeemen in Arkansas, Maryland, and Mississippi; black advisory committees; and summer hires as proof of civil rights progress. It claimed falsely that placing blacks on the ballot "succeeded in increasing the number of Negro community committeemen." The FES smoothed over compliance bumps and claimed far more integration than existed. The FHA went over the top in claiming "real progress" and predicting "even greater accomplishments as the program gathers greater momentum and public support." None of these reports mentioned continuing problems that plagued the USDA.[38]

Such evasion and incompetence came to the attention of the Civil Rights Commission's Georgia advisory committee, which held hearings in Macon in 1966 to follow up on the commission's 1965 report. It concluded that African Americans "receive substantially less than is constitutionally guaranteed them as Americans." From early school years, black children were denied services and opportunities and later denied participation in programs and jobs. "To the day they retire or die," the report stressed, "Negro farmers in Georgia experience second-class citizenship unknown to whites." Black farmers were addressed as "boy" or "girl" and treated by USDA agencies "with noxious difference." Black farmers throughout the state "feel they are being phased out as farmers in Georgia."[39]

Instead of holding state and county officials accountable for continuing discrimination, Horace Godfrey launched a public relations campaign in January 1966. To ensure positive treatment in the press, he notified top ASCS administrators that in the future the agency would issue regular press releases "on activities generally included under civil rights and equal opportunities." He added that offices need not bother "unless there is something positive to report" and listed several examples of "success types," including a young black man who worked during the summer "and as a result was able to return to college in the fall and complete the work for a college degree" and a black farmer who participated in the cotton-loan program and sold his cotton for a profit. Such examples omitted that blacks had only recently been hired for summer work and that, because of discrimination, black farmers had been denied information on most USDA programs. A month later, another memo reminded state ASCS committees to send material, for "this request gives all of us a wonderful opportunity of making known the results of your tireless efforts in administering the ASCS policies on civil rights."[40] Burying discrimination under success-story headstones complemented the passive nullification tactic that claimed USDA compliance in civil rights matters. Improving public relations became more important than protecting minorities.

In January 1966, as Secretary Freeman prepared for a Capitol Hill hearing, Godfrey provided him with a remarkable briefing paper defending ASCS committees and offering a dubious justification for denying county committee employees civil service status. "We believe that placing county committee employees under the Civil Service system," he imaginatively theorized, "would destroy the local control necessary for effective county committee operations." The lack of federal employment guidelines outside the civil service system provided cover for continuing discrimination, as Godfrey well understood. Later in the year, Kenneth Birkhead reminded an undersecretary of agriculture of approaching appointment deadlines to state ASCS committees and stressed that the secretary's Citizens Advisory Committee was keen on black appointments. Failing to appoint African Americans, he cautioned, "will be another clear evidence that the Department is not really interested in civil rights." As if to illustrate disinterest, the USDA director of personnel ignored Birkhead's suggestion, as, evidently, did the secretary.[41]

By 1966, African Americans were realizing benefits from the 1964 Civil Rights Act and the Voting Rights Act a year later. In January, President Johnson appointed Robert C. Weaver as secretary of housing and urban development, making him the first black cabinet member, and Edward Brooke that autumn was elected the first black U.S. senator. Still, violence continued, and in 1966, Mississippi civil rights leader Vernon Dahmer was killed when his house was firebombed. In May, Stokely Carmichael was elected chairman of SNCC. During the summer, riots exploded in Cleveland, Lansing, Waukegan, and Atlanta, and antiwar and antidraft protests generated increasing opposition to the Vietnam War. In the rural South, African American farmers continued their efforts to win seats on county ASCS committees, often in obscurity.

Even as Seabron and his staff worked on ASCS election issues in the South, Aaron Henry, a member of Secretary Freeman's Citizens Advisory Committee, alerted President Johnson in February 1966 to the deteriorating conditions among black farmers. Henry had emerged as an outspoken civil rights advocate. After service in World War II, he went back home to Clarksdale and opened a drugstore. In 1960, he was elected Mississippi NAACP president, but he had already surpassed its cautious civil rights approach. "Here by a combination of Agriculture Legislation, Automation, and Racial prejudice," he explained to the president, "Negro farmers by the hundreds are being told that there will be no work for them on the plantations this year." He blamed a 35 percent reduction in cotton acreage. Henry suggested that Johnson distribute surplus federal and state property, make surplus food available, and initiate jobs. "I do have to remind you," he stressed, "that to take these steps will require courage and determination that in many instances will upset the local political power structure."[42] Johnson had already refused to anger white Democrats when he humiliated the MFDP at Atlantic City, as Henry, one of the MFDP leaders, well knew.

Seabron visited Henry in Clarksdale on February 7 and met with forty farmers at Haven Methodist Church across the street from Henry's drugstore. Some of the larger operators seemed to be "getting on" but "complained of unfairness in obtaining true weight for their cotton at the gin." Many renters and sharecroppers dejectedly announced that they would not be farming in 1966 because of mechanization and the 35 percent cotton-acreage reduction. "One reported that his boss in Quitman County had pushed all of the shanties on the plantation in one heap and set them on fire." Some owners and tenants had bought farming equipment, but due to distressing economic conditions, they were facing repossession. "Maybe the Department should declare a one-year moratorium on FHA payment for farmers caught in this situation," Seabron suggested. Both Henry and Seabron saw the dire consequences of the transformation that was decimating rural life. "Generally speaking," Seabron stressed, "their stories were of despair, frustration, and uncertainty as to their future and their children's."[43]

In May 1966, Jac Wasserman appealed to deputy ASCS administrator

*Aaron Henry, 1971.*
*Photograph by*
*Pete Daniel.*

Ray Fitzgerald to implement new regulations for ASCS community elections. Wasserman observed that many sharecroppers and tenants as well as farm owners did not receive ballots (in the margin, Fitzgerald wrote, "This is being done"). Wasserman urged the ASCS to post a list of eligible voters ("no" in the margin) and clarify how many candidates a farmer could vote for ("we do provide this" in the margin). Wasserman suggested seven changes, but the final ASCS reply amplified Fitzgerald's marginal comments. None of the changes Fitzgerald proposed would challenge the power of county committees to manipulate elections.[44]

Wasserman shared Fitzgerald's suggested changes in ASCS election procedures with Louisiana CORE worker John Zippert, who wrote to Fitzgerald that they were "wholly unacceptable to me and to the many Negro farmers I am working with." He complained that the 1965 election was "rigged" and that "hundreds of eligible voters were prevented and defrauded from voting." Zippert ridiculed Fitzgerald's claim that the ASCS would attempt to gather names and addresses of eligible voters while simultaneously stating that it would be excessively costly to compile such a list. He asked why some 3,000 unverified sharecroppers were compelled to get ballots at the ASCS office while landowners received theirs in the mail. "It is very difficult to understand just exactly how you are going to run any sort of election much less a fair one," Zippert mused, "not really knowing who is eligible to vote." He suggested that "these anachronistic regulations" be revised to ensure fair elections "over the South but especially in Louisiana." On a larger scale, Zippert added, "the present system of ASC elections [has] perpetuated in office a group of people totally unresponsive to the needs of the small farmer and the sharecropper." Fitzgerald replied with the usual empty assurances.[45]

Meanwhile, analysis of the 1965 Lowndes County, Alabama, election dragged on until May 1966, when William Seabron caustically reminded the DASCO office of the issue. "Because of the blatancy of the form of discrimination and the seemingly total lack of good faith on the part of the County Committee, I can see no other recourse than to void the election and schedule new elections as soon as possible." The ASCS had tested Seabron's patience by seizing on a preliminary OIG report, misconstruing it, and announcing that the election had been proper. SNCC's Stokely Carmichael was not assuaged. "We are sick and tired of the tricks that this racist government—from federal to local—attempts to play on black people," he complained in July 1966. He insisted that the federal government monitor the Lowndes County ASCS election that fall, for if left to local authorities, it "will be a fraud." Carmichael's letter also reflected concern that Washington's focus roamed far from southern cotton fields. "If the government can spend billions of dollars to kill people in Vietnam to assure free elections," he wrote, "then they had better spend some of those dollars to assure free elections in the Lowndes County ASCS." He added that he was also posting his letter as a press release. A U.S. district court extended the dates for mailing Lowndes County ballots, and Ray Fitzgerald assigned staff from both Washington and Montgomery to monitor the election. It was mid-August 1966 before Seabron's much-vetted letter assured Carmichael that federal officials would ensure a fair election.[46]

The USDA's unresponsiveness to discrimination prompted the Lawyers Constitutional Defense Committee (LCDC), an arm of the American Civil Liberties Union founded in 1964 to provide legal assistance for promoting civil rights issues in the South, to file suit in behalf of thirty-seven black farmers from eleven Alabama black belt counties. Donald A. Jelinek, an attorney for the LCDC, prepared the case, *William v. Freeman et al.*, which listed discriminatory practices that included denying African Americans the vote, conspiracy among large landowners, nominating blacks without their consent, violence, coercion, and failing to furnish information. Jelinek had given up a Wall Street legal position to join the civil rights movement, first in Mississippi when Aviva Futorian convinced him to remain in the South. The suit asked for postponement of the August 16 election in order to correct problems resulting from the 1965 election, and then it requested two elections, the first a re-vote in Lowndes County for 1965 and then the regular 1966 vote adequately supervised by federal officials.[47]

Thirty farmers involved in the suit traveled to Washington for a hearing, and Peter Agee, a fifty-year-old farmer and sculptor from Marengo County, testified that a white man running for the ASCS committee offered him thirty acres and some cows if he would get out of the race. The delegation wanted postponement of the election so they could organize a challenge. The impatient judge hurried the hearing but agreed to move the election from August 15 to September 15. The Saturday after Agee returned to Marengo County, white men fired shots at his store and he fled the county. Afterward, some white men told his sister there was a body out by the barn and cruelly suggested, "It might be your brother."[48]

The civil rights landscape was shifting as Stokely Carmichael led SNCC further away from old-line civil rights organizations and their white allies and focused on color rather than class. He articulated SNCC's frustrations with the Johnson administration's focus on war at the expense of civil rights and the president's shabby treatment of the MFDP at Atlantic City. After James Meredith was wounded in early June 1966 during his march across Mississippi, SNCC repudiated the NAACP's Roy Wilkins, criticized Martin Luther King Jr., and vowed to complete the march. In mid-June, a month before he wrote his letter challenging the Lowndes County ASCS election, Carmichael was released after what he announced was his twenty-seventh arrest, which occurred in the spring of 1961 when, as one of the freedom riders, he entered a white waiting room in Jackson and then spent forty-nine unpleasant days in Parchman Penitentiary. He spoke to a crowd of supporters in Greenwood, Mississippi, and began chanting "Black Power," answered by the crowd's roaring echoes. The slogan reached deep into African American frustrations and chal-

lenged not only the white power that had erected segregation and refused to tear it down but also civil rights leaders who shied away from
Carmichael's rhetoric. The concept of Black Power sent chills down the
spines of whites, who imagined in Carmichael and SNCC the ghosts of re-
volting slaves and saw visions of burning cities. As the term spun around
the country, it became encrusted with self-serving interpretations, for
some, a term of pride and hope, and for others, an incitement to violence.
Carmichael became a celebrity, a status most SNCC leaders had shunned,
and his belligerent words alienated many of the increasingly urban staff.

John Lewis, Charles Sherrod, and Julian Bond resigned from SNCC in
1966, and in the spring of 1967, the last whites departed. For whites, leaving
SNCC proved emotional. Jane Stembridge recalled that even during the
most dangerous times, the staff "talked, refused to talk, laughed, cried,
got mad, slammed doors, got drunk, whatever." The next day, they would
be out canvassing people to vote. "Leaving SNCC was one of the most
painful things that ever happened to me," Dorothy M. Zellner remembered. "It meant not only the loss of what I thought was my life's work,
but it was also the end of my five-year proud association with SNCC, the
most creative, funny, innovative, daring, fearless group of people I ever
met."[49]

Just before the 1966 ASCS elections, Virgil Dimery, writing for the Wil-
liamsburg County Voters League, a South Carolina citizens group with a
broad agenda, appealed to Secretary Freeman to ensure fair ASCS elections in 1966. The year before, Dimery explained, the group had "invested
hours of time and a great deal of energy in order to insure more equal
and fair administration of the programs of the Department of Agriculture
in this county." The ASCS "thwarted" its efforts. "The local director, it is
reported, covered the countryside inviting Negro farmers to place their
names on the ballot." Some thought this meant they were already on the
committee. In Dimery's opinion, this was "a deliberate attempt" to prevent blacks from being elected. One African American was elected to a
community committee. Dimery asked Freeman to investigate the irregularities from 1965 and assure blacks of a fair 1966 election. Fitzgerald told
Seabron that an OIG investigation would be requested and that Dimery
would be informed of the changes in the ASCS election rules, including
adding black candidates, the very thing he had complained of.[50]

The NSF's Fay Bennett, of course, looked beyond the ASCS elections
to the changing structure of farming. She sent Fitzgerald a report, "The
Condition of Farm Workers and Small Farmers in 1965." Predictably,
he evaded the report's implications that government policy displaced

farmers and claimed that the decline of black farmers "may be accounted for by mechanization and the growing demand for and use of synthetic products which have taken the place of cotton." Bennett's report also revealed that African Americans elected to community or county committees were not informed of meetings and thus were excluded from even a modicum of input. "This information has not previously come to our attention," Fitzgerald lamely replied, and he asked for specific examples. He impudently assured Bennett that "all programs are administered equitably and without regard to race, creed, color or national origin."[51]

The ASCS had moved haltingly to hire African Americans, and Fitzgerald reported to Horace Godfrey in June 1966 that he was "continuing to conduct a vigorous program to employ minority group members." He mentioned a handful of positions held by African Americans in ASCS offices, boasted of temporary hires in state and county offices in 1965, and predicted that many temporary jobs would go to African Americans in 1966. The Mississippi training program, he insisted, had great potential, and Arkansas, Louisiana, Alabama, Georgia, North Carolina, South Carolina, and Tennessee were instituting similar programs. "A few weeks ago," he announced, "I employed a female Negro college graduate to work in my outer office. She is extremely capable and is a definite asset to the DASCO staff."[52] Fitzgerald's tokenism added little to the unbalanced white/nonwhite equation.

As COFO merged into the MFDP and the heat dissipated from SNCC throughout the South, the NSF and CORE continued efforts to elect blacks to ASCS committees. On July 20, 1966, Jac Wasserman mentioned ASCS elections at the Scholarship, Education and Defense Fund's meeting in Frogmore, South Carolina. "Most of the people there," he recounted, "were also not aware of the forthcoming elections in their states and had not made any plans to mobilize the communities to participate." He handed out manuals and other literature and offered support. Wasserman praised the NSF's Mike Kenney and James Mays for their work among black farmers. Kenney had given up his SNCC work in Holmes County, Mississippi, to become Alabama field coordinator for the NSF. Wasserman spoke at several locations throughout the South in late July and early August. Organizing the ASCS vote strained the NSF staff, and Kenney enlisted SCLC staff to help. SNCC workers, he sadly observed, "were at best lukewarm to the campaign."[53]

Even as SNCC slid further into internal chaos and controversy, Lowndes County farmers prepared to vote again for county ASCS committeemen. The NSF through its Rural Advancement Fund devoted substantial resources to the 1966 ASCS elections and contributed to transportation and food costs for black farmers. In Alabama, field representative Lewis Black focused on twenty-five counties and hoped to enlist 150 volunteers. The NSF also helped prepare imaginative and easily understood literature to circulate among black farmers. "This whole ASCS campaign was on NSF *initiative*," Mike Kenney stressed, adding that the SCLC also did excellent work in several counties. The Alabama Council on Human Relations hired Lewis Black, who had worked on a rural-advancement project for the NSF. In the summer and fall of 1966, Black used a $1,650 Southern Regional Council grant to organize black farmers for the ASCS vote. Still, no African Americans were elected to county committees, although eighteen won community committee seats and seventy-six were elected as alternates.[54]

No matter what reform pledges streamed from ASCS headquarters in Washington, county offices continued business as usual. In early August, Ralph S. Tyler III attended an ASCS appeal hearing in Greene County, Alabama. Two black farmers had been nominated to run for community committees, but the county ASCS committee rejected their petitions. "The treatment which the two men received was a disgrace to your department," Tyler wrote to Secretary Orville Freeman. "The county committee, Mr. James A. Smith as county office manager, treated the two Negroes with paternalism, but not respect," he wrote, and did not use courtesy titles, addressing the men by their first names instead. In Tyler's estimation, the treatment was "indicative of the way in which your organization and other federal organizations participate in perpetuating the feeling that only white men are deserving of respect."[55]

Lewis Black continued his rural work and was instrumental in setting up the Southwest Alabama Farmers Cooperative Association (SWAFCA) and gaining support from the Office of Economic Opportunity. In 1967, as word of a possible half-million-dollar OEO grant to SWAFCA spread across Alabama, Selma mayor Joe T. Smitherman organized a strong coalition to strangle the organization before rural blacks realized its potential political strength. The historian Susan Youngblood Ashmore unraveled the complex efforts that Alabama whites and some jealous Selma blacks exerted to crush the fledgling black farmer organization, which they portrayed as subversive, pro-union, and incompetent but in fact feared as a powerful force they could not control. A successful co-op would encourage black farmers to stay on the land and allow them to exert political muscle. In May 1967, the OEO approved a $399,967 grant, but Governor Lurleen Wallace vetoed it; in July, OEO director Sargent

Shriver overrode the veto. SWAFCA hired a professional staff, enticed reluctant USDA agencies to assist members, and marketed vegetable crops at much higher prices, but the sniping from its enemies, including Alabama FHA head Red Bamberg, eroded its chances for success. After a decade of hard work, SWAFCA went out of business in January 1981. The NSF's Fay Bennett reported in July 1969 that 17,000 rural southerners belonged to 75 cooperatives and asked for federal support through long-term low-interest loans, direct payments, and access to land.[56] It seemed that no matter how hard black farmers attempted to stay on the land, potent opposition undermined them.

As 1966 ASCS elections approached, Mike Kenney appealed to Ray Fitzgerald for lists of eligible voters that could be circulated among farmers prior to the election. It was crucial, he reminded Fitzgerald, that African American farmers know if they were eligible, and posting a list at the ASCS office was not sufficient since farmers would have to make a trip to town to peruse it. Fitzgerald stubbornly replied that posting one list and making another list available for copying should suffice. Kenney, not assuaged, stressed that the point of his earlier letter was that he did not find that arrangement adequate. "Agency policy does not permit supplying lists to you without cost," Fitzgerald perversely retorted, dredging up a rule that prohibited providing the list "to non-governmental organizations or persons," presumably including farmers whose names might or might not be on the list.[57]

Despite his hard work, by September Kenney was discouraged about the coming ASCS elections. Statewide meetings were planned for Louisiana and Arkansas, but there was little enthusiasm, and the NSF had waited too long to organize in Florida and South Carolina. Leaders in Alabama predicted "moderate rather than big success." Kenney summed up for Fay Bennett the worsening problems facing the ASCS campaign in Mississippi. "It became clear from our discussion that there is too much tension and distrust within the movement itself in Mississippi for us to try to set up a temporary ad hoc committee to work on ASCS," he admitted. Still, he argued, James Mays and several other staff could put together a campaign, handle the WATS line, keep records on each county, chronicle complaints, monitor deadlines, and answer questions. "It is almost a sure bet that Arkansas will not be conducting an ASCS campaign," he dejectedly reported, "and the Louisiana operation will be much smaller than we had anticipated." On October 19, Kenney wrote to the file that no blacks were elected in Georgia and added that one of the workers admitted that he could not get people to work together.[58] Many of the SNCC foot soldiers who marched along the back roads enlisting farmers for voter registration and ASCS elections had retired from battle, and ASCS strategic inefficiency in providing accurate maps, lists of voters, and ballot counts had ground down earlier enthusiasm.

In early November 1966, Kenney visited Mississippi to survey ASCS election preparations and meet with the Hinds County Farmer's Association. "The ASCS campaign is proceeding slowly and with little chance of marked success," he dejectedly reported. In Holmes County, he talked with Edgar Love, who explained that the enthusiasm of earlier years had disappeared. "They are doing very little this year," he said of the local farmers, for they had lost interest in ASCS elections. County committees continued to hack away at black farmers' benefits. Howard Taft Bailey, one of the most prosperous and active black farmers in Holmes County, reported that the ASCS committee had "grossly underestimated" black farmers' cotton yields, and he had taken thirty farmers to the office to get their yield figures raised. Kenney suggested that the NSF launch its own investigation of the Holmes County ASCS office. On a brighter note, he learned that Ralthus Hayes and nine other farmers had banded together to buy a two-row cotton picker. Wainwright Blease, director of the ASCS's south-central area, toured ten Mississippi counties in mid-October and talked with ASCS fieldmen about publicity, meetings, and election materials. "They also reported that no problems had been encountered and, as far as they knew," Blease sighed with relief, "no civil rights organization is active in ASCS elections at this time."[59]

Even as the election process unfolded, SNCC's Aviva Futorian complained on October 24 from Mississippi about other ASCS issues. In Benton County, she fretted, the African Americans who were hired for temporary jobs were schoolteachers who had "nothing to do with the movement" and were considered safe because "their jobs are dependent upon the will of the white power structure." She dismissed the ASCS claim that other qualified candidates were unavailable. "This seems to be a flagrant example of Mississippi officials using federal money to reward 'good niggers,'" she wrote, at least that was the way most blacks in Benton County interpreted the hiring practices. ASCS officials had also failed to explain a key form to black farmers, she charged, and she distrusted any investigation by white ASCS officials. "In fact," she suggested, "I would request that any further investigation regarding these sharecroppers be stopped if it is to be conducted by white Mississippians, as it will only increase these sharecroppers' problems rather than alleviate them." When Futorian left SNCC, she went to law school and then worked as a legis lative assistant for Congresswoman Elizabeth Holtzman and was active in women's rights and prison reform. It had taken William Seabron from May to September to dismiss an earlier complaint from Futorian, and it took two months for him to dismiss her October 24 letter.[60] Why it took months for county and state ASCS officials in Mississippi to compile a report and send it to Washington, for DASCO to further massage it, and for Seabron to sign off remains a mystery, especially considering that all levels agreed beforehand that there were no violations. This pattern occurred over and over; the ASCS never admitted to discrimination, but it had to work at it.

ASCS elections in 1966 followed the blueprint of previous years. Eighteen Louisiana African American farmers who ran for community committee seats in St. Landry Parish accused the ASCS committee of "a deliberate attempt to divide the strong Negro vote and defraud many of our poorest Negro farmers of the chance to vote." Several communities put up candidates by petition, but the ASCS office placed other African American candidates on the ballot, which would "significantly split the Negro vote and make it impossible for anyone of us to win." Leaders had demonstrated how to sign the ballots and had explained eligibility, but choosing among dozens of candidates "is too great a burden to bear since it undermines the whole democratic character of the election." Although ASCS committeemen were supposedly neutral, they canvassed the parish for black candidates and "used falsehood, intimidation and subterfuge to gather the signatures of additional Negro candidates for the ballot." Because of the committee's power, the candidates stressed, a farmer would think twice before refusing to run. Sharecroppers' wives did not receive ballots, nor did some sharecroppers. William Seabron dismissed the well-reasoned complaint as coming from "disgruntled Negro farmers" and asked for a draft letter explaining the nominating procedure that allowed the white ASCS committees to place black candidates on the ballot.[61] His passivity gave the green light to county and state ASCS officials to crush black opposition.

In Alabama, the efforts of the NSF and other civil rights groups yielded poor results despite publicity from its lawsuit and $1,600 spent in the final month before the election. Only eighteen black community committeemen were elected, while the year before, without such focus, eleven had been elected, and, of course, there was no chance to elect county committeemen. Civil rights worker Shirley Mesher complained that Dallas County ASCS officials were "cold" and "caustic" and that black farmers feared powerful community and county ASCS committeemen. Better-informed black farmers complained to Washington and received "letters in return referring them back to the local officials, the very source of their initial dissatisfaction," she wrote. "Even the staunchest eventually lose heart." Counties in the South with black majorities were decreasing, and some black farmers even questioned whether winning ASCS elections in all of the counties with black majorities was worth the effort.[62]

Shirley Mesher had moved from California to Selma in 1965 to work for the SCLC. Her clear thinking and activism endeared her to poor farmers but alarmed planters. Mesher grew progressively acerbic as the USDA ignored not only black farmers but also the forces of transformation. "Investigations have come and gone," she wrote, probably in the fall of 1966. "We have heard nothing. We have seen nothing." Plantation owners, she understood, abused workers with impunity for they realized that "the government will do nothing." Cotton was being replaced by cattle, workers were being replaced by machines, and no one was preparing poor farmers for this change. The stacks of documentation she had gathered were ignored. "Winter is coming," she wrote. "It will be very cold." Many former sharecroppers had no housing, and in the spring, they would have no work. Mesher threw her energy into founding the SWAFCA.[63]

Civil rights workers in other states also became discouraged. On November 2, Vincent O'Connor, a white SNCC staff member from California, wrote from Jefferson County, Arkansas, that although black farmers had met and nominated candidates, the county committee had added more. He insisted that packing the ballot meant that no fair election could be held, and he recommended that the election be rescheduled.[64] O'Connor worked hard in Jefferson County, but in the end, the county committee added eight additional blacks to the five placed on the ballot by petition. All of the candidates by petition had signed the proper forms, and O'Connor had notified both state and federal administrators. He had resigned from SNCC in August 1966, he explained, "because of its financial irresponsibility toward its workers and the continuing drift of its policies towards violence, separatism, and the eventual destruction of the American left." Still, he signed his letter to Fitzgerald as if he were affiliated with SNCC, and, he seethed, Fitzgerald "had the gall to suggest that I had chosen the candidates or that SNCC had." Signing as a SNCC worker was "the only means I have to identify myself to some of these federal bastards," he complained to Donald Jelinek, whom he asked for advice on filing suit against the ASCS committee. "They need to be challenged," O'Connor wrote, because whites assumed they could exclude black farmers.[65]

In a November 28 memo to Fitzgerald, Wainwright Blease made an oblique reference to "non ASCS organizations" that would "lead to influence not in the best interest of all farmers involved," a reference to Vincent O'Connor's activities. He had informed O'Connor that "while there may be an instance cited where one may suspect chicanery on the part of the nominating committee, redress for this type of grievance may best be expressed through nomination by petition," the very strategy that black farmers had attempted.[66] Basically Blease repudiated civil rights activists as not in the best interests of black farmers and winked at shady county-committee tactics. Such bureaucratic subterfuge increasingly typified USDA correspondence.

John B. Vance, director of the ASCS's southeast area, interpreted African Americans' slight gains in Alabama as a sign of progress. In 1966, the twenty-two blacks elected to community committees doubled the 1965 figure, and alternates increased from eighty-five to ninety-one.[67] The results, of course, epitomized ASCS tokenism. A few black farmers on a community committee could not hope to elect a black committeeman or in any way influence ASCS policy. After three years, the combined forces of SNCC, COFO, CORE, the SCLC, the NSF, and the U.S. Commission on Civil Rights had not cracked the white monopoly of county ASCS committees. Although 85 (434 alternates) African Americans won regular seats on community committees in 1965 and 113 (430 alternates) in 1966, no black farmer was elected to a county committee, except for 7 alternates.[68] Reassurances from Washington promising fair elections were diluted by the time they arrived in southern counties, and cunning and often illegal strategies went unadmitted and unpunished. White USDA employees pretended to carry out civil rights laws, all the while undermining equal rights and rigging the flawed committee system. As Thomas Hughes had suggested, USDA staff marginalized COFO and SNCC activists who pushed for more federal action, and this proved fatal to black farmers who remained outside the decision-making structure.

*Like now they trying to get rid of all proof*
*that black people ever farmed this land with plows and mules—*
*like if they had nothing from the starten but motor machines.*
—*Ernest Gaines,* A Gathering of Old Men

# 5

# DISSOLUTION

In an undated letter evidently written in December 1966, Geer Morton, who worked with the Hinds County, Mississippi, Community Development Agency, shared with the NSF's Mike Kenney his despair over the disintegration of rural life. "Look, Mike," he insisted, "things are getting bad here in a hurry. Tenants, sharecroppers, and renters are getting tossed off the land right and left." Small farmers were leaving agriculture, he continued, and larger farmers were buying machines. "But the immediate reason for throwing the people off the land," according to Morton's analysis, "is their increasing activity and interest in sharing in federal programs they're entitled to." In one Hinds County community, blacks won all five seats in the ASCS election. "This has made the local whites pretty mad," he warned. Black tenants and sharecroppers began demanding receipts from planters and furnishing merchants and refusing to sign away their share of USDA payments. A note of urgency ran through Morton's letter; he insisted that "things are going to hell in a basket." The county farmers association was crumbling, leaders were searching for Head Start jobs, and others were losing interest. Morton complained of the "inscrutably complicated" cotton program and the tangle of USDA policies. He insisted that Kenney send NSF staff to work on these issues "for a good month or so in order to get some answers and stop fucking around (by that I mean all of us)." Kenney, however, was fighting a prostate infection, and, ill and exhausted, he asked for a leave that in May ended with his resignation from the NSF.[1]

In February 1967, Morton and another Community Development Agency staffer chronicled for William Seabron a familiar litany of ASCS committee abuses, including inaccurate community maps, confusion regarding petitions, and gerrymandering, and they argued that these abuses suggested "a concerted and deliberate effort" to deny blacks ASCS representation. Blacks were a majority of eligible farmers in Hinds County, but they had won only a handful of seats. Whites interpreted black farmers' voting in federal ASCS elections as "another facet of civil rights activity," and they "discourage[d] participation with open hostility." The ASCS office staff treated African Americans "with varying amounts of hostility," usually "with brusqueness and disdain." The office employed one African American clerk, but she did not work at the counter, which was controlled by "elderly white women." The staffers described a dire situation. "Unless solutions to these problems can be found, the Negro's right to participate in rural life will be lost."[2]

The NSF strategy to contest ASCS elections in Hinds County had been flawed, and Morton admitted that the NSF had organized too late. After confusion over petitions for black candidates, two resigned but remained on the ballot. Morton suspected that they withdrew because they preferred not to be implicated in the civil rights movement and hear, in Morton's words, "Now Willie you all know you shouldn't be fooling around *those* kinds of people." Morton also learned that black farmers often split their votes. "Something like two votes for friends, one vote for a 'good' white and the other votes going to whites owed money."[3] Such a voting pattern, if widespread, would drain votes from the numerous black candidates that county committees placed on the ballot and favor the few whites.

While African Americans endured fraudulent elections and persistent discrimination, Secretary Orville Freeman's staff fed him only positive information, prompting him to amplify incremental civil rights gains. In September 1966, in reply to Michigan congressman Charles C. Diggs's inquiry about ASCS elections, Freeman beamed that ninety-six black farmers were elected to Alabama community committees, failing to clarify that none were elected to powerful county committees or that honest elections would have produced dozens, perhaps hundreds, of black county committeemen. "During this same period," Freeman continued, ASCS "county committees employed 20 full time and 204 temporary Negro employees." Freeman's claims implied far more significance than warranted, for they came amid ASCS chicanery and theft. Some NSF staff understood that Freeman was isolated from civil rights issues and that the well-meaning William Seabron lacked access to the secretary.[4] Whatever the truth, Freeman sang frivolous cover songs and tuned out the world of rural southern blues. By citing flawed statistics and not demanding ASCS reforms, he was complicit with the department's discrimination.

The unlikelihood of reforming county ASCS elections was the result of both southern whites' intransigence and Washington's feckless acquiescence. On February 21, 1967, a hastily called meeting by Seabron to discuss ASCS election reform brought together ASCS officials, Seabron's office staff, William Payne from the U.S. Commission on Civil Rights, Francis Pohlhaus from the NAACP, Clay Cochran from the secretary's Citizens Advisory Committee on Civil Rights, and Marvin Caplan from the Leadership Conference on Civil Rights. Conspicuously missing were representatives from CORE, SNCC, and the MFDP and, of course, black farmers, the people most invested in reforming elections. Most suggestions were structural and ignored the power of county committees and office managers. The ASCS staff fought off proposals to reform county committees, predictably insisting on the sanctity of local farmers electing their own representatives. That Freeman had appointed only five black state committeemen drew sharp criticism. ASCS representatives expressed shock that those outside the USDA "measured our accomplishments strictly on the basis of how many Negroes were elected to county and community committees (not alternates)." William Payne, who came from a Texas cotton-growing family and later worked for the USDA Office of Advocacy and Enterprise, suggested that Secretary Freeman appoint county committees. "He charged," one ASCS staffer smarted, "that we were more interested in animals than in people" and accused the USDA of being beholden to Congress rather than supporting the president's agenda. Payne's "genuinely intemperate" remarks did not sit well with USDA bureaucrats. Seabron courted trouble for even holding the meeting and justified it to Freeman by arguing the wisdom of allowing outside critics to have a say.[5]

More and more, the U.S. Commission on Civil Rights and the NSF distrusted USDA statements and promises. Isolated from civil rights advocates, Secretary Freeman seemed to parrot whatever his staff told him, for example, announcing incorrectly that cooperatives were being integrated. Seabron reported not to Freeman but to Assistant Secretary Kenneth Birkhead, and USDA agencies submitted dubious reports that the secretary proclaimed as examples of civil rights "progress." It was smoke and mirrors. The NSF's Jac Wasserman had acquired questionable

ASCS compliance reports that were withheld from the commission, and he hoped to publish a report on USDA discrimination. "Coupled with our own misgivings about the credibility of USDA's equal opportunity program," the commission's Gilbert Ware wrote in January 1967, "they certainly merit investigation." He also suggested updating the commission's 1965 report on USDA programs. The commission's mandate was staggering, for it surveyed the entire federal government for civil rights issues, and monitoring the USDA, a bureaucracy practiced in sleight, consumed tremendous resources.[6]

After offering Freeman meek suggestions for ASCS election reform in March 1967, Godfrey sent the secretary a self-serving memorandum falsely claiming that blacks had been unsuccessful in gaining victories in ASCS elections because "the civil rights workers worked primarily with sharecroppers and tenants" and were unable to "obtain much support from the Negro landowners." Karel M. Weissberg's study of Panola County, Mississippi, showed the opposite. Landowners were at the forefront of ASCS election activity, and civil rights workers canvassed indiscriminately. It was easier to misrepresent civil rights strategy than to deal with ASCS chicanery, intimidation, and stonewalling or to explain why black landowners would vote for whites or why complaints of violations came from black farmers of all tenures. It was also unconscionable and inaccurate to fault COFO and SNCC for focusing on sharecroppers when they carried out educational work that the ASCS and the Extension Service had neglected. As civil rights activists moved on, so did the memory of their work, and USDA leaders erased the SNCC and COFO legacy.[7] It was easier to fake compliance without the young, sharp eyes of civil rights workers.

The issue of civil rights activity arose again in March 1967 when the U.S. Commission on Civil Rights requested information on twenty-six Alabama counties. Dissatisfied with the ASCS response, the commission renewed its request in April. "Indicate whether or not civil rights activity occurred with regard to elections in the years indicated, the nature and extent of such activity," it queried, and whether the activity "served to increase the number of Negro committeemen." Godfrey had earlier responded that he was not in a position to supply such information, but after the commission insisted, he bent it to the ASCS's advantage. Wainwright Blease, director of the south-central area, supplied statistics from six Louisiana parishes and six Mississippi counties that showed that more black committeemen had been elected in counties and parishes without civil rights activity. Whatever the accuracy of Blease's analysis, before the

1964 SNCC and COFO campaign, African American farmers had barely participated in the elections, and the increase in black farmer participation was directly attributable to civil rights action. Godfrey's stonewalling reflected his apprehension that an honest report on civil rights activity would reveal the significant role of civil rights workers in educating black farmers of their rights.[8] ASCS leaders were eager to erase SNCC's historical role in challenging ASCS elections.

The commission renewed its request for a complete record of complaints and was adamant that the ASCS provide information on black farmer participation in price-support payments and conservation programs "by race, kind of operator and economic class of recipient." Godfrey refused the request, claiming that such information would "entail considerable cost and considerable work on the part of our county offices." The commission also requested a breakdown of state committeemen by "race, kind of operator, and size of farm." The ASCS offered dribbles of information to mollify the commission.[9] A full breakdown of price-support payments would expose how elite control of county committees benefited ASCS committee members and wealthy farmers. The commission was probing the center of ASCS economic power, and Godfrey surely understood that its report, even without the guarded information, would not add to his glory.

In the spring of 1967, Seabron added to his staff several civil rights specialists who would assist state and county offices with civil rights issues. "Since your State and county office personnel are often closer to troublesome situations," Wainwright Blease's memorandum pointed out, "they are in a position to detect apparent trouble spots where the services of the Department's Field Civil Rights Specialists could assist in alleviating the situation." The specialists were available "upon request." Given the level of denial among state and county ASCS personnel, it was unlikely they would request fieldworkers, an admission of trouble. Meanwhile, ASCS offices, under pressure from Godfrey to hire African Americans, reported that in Arkansas eighteen blacks and twelve whites filled thirty vacancies for permanent positions. In Louisiana, ten blacks and seven whites filled seventeen vacancies.[10]

On May 26 and 27, 1967, the U.S. Commission on Civil Rights held a meeting in Selma, Alabama, to investigate the ASCS. John B. Vance, director of the ASCS's southeast area, whined that the commissioners favored civil rights workers at the expense of ASCS personnel. ASCS representatives were "harshly treated" with "antagonistic interrogation and cross examination," he complained, while civil rights workers "were treated cordially and without interrogation and without being asked to in any way support the very serious charges they made against ASCS." Vance might have reflected on how similar this treatment was, if true, to the treatment ASCS employees visited upon black farmers. He listed Donald Jelinek (whom he mislabeled as from SNCC, not the Southern Rural Research Project), Shirley Mesher (a civil rights worker), and Lewis Black and Robert Valder (from the Alabama Council on Human Relations) as recipients of favoritism. Vance was appalled that the commission judged ASCS civil rights efforts a failure. The tone of Vance's memorandum reveals how uncomfortable USDA officials were with the aggressive questioning of astute commission members and the testimony of dedicated civil rights workers. In Vance's estimation, the representatives from the state ASCS committee and from several counties were victims. There was nothing in his recriminatory memo that acknowledged that there was any substance to the charges against the ASCS.[11]

Reporter Beth Wilcox from the *Southern Courier*, a nonprofit weekly newspaper published in Montgomery, interpreted the hearings quite differently. In a thoughtful letter to Horace Godfrey, she asked that he take a hard look at the Alabama ASCS staff and consider hiring replacements for state office supervisor Jack M. Bridges and southeast-area director John B. Vance. Wilcox admitted that until a year earlier, she did not even know what "ASCS" stood for. Still, she had come to realize that reform would come only when the old guard was replaced with a staff that would comply with civil rights directives. She reminded Godfrey that despite his order that six county managers should attend the hearing, only two showed up, "sorry proof of how directives are followed in the state of Alabama." Those who represented county ASCS offices, she observed, pleaded ignorance of discrimination and falsified compliance reports, while higher-ups had no knowledge of county operations. "The worst bureaucratic appointment could not be worse than the people elected county manager in the black belt counties," she judged. "Perhaps elections should be seriously considered to be the most open avenue for discrimination in ASCS," she bitingly suggested, "and therefore the abolishment of elections be considered." Godfrey made a polite reply, but, of course, he was not willing to consider Wilcox's penetrating analysis or wise suggestions. He continued instead to tolerate the ignorance and hubris of a mediocre staff that nullified civil rights orders.[12]

Victor B. Phillips, the ASCS's civil rights coordinator, realized that the agency failed to communicate with most farmers, and he was determined to reach "disadvantaged people." Most farmers did not read the thousands of technical bulletins put out by USDA agencies, he suspected, so, taking a page from the SNCC and NSF strategy book, he proposed that bulletins should use art to tell "a picture story." In addition, the ASCS should sponsor face-to-face talks with farmers. He belatedly suggested that pastors and civil rights workers could be useful in educating farmers about the ASCS's programs. Phillips did not define what he considered "disadvantaged people," but from the context of the memorandum, he apparently meant poor African American farmers. In June 1967, Phillips learned that Jac Wasserman was leaving his position as NSF field director and wrote him that it would "be like losing an old friend when you leave." He reported to Wasserman the new ASCS publications and revisions in ASCS election rules.[13] Civil rights activists had given the ASCS a wake-up call that it had neglected a large part of its constituency, and although the ASCS was promising to do better, revised publications would not prevent African American farmers from dwindling away.

The NSF's James Mays traveled to Batesville, Mississippi, to help with preparations for a visit by Secretary Freeman. On Monday, June 26, 1967, Freeman met with local leaders in Clarksdale and discussed the food stamp program. Robert Miles, a leader of the West Batesville Coopera-tive, civil rights activist, and prosperous farmer, attended the Clarksdale meeting and then traveled to Batesville along with county agents and an FHA supervisor. Miles had gone to Marks with Mississippi agriculture officials before heading to Batesville, leaving him no time to check last-minute plans for the secretary's visit to the cooperative. Mays suggested that Miles "was manipulated in a very shrewd manner" to prevent his dis-cussing civil rights. The sheriff led Freeman onto the cooperative grounds (the same sheriff, Mays helpfully pointed out, "who proved useless when the night riders shot into the home of Mr. Miles"). Freeman's glib re-marks about truck crops, Minnesota agriculture, recreational loans, and cooperatives ignored evicted and endangered farmers. He mentioned nothing about ASCS elections or FHA bias but instead praised as patri-ots large farmers who received $100,000 subsidy checks, explaining that unless production was controlled by such expensive USDA programs, "farmers would produce so much we would have an economic chaos." Subsidy checks might also convince restless southern planters to remain loyal to the Democratic Party. Freeman then got a haircut at a shop built with an FHA loan, toured a house financed with FHA funds, and attended a groundbreaking ceremony for a plywood company in Oxford.[14]

The secretary's defense of planters did not go down well with black farmers. James Mays offered a brilliant suggestion: let the big farmers

grow crops that were not in surplus and give the small farmers large subsidies to grow cash crops. That would get rid of the surplus and keep
small farmers on the land. Two observers, Mays reported, scoffed at Freeman's claims for African American progress. Mays added, "Mr. Freeman's
visit must have been a real joke as far as listening and learning is concerned." Reporters took the opportunity to grill William Seabron about
unfair ASCS elections, and while admitting the positive role of civil rights
activists, he inverted the credit, claiming that before Freeman became
secretary, "Negro participation in ASCS elections was practically nil." Of
course, black farmers' interest in ASCS elections began three years after
Freeman became secretary only because of SNCC's initiatives. Seabron
conceded that county committees nominated an excessive number of
African Americans to confuse voters. The ASCS "came out fairly well," he
confided to Freeman later.[15]

As these issues consumed African American farmers, their supporters,
and the USDA staff, events moved toward a violent summer. As the Vietnam War escalated and cost more American lives, the draft call increased
and fueled even more opposition to what was becoming an extremely
unpopular war. Martin Luther King Jr. denounced the war in April, and in
October, some 100,000 people demonstrated in the nation's capital. Most
upsetting, during the summer, race riots exploded in Roxbury, Buffalo,
Newark, Cairo, Memphis, Detroit, Milwaukee, and Washington. Attention
strayed from civil rights, and farmers faded even deeper into the background.

ASCS officials meanwhile outlined numerous steps to be taken before the 1967 elections in Arkansas, Louisiana, and Mississippi. The ASCS
would make newspaper and radio announcements, hold meetings, provide pamphlets and accurate maps, distribute posters, and circulate information on eligibility. In the southeast area, the ASCS emphasized black
participation.[16] When copies of the posters were shown at a meeting of
the secretary's Citizens Advisory Committee on Civil Rights, Aaron Henry
"remarked that he had not seen any of them posted in Mississippi." Ray
Fitzgerald slyly offered the possibility that posters were not displayed in
the intended areas. Henry suggested placing posters where black people
congregated, that is, in the African American community. It remained a
mystery whether or not the posters had been put up at all.[17]

In Macon County, Alabama, ASCS elections in 1966 and 1967 generated enormous tension and challenged the ASCS's ability to gild its mis-
deeds. It had been nearly a century since Booker T. Washington arrived

in Tuskegee in 1881 to found Tuskegee Institute and build the school with black ingenuity and sweat and white money. Tuskegee Institute prospered in large part because Washington perfected a high-wire act along the volatile color line. By the mid-1960s, Tuskegee had matured into a university and had left behind most courses based on trades and agriculture. In an effort to ease tensions in 1966, ASCS officials decided to count votes in the Tuskegee National Guard Auditorium with local and state observers present. In late October, three civil rights workers accompanied Alabama Council on Human Relations' Robert Valder to Tuskegee, where they joined other public witnesses, including the press, in monitoring the counting. According to Philip C. Beach, deputy director of the ASCS's southeast area, the civil rights workers "were very disruptive and highly critical of ASCS and our election procedures. There were constant minor disruptions but no major incidents." Three African Americans were elected to community committees and seven were elected as alternates. Beach observed that the percentage of black farmers in Macon County had steadily decreased, from 73 percent in 1959 to 60 percent in 1964 to 50 percent in 1966. He did not address how a majority of black farmers had not been able to elect a county committeeman.[18]

A year later, on September 25, 1967, Beach again witnessed the Macon County vote counting and reported some twenty people present, including a contingent of civil rights workers who remained throughout the day. The count was orderly, he wrote, except for a dispute over thirteen ballots that the county committee disqualified. After the counting ended at 6:30 P.M., a despondent Robert Valder was at a loss to explain how the civil rights activists' hard work for three months had resulted in only four black community committeemen. "If you people in Washington are serious about electing Negroes to county committees in the South," he told Beach, "you will first have to figure out how to do it and then write the regulations accordingly." Valder had decided to appeal the election based in part on the issue of wife voting.[19]

Ellis Hall, a Tuskegee Institute veterinarian active in the campaign, observed that the black vote was split and that whites recruited well. Leary Whatley, the county ASCS office manager, defended the elections as fair and insisted that he encouraged all farmers to vote. Reporter Mary Ellen Gale gave a graphic account of how the white vote increased. In 1966, African American Modichia Sims garnered 107 votes, most in the Little Teas–Society Hill community, but in 1967, P. T. Godfrey, a white candidate, won with 156 votes while Sims received 139. Civil rights activists worked hard in four communities, but only in Fort Davis–Hardaway did black voters follow the ticket. "The white farmers, meanwhile, voted in a bloc," Gale concluded.[20]

Four days after the election, Ray Fitzgerald met with Beach and a representative from the OIG to discuss "the possible forged ASCS election certifications" in Macon County. Instead of asking black farmers about the contested signatures or, more to the point, investigating claims of fraud in the ASCS office, Fitzgerald secured an expert for a "signature comparison." Addressing another complaint, the Alabama state ASCS committee sent a fieldman to check on the issue of wife eligibility, and he learned that information on how wives could secure ballots had not been distributed in newspapers, in pamphlets, or on the radio. It had been on the agenda at several meetings, and the ASCS office manager informed those who came by his office.[21]

A hearing on November 9, 1967, in Montgomery pitted southern, state, and county ASCS officials against defeated Macon County farmers and their supporters. Donald Jelinek of the Southern Rural Research Project represented African American farmers, and state ASCS office supervisor Jack M. Bridges chaired the hearing. When Jelinek asked for information from the county files, the ASCS staff balked and demanded the request in writing. Tired of the stonewalling, Jelinek focused on the issues and accused the Macon County ASCS committee of negligence, incompetence, and fraud. Three white community committee members had nominated themselves and then added five or six black nominees. Suggesting factiously that perhaps he had a "mental block," Jelinek stated, "I am unable to think of any reason that they would have added those extra names except for the sole reason of splitting that vote." He reminded the ASCS staff that in Lowndes County the 1966 election was voided for basically the same reason. Jelinek produced evidence that some sixty new voters— wives, brothers, and sons of white owners—were added to the eligible list just before the election. He castigated the ASCS committee for its punitive attitude toward 104 farmers whose ballots were rejected because they signed with an X or needed help signing. "The ASCS election is not a grammar school exam where people are punished for making mistakes," he lectured. Rather than confusing black farmers, he insisted, "it is incumbent upon the ASCS to protect the rights of its voters."[22]

In a phone conversation with Beach, Leary Whatley portrayed his Macon County office as an efficient machine. "There is not and never has been a voter list set-up by race," he stated categorically. There were, he hedged, "a number of code marks" for different purposes that accomplished the same thing. "Some of these were to indicate to best knowledge of the county office people which were White and Non-White farms for the purpose of making reports by race as required by Washington and the State office." All of the certifications, he insisted, had been received before the cutoff date. While the office record-keeping methodology showed inefficiency and carelessness, Whatley denied point by point that the office had done anything to alter the election.[23]

James H. M. Henderson challenged the ASCS handwriting expert who claimed that on some ballots signatures were forged. He made his own investigation of the contested signatures and discovered that five of the six contested certifications were indeed signed by someone other than the person named. Fred Simmons's daughter-in-law signed; a neighbor signed Frank Martin's certification; the wives of Billie Johnson, Frank Martin, and Elisha Peterson signed; and Charlie Jackson personally signed his certification. Henderson concluded that "they all felt that they were not committing a crime, nor invalidating their ballot." He also had serious doubts about the USDA handwriting expert: "I must conclude that the honesty and innocence of my fellow neighbors holds higher esteem and integrity than that of the so-called handwriting expert." Among black farmers, he stressed, voting was regarded as a privilege, more so than among "those of us who take the matter as pure routine and matter-of-fact." The same day that Henderson wrote this letter, Bridges certified that the Macon County election was held according to regulations. Macon County black farmers appealed. Donald Jelinek did not expect to win on appeal to the state committee or to Washington, for that would constitute "appealing to the very people we're accusing." He was ready to go all the way to the U.S. Supreme Court if black farmers desired.[24]

Meanwhile, Ray Fitzgerald's office had been busy constructing a reply to Jelinek's requests for more information. On December 27, he sent the boilerplate, which proposed charging Jelinek for the information, to the chairman of the Alabama state ASCS committee. If Jelinek wanted a list of all farmers who participated in ASCS programs, it would cost him six cents per name. If he wanted envelopes to compare handwriting, the charge would be twenty-five cents per copy. If he wanted public documents from the ASCS, he would have to pay dearly. Fitzgerald's reply followed a phone conversation with Leary Whatley on December 7 that dressed the flawed Macon County ASCS office in stylish bureaucratic clothing.[25] Evidently, civil rights activity and challenges to elections had alerted Whatley to possible charges of bias, so he sought to cover his tracks.

Hosea Guice, one of the men who ran for the community committee and challenged the result, gave a succinct account of county ASCS committee power. The committee assigned acreage allotments and calculated projected yields, both vital in farming operations. Although Guice produced 500 pounds of cotton per acre, the committee projected only 290 pounds. He appealed, and after making a number of trips to the ASCS office to document his case, he won the appeal and his projected yield for 1969 was raised to 500 pounds. Three other African American farmers also successfully appealed. Guice argued that racism played a large part in the committee's lowballing of black farmers' yields, explaining that "there's only so much (federal) money allotted to Macon County. If they cut down the Negroes, they can raise the whites." Roosevelt Harris, who also successfully appealed, was proud of the pile of documents he provided to prove his case. He noted that some black farmers were hesitant to visit the ASCS office because they were treated rudely. "A man went in there," he had heard, "and the lady told him he might get his yield lowered if he went on asking questions." Not surprisingly, Leary Whatley denied any discrimination.[26] Guice's analysis epitomized the way whites controlled ASCS committees, penalizing black farmers and rewarding whites.

In February 1968, the USDA's John Slusser and other Washington staffers visited Macon County and brought the county committee and black farmers together to discuss ASCS programs. "We had a frank discussion of the issues," James Henderson said of the meeting of six black farmers, two federal officials, and the county committee. The committee even promised to open its records but then hedged on a date. Troup R. Cunningham, a member of the county ASCS committee who owned 6,816 acres, said the meeting cleared up issues. The committee composition personified the elite lock on ASCS policy. Jim Weldon, the chairman, owned a 457-acre farm, and the third member, B. M. Segrest, owned 176 acres. Federal officials investigated committee records and found irregularities in both black and white projected yields.[27] The meeting at last brought black farmers into the discussion about Macon County policy, a rare and unreplicated moment.

Macon County black farmers lost their appeal for a new election for, unsurprisingly, the Washington office supported the Alabama staff. Robert Valder derided the decision as "three pages of nothing" and faulted the report for denying everything without paying attention to the validity of the charges. Ray Fitzgerald insisted that all eligible voters received ballots while admitting the difficulty of keeping a current list since farmers often moved. Instead of faulting the Macon County office for failing to keep an accurate list, he blamed black farmers for not reporting they were missing from the list. Fitzgerald denied that the office solicited white votes at the last minute, supported the handwriting expert's conclusions, and stated that the committee's nomination of additional black candidates did not violate ASCS rules. As Donald Jelinek had surmised, they were not going to win on an appeal within the ASCS. Before the 1968 election, Fitzgerald assured Henderson and other black farmers that the ASCS elections would be conducted according to regulations and addressed some of the complaints from the 1967 election.[28] Surprisingly, in the 1968 Macon County ASCS vote, an African American was elected chairman of the county committee.[29] Macon County had become too hot for the ASCS committee to continue its flagrant discrimination.

In Dallas County in 1967, African Americans lost the one alternate community committeeman elected the year before, Marion Green. He did not attend vote counting but stayed at home and picked cotton. In Lowndes County, three black farmers were elected to the Community C committee. R. C. Mays explained how he won re-election. "In this community, most of us own our own farms," he pointed out. "In other communities, they are tenants—the man says, 'Vote for me or you're off the land.'" Because fifteen white men were elected in other communities, Mays and Eugene Peoples, also re-elected in Community C, did not expect to place a black person on the county committee, but Mays observed that manners in the ASCS office had improved. He looked back over a wet spring and dismal stand of cotton. "I planted cotton three times," he complained, "and I still don't have a crop at all."[30]

In Mississippi, much of the vitality had leaked out of the ASCS challenge when COFO broke up and movement veterans left. Labor-intensive canvassing required legwork and patience, and although the NSF and other civil rights groups continued to organize black farmers, they were often outflanked by ASCS bureaucrats with their unfaithful promises. The Holmes County ASCS committee placed three African Americans on the ballot in the Acona community, resulting in the election of one committee member and two alternates. Jessie Williams Jr. and Lee H. Lewis, both active in the civil rights movement, had won in another community in 1966, but both lost in a close 1967 election. Robert B. Fitzpatrick of the Lawyer's Committee for Civil Rights Under Law found ample issues to contest. Holmes County ASCS office manager Will P. McWilliams Jr. arbitrarily ruled on a signing technicality and then months later admitted that the issue did not apply to ASCS elections. Fitzpatrick insisted that the excluded ballots could have changed the results, and he asked for the disputed envelopes if they survived and for a new election if not.[31]

In Kemper County, Mississippi, however, organizers held workshops and observed the counting carefully to monitor how sixteen black farmers fared. They all lost. One person reported that a landlord made his tenants mark their ballots as he watched. The fifteen black candidates in Grenada County also lost. The ASCS dismissed complaints from Yazoo County. "Across the state," reported Estelle Fine in the *Southern Courier*, "there were reports of votes cast by white absentee landlords, by school districts that own land, and by hunting clubs." White farmers spent a substantial amount of money in one county to get their wives' names placed on deeds so they could vote.[32] William Seabron disputed Jan Maedke's complaints of rule violations in Carroll County, Mississippi. "These findings would not have changed the outcome of the election," he ruled, "and action to correct these errors has now been taken." He dismissed Maedke's complaints, as if breaking the rules was acceptable so long as the outcome, white control, was not threatened.[33]

Seabron wrestled with dubious drafts that crossed his desk from the ASCS. He reminded Ray Fitzgerald that the information sent from his office came from other USDA agencies through Fitzgerald. "I rely on you folks to prepare letters, in technical matters, that we can all live with without consequent or subsequent embarrassment," he wrote. By the time drafts reached Seabron, of course, the bureaucracy had bled the complaints, massaged inspector general reports, and exonerated ASCS employees of wrongdoing.[34] Sadly, Seabron seemed incapable of demanding honest replies from Fitzgerald and his staff.

In 1967, results in ASCS elections revealed that 4 African Americans were elected as alternate county committeemen, 102 as community committeemen, and 353 as alternate community committeemen. These successful candidates were among 6,211 placed on the ballot by county committees and 133 by petition, adding validity to charges that white committeemen put blacks on the ballot to confuse voters. In Alabama, the committees placed 829 blacks on the ballot to compete with 67 by petition, and in Mississippi, the figures were 937 and 26, respectively. The statistics for ASCS employment in Alabama showed only 24 African Americans employed full-time, an increase of 1 over 1966. Georgia's numbers increased from 15 to 29, and Mississippi's from 32 to 56, probably because of internship programs. In fourteen southern states (including Oklahoma and Texas), there were 305 full-time black employees.[35]

Even such puny advances did not always go down well among whites, as evidenced by an incident in Calhoun County, Mississippi, concerning Grady Ford, a farmer and compliance reporter trainee. At 1:30 A.M. on Sunday morning January 23, 1968, someone fired a shotgun blast into Ford's house. He immediately contacted the sheriff, who alerted the FBI "on the basis that possible violation of civil rights laws is involved." Ford had recently secured an FHA loan and built a new home. State executive director C. W. Sullivan boasted that Ford was well respected, "and there is no indication he has been actively associated with any civil rights movement," implying that such activity might attract a shotgun blast. Sullivan doubted that the attack had anything to do with Ford's ASCS position.[36] The fact that Ford had a new house, had received federal funds to build it, and was training as a compliance reporter for the ASCS might well have rubbed some Calhoun County residents the wrong way, even if he was not involved in the movement.

Reporter Nick Kotz's series in the *Des Moines Register* in February 1968 explored the vast economic distance between Mississippi planter and U.S. senator James O. Eastland and his cook, Irene Taylor. As in many other black Delta families, Taylor's eight children crowded together two or three to a bed in a sagging house with a pump for water and a leaning outhouse out back. The $1.00 an hour minimum wage had taken effect, but as Taylor explained, "A dollar a day ain't worth nothing when, maybe they give you one day, and then you're off for two weeks and like that." Four of her children lacked shoes and could not attend school. Taylor had cared for Eastland's mother before she died and for the senator when he was ill, but she complained bitterly that he refused to give her Christmas money. In 1967, the USDA provided $446,000 for food programs in Mississippi, Kotz reported, and paid cotton farmers $10.2 million in subsidies. Only 80 of the original 200 families were left on the Eastland plantation, and Kotz found that 40 tractor drivers operating 25 tractors, 8 cultivators, 8 planters, and 8 pickers did all the labor. Kotz interviewed Senator Eastland in the study of his Sunflower County plantation home, and when the senator briefly left the room, Kotz took down a volume in a set by Lothrop Stoddard and found many of the pages unseparated, indicating that the book had not been read. When Eastland returned, Kotz complimented him on the fine volumes, and Eastland replied, "Yes, I read that constantly." The reporter made a note that his first impression of Eastland was correct: "He is a petty man." Later in the year, Kotz won a Pulitzer Prize for a *Washington Post* series on meat-packing plants. William Seabron compiled a four-page response to Kotz's articles and saw them as a lever to pry action from Secretary Freeman.[37]

The secretary's Citizens Advisory Committee, meeting on March 14–15, 1968, in Tampa, Florida, suggested that an African American or minority-group representative be included on all county committees in predominately black counties. ASCS election procedures were flawed, it argued, and the rules worked against African Americans. When the issue arose of notifying all candidates of election vote counts, USDA personnel at the meeting stubbornly echoed Ray Fitzgerald, insisting that anyone interested could get the results at the county ASCS office. One committee member scoffed and observed that black farmers "fear reprisal" and would not go to the office. Aaron Henry asked who had selected members to serve on the African American state advisory committees, noting that in Mississippi they were "'safe' Negroes acceptable to the State office." Victor Phillips traced the nominations from the state ASCS committees to area directors and on to Ray Fitzgerald, who "clears them through Mr. Seabron's office." Seabron reiterated that he was consulted as "an afterthought" and usually had only twenty-four hours to comment, and he admitted that he sometimes "approves appointees in the dark because he does not know them." G. T. Dowdy, an educator; J. H. Glanton, a farmer who was also in real estate; and Bishop William M. Smith composed the Alabama ASCS advisory committee.[38]

Even as civil rights initiatives faded in the southern countryside, significant and tragic events occupied the national stage. In Memphis to support a strike by garbage workers, on April 4, 1968, Martin Luther King Jr. was assassinated at the Lorraine Motel and rioting spread across the country. President Lyndon B. Johnson had already announced that he would not stand for re-election that fall. On the night of June 4, Senator Robert Kennedy was assassinated after winning the Democratic primary in California. In late August, the Democratic National Convention nominated Hubert H. Humphrey for president amid violent demonstrations in Chicago. Antiwar sentiment drowned out civil rights initiatives. Meanwhile, African American farmers in the South watched support from civil rights organizations fade, leaving them at the mercy of local committees and the USDA.

Criticism from civil rights workers and fieldwork by Seabron's staff generated suggestions for changes in county ASCS elections that eventually went to Secretary Freeman, who in May 1968 sent them to Horace

Godfrey's desk for comment. Seabron recommended that county committees sponsor open houses to bring black and white farmers together
to discuss problems across the color line, and Godfrey indicated that just
such a meeting was scheduled in Macon County, Alabama. Taking a page
from the Extension Service, Godfrey hinted that in counties with a large
percentage of black farmers, the ASCS might employ a "county program
specialist" to interpret ASCS regulations. Seabron also proposed "spon-
soring the cooperative gathering of small cotton allotments into a single
piece of ground with each farmer exchanging his allotment for shares in
the producing cooperatives." Godfrey was open to the ideas and made
several suggestions on implementing them.[39]

In June, Victor Phillips, who had become the Equal Employment Opportunity program coordinator for the USDA, reported on the sixty-eight
equal opportunity plans submitted by ASCS offices. After deciding that
"most of the EEO Plans are well-developed," he observed that there were
still few minority-group employees, and he focused on "the low proportion and in many cases no women employed in sub-professional, professional and technical positions, especially in State offices."[40] Tokenism
was not making a big dent in southern ASCS offices, but increasingly the
focus of discrimination included women. Employing blacks and white
women in administrative positions would change office culture, and the
assumptions about black farmers and what was due to them could likewise change.

In mid-June, the secretary's Citizens Advisory Committee on Civil
Rights met in Memphis. In his report, John B. Vance, director of the
ASCS's southeast area, expressed relief that Vivian Henderson, who he
claimed had "been rather discourteous to ASCS representatives" at the
Florida meeting, was not present. One committee member, upset at the
lack of black county agents, "asserted that in his judgment Extension Service should be abolished." The Arkansas extension director drew a re-
buke from Aaron Henry "for his mispronunciation of the word 'Negro.'"
Henry's patience was running thin, and he suggested that a suit should
be brought against the USDA to push civil rights enforcement. When John
Gammon Jr., an Arkansas state committeeman, suggested that blacks
were not elected to county ASCS committees because "one Negro would
not vote for another," F. B. Pierce, a black representative of the Arkansas Farm Bureau, agreed. "Committee member Henry asked Gammons
in so many words," Vance reported, "if he had not been 'brain washed'
by the Department of Agriculture." Advisory committee member Clay
Cochran suggested that the USDA be abolished, arguing that it "was the

least popular of all the Departments of the Government with the possible exception of the Pentagon." Despite sharp questioning and imaginative suggestions, Vance allowed that ASCS witnesses "were treated most courteously, were not harassed in any respect." Obviously the advisory committee had seen enough to conclude that civil rights were not high on the USDA priority chart, and its recommendations showed impatience with stonewalling and broken promises. Because committee member Aaron Henry often raised troubling questions about USDA policy, his role came to the attention of Horace Godfrey, who on June 24, 1968, advised Thomas Hughes, "there is considerable criticism throughout the mid-South as a result of Mr. Henry's membership on the Secretary's Advisory Committee."[41]

The ASCS also faced discrimination complaints from other rural minorities. When Horace Godfrey responded to William Seabron's memorandum about the Citizens Advisory Committee meeting, he summarized other minorities on county committees. An African American served on the county committees in Massachusetts, Pennsylvania, and Indiana, and "34 Spanish Americans, 22 of Oriental ancestry, and 23 Indians" also served on county committees. There were 134 African Americans on community committees, 73 Hispanics, 45 Asian Americans, and 122 Native Americans. His list did not include any women. Godfrey made an effort to justify the existing election system, reviewing its New Deal origins and insisting, despite evidence to the contrary, "that the cause of civil rights will be served better through use of the ballot in free elections than through abandonment of the ballot." His boilerplate justifications failed to address the issues that had prevented electoral success.[42] If one ignored the racism of white farmers, county committeemen and administrators, and state and federal ASCS staffs, Godfrey's analysis made sense. The figures that he cited, however, vividly illustrated how little ASCS elections had done to promote diversity on county committees.

In July 1968, Godfrey responded to Seabron's request for commentary on the commission's report on twenty-six Alabama counties, and even he could not put a gloss on the ASCS civil rights record. There were no black professional employees in the state ASCS office and only two clerical workers, and in the twenty-six counties, the number of full-time black positions had crept from seven to nine, but the number of temporary positions had declined. Arguing that it could not find qualified applicants, the Alabama ASCS began a training program, and while Godfrey observed that trainees learned basic knowledge and skills, the ASCS obviously learned whether they harbored civil rights ideas or were not team

players. Godfrey went to some length to defend his state advisory appointees, claiming that all-black committees were established "to provide a practical and immediate administrative mechanism for obtaining
advice and suggestions to solve civil rights problems." He insisted that
"Negro educators, businessmen, ministers and farmers" were best suited
for this task while weakly admitting that they had no "authority for policy
decisions." He did not explain how educators, businessmen, and ministers could bring anything to the examination of agricultural issues and
furnished no examples of their counsel. In early 1968, a rumor circulated
that Godfrey would resign his ASCS position to run for Congress, and
a staff member warned Senator Richard Russell that both men in line
for his position "are from areas of the country other than the South and
would be much more difficult to work with in connection with the Department's program."[43]

At the end of July, Godfrey reacted to the recommendation from the
secretary's Citizens Advisory Committee "that ways and means be found
to guarantee that at least one member of each county committee shall
be a Negro or a member of some other minority group that is present
in the area in question." Godfrey wrote to Seabron, "The word 'guarantee' presents a serious problem." He reiterated the sanctity of the election process and again insisted that the ballot better protected black
farmers' interests. He deftly sidestepped the long train of violations in
ASCS elections and claimed that "the way is completely clear for free and
open elections." Two weeks later, he pleaded, "We have made what we
feel are exhaustive efforts to encourage Negro participation in the farmer
committee elections within the legal authority under which the committee system exists."[44] By flooding ASCS committees with civil rights pro-
nouncements, listing numerous new rules, and pledging commitment
to civil rights, Godfrey papered over county committees' enduring bias,
which would never yield to black members without force. Godfrey and his
staff preferred to ignore the role of civil rights activism and blame black
farmers for failing to win elections.

At the U.S. Commission on Civil Rights hearings in Montgomery in the
spring of 1968, C. H. Erskine Smith, a Birmingham attorney and chairman of the commission's state advisory committee, reported on issues
facing African Americans. The level of violence overall, he contended,
had decreased, not because whites had mellowed but because the civil
rights movement had cooled as "civil rights workers, the 'outside agita-
tors,' SCLC, SNCC, and the others" departed. The higher level of violence
may well have been a barometer of civil rights effectiveness, as sad as that

reading might be. Smith's statement implied that SNCC and the SCLC's departure had a healing effect, but without pressure, county USDA agencies stubbornly refused to hire black workers, explain programs, administer loans, or show civility.[45] Already SNCC's invaluable contributions were being written out of the story, and in rural areas, the situation suspiciously resembled the antemovement status quo.

The ASCS bureaucratic engine had been tuned up to run on fuel composed of platitudes and promises. In September 1968, William Seabron brushed aside a stinging indictment of ASCS discrimination in twenty-six Alabama counties by the commission's Carol B. Kummerfeld, director of federal programs. Her definitive list of instances of ASCS discrimination covered employment, committee representation, absolute white control of county and state offices, insubordinate county and state executives, communication, voting irregularities, and the eviction of tenants who refused to sign over ASCS checks to landlords. Seabron's twenty-one-page reply featured confession, denial, and substantial boilerplate recycled from earlier replies to critics. Victor Phillips responded to Kummerfeld's November complaint that "the Marshall County, Mississippi, ASCS Committee is set up on a basis which discriminates against poor people, tenant farmers, Negroes and all other persons except landlords and large landowners." Unimaginatively, Phillips recited a litany of half-truths, and based on his reasoning, it would have been impossible for the white elite of Marshall County to discriminate. Phillips added that the secretary of agriculture appointed an African American to the Mississippi state committee and there was also the all-black advisory committee. All ASCS operations were governed by regulations, he insisted, and thus there could be no racial discrimination.[46]

As the Johnson administration wound down, Secretary Freeman sent William Seabron a farewell note. The secretary thanked Seabron for his "friendship and dedicated service" during "these hectic rewarding years." Seabron had done well at "a difficult often unrewarding job," Freeman observed. He conceded that they might have "moved faster but we have come a good way and the momentum is building." Seabron replied that he had established "warm and trusting relationships with civil rights leadership (white and black) across the country." Despite the intense friction with bureaucrats, misdirected memos, and impatience with the tortured replies he signed, Seabron wrote that the "top staff have been cordial and friendly, particularly when they are equated against the assignment I've had." Like Freeman, he concluded, "there is still much to be done." Horace Godfrey simply wrote to Seabron, "I want to let you know what a pleasure it has been to be associated with you during the past several years." The innocuous farewells grated against the USDA's continuing discrimination, and they ignored the ghastly farm-population decline during the Kennedy-Johnson years. In the decade since 1959, the number of black farmers had decreased from 273,000 to 87,000, and overall the country's farmers decreased from 3.7 million to 2.7 million. Freeman might have clarified the kind of momentum he had in mind. Fewer and fewer people knew or cared about what was happening in rural America.[47] For a brief moment, COFO, SNCC, CORE, local NAACP activists, and dedicated black farmers had challenged the nerve center of county committees, which dispersed millions of dollars. Despite a few election victories, elite white control of county ASCS committees remained inviolate as the Democrats moved on.

Maxine W. Lacy from Compton, Kentucky, penned one of the first letters to the Nixon administration's ASCS office. Her biting and astute analysis dissected the ASCS election procedure. "The incumbent not only has the ballots printed in abundance," she wrote in mock admiration, "but is allowed to run for office and receive these voted ballots by mail and keep these ballots for near two weeks then he, the incumbent, is allowed to count the ballots and declare himself the winner." She gently raised questions about this marvelous process. "Now knowing human nature," she ventured, "do you believe these voted ballots are ever changed or spoiled?" It was true, she admitted, that the public could watch the count if they could crowd into a tiny room. A voter lost control of the ballot unless he or she guarded it "with a shot gun for two weeks to see that it is counted." The other option was to "smile and say the election laws are fair, lick a dozen pair of boots, hold our little hand out for a hand out and fight the wars and pay the taxes. I'm afraid my political disposition has turned sour." William Galbraith, the new DASCO, wasted a page and a half of drivel evading the implications of her inquiry.[48] There is no evidence to indicate whether Maxine Lacy was black or white, but elite discrimination cut by class and gender as well as by color.

In a 1968 interview, Earl Butz, who would become best known for urging farmers to expand operations and plant fencerow to fencerow, condescended to small farmers and ridiculed idealists who opposed farm programs that favored the wealthy. He boldly stated that "the trend in modern agriculture toward bigness, toward well-capitalized and heavily mechanized units, is inevitable." It was also inevitable that "the little fellow either has to make quite severe adjustments, or sometimes adjust out to other work." When Richard Nixon became president, Butz joked, he "probably didn't know the difference between a pig and a soybean."[49] Butz personified the USDA's callous disregard for small farmers, who were rapidly disappearing, and he realized that the executive office knew or cared little about agriculture. Nothing stood in the way of agrigovernment.

Despite ASCS rhetoric proclaiming fair voting, the number of minority members of county ASCS committees had been decreasing since 1967. A memo from USDA civil rights staffer Percy R. Luney in 1971 expressed surprise that declines continued despite annual changes in election machinery. He proposed six topics for study and implied that research rather than action could solve the problem. "We believe that this is the kind of research that could be conducted at an 1890 institution [i.e., an African American land-grant college]," Luney said, and he suggested a $5,000 to $8,000 budget.[50] Material in the civil rights office offered compelling evidence of the reason African Americans were not elected, but a study of such material could only inform the office of what it already knew, though it would, of course, postpone any action.

The cumbersome bureaucratic equal rights machinery was used to mask continuing discrimination rather than ending it. A 1973 U.S. Commission on Civil Rights report tallied 23,000 on-site USDA compliance reviews in 1972 that covered 213,000 Title VI recipients, and all were in compliance. The report concluded that such unlikely statistics cast "serious doubt" on the veracity of USDA reviews. When investigations did follow complaints, the USDA "permits inordinate delays between the time investigative reports of Title VI complaints are prepared and the time the cases are closed."[51] Nullification by false compliance reports became a sharp USDA tool.

With its compliance reports in order, the ASCS had no compelling reason to attack the discrimination embedded in its programs. In 1986, Secretary of Agriculture Richard Lyng, like Orville Freeman before him, insisted that the ASCS must change. Ward Sinclair, the *Washington Post*'s astute agriculture reporter, chronicled how little ever changed in the USDA. One of the congressional oversight committees investigating Florida and Arkansas found numerous incidents of discrimination. In Florida, none of the forty-one county ASCS offices was headed by an African American, Sinclair discovered, and only a GS-4 mailroom worker, who was trained for a county director position, served in the state headquarters. The Arkansas ASCS employed only twelve blacks full-time. Insensitivity thrived in this atmosphere, with some managers referring to black males as "boys" and black women as "negresses." Because of bud-

get cuts by the Reagan administration, there were fewer compliance reports, and ASCS administrators argued that classes on sensitivity sub-
stituted for compliance. The ASCS's Richard W. Goldberg stated that the
agency "looks pretty good compared to many other federal agencies." Yet
interviews with black ASCS employees revealed that discrimination continued in part because of the political nature of the ASCS. The party in
power appointed state directors and state committees, while the county
committee hired a staff director to do its bidding. Not surprisingly, as
of July 31, 1987, only 33 of 2,520 county directors in the nation (1.3 percent) were black. "Most of the offices are staffed by white women," one
of the few black county directors explained to Sinclair, "and the white-
dominated committees just don't want to put a black man in charge." "It's
a good-old-boy buddy system," he judged.[52]

*'Cause with them what you always*
*have to be listening for is the opposite of what they say.*
—*Thomas Pynchon,* Against the Day

*We are tied down to a language which makes*
*up in obscurity what it lacks in style.*
—*Tom Stoppard,* Rosencrantz and Guildenstern Are Dead

# DUALITY

Farmers have always studied the soil and the sky for signs that might increase their bounty, and some have vested almanacs, treasuries of agricultural observation and practice, with almost biblical authority. Before formal agricultural education emerged, farmers often exchanged insights on, among other topics, plant varieties, fertilizer, labor needs, and, of course, the weather. By the early nineteenth century, agricultural education emerged as a counter to almanacs and everyday practices, and spirited debates consumed pages of agricultural journals. Many farmers shunned what they judged as meddlesome notions divorced from rural experience, but schools expanded agricultural offerings and insisted on the benefits of academic study. Congress formalized federal interest in agriculture by creating the Department of Agriculture and land-grant colleges during the Civil War, passing the Hatch Act in 1887, and establishing the Federal Extension Service in 1914. Former slaves had no voice in agricultural policy, nor did they share equally the fruits of education and research. Since white schools in the South would not admit black students, Congress funded African American land-grant schools in 1890. From the beginning, these "1890 schools" were underfunded and, like many products of segregation, were demeaning tokens created to suggest equal treatment of African Americans. Similar to other institutions during the Jim Crow era, however, African American land-grant schools constructed autonomous operations that imaginatively utilized stingy budgets.

Since most federal activity targeted white farmers, African Americans took their own initiative in improving rural life. Since its founding during Reconstruction, Hampton Institute focused on teaching agriculture and other useful trades. Booker T. Washington, Hampton's most notable graduate, founded Tuskegee Institute in 1881 and played an active role both in promoting agricultural education and in spreading knowledge to farmers. Using student labor to construct the school and tend its fields and herds, he built a strong agriculture department that employed, among other notables, George Washington Carver. In 1892, Washington founded the Tuskegee Negro Conference, which provided an opportunity for area farmers to hear speakers address rural problems and encourage better hygiene, housing, habits, and thrift. Tuskegee's Jesup Wagon, funded by New York banker Morris K. Jesup, took agricultural education to farmers and recruited for Tuskegee's weekly farmers' institutes and short courses. Tuskegee graduates planted education seeds throughout the South and even in Africa, and schools sprouted and spread agricultural knowledge that helped farmers to prosper and buy land.

As the boll weevil spread east from Texas in the early part of the century, several strands of rural education converged. Seaman A. Knapp, who had begun demonstrations on boll weevil control in Texas in 1903, advocated demonstrations across the South both for white and for black farmers. Tuskegee's Thomas M. Campbell, with the support of the General Education Board, became the first black demonstration agent, and by 1911, there were twenty-three African American agents among the hundreds of white agents. Campbell grew up in rural Georgia and left in 1898 at age fourteen for Tuskegee Institute, walking and working his way there much as Booker T. Washington had traveled to Hampton Institute thirty years earlier seeking his education. Women's home-demonstration work evolved slowly with often-heroic efforts to convince male administrators of its importance. Dedicated black women overcame daunting obstacles in carrying their message to farmwomen. When Congress passed the Smith-Lever Act in 1914, federal funds went to states for extension work, and a separate and unequal Negro Extension Service worked under white administrators. Alabama Polytechnic Institute (later Auburn University) administered the state program and split about 30 percent of the funds between Tuskegee Institute and the Agricultural and Mechanical College for Negroes in northeastern Alabama, the official 1890 school. Paradoxically, neither Tuskegee Institute nor Hampton Institute was a land-grant school despite their important role in agricultural education and extension.[1] The Alabama pattern spread across the South, and the 1890 land-grant schools became the headquarters for black extension, always under the watchful eye of whites at the segregated white land-grant schools created in 1862.

Both the Farmers Home Administration and the Agricultural Stabilization and Conservation Service had straightforward organizational structures, but the organization of the Federal Extension Service was convoluted, esoteric, territorial, and discriminatory, and it ultimately grew into an octopuslike creature with tentacles extending from Washington to land-grant schools and on into the most remote communities. The Washington office reached into state agriculture departments, black and white land-grant schools, and counties, where extension agents wielded enormous power. The FES budget drew from federal, state, and county sources, and, chameleonlike, the agency changed its affiliation as the situation demanded. From the beginning, extension programs were tailored to the needs of educated and prosperous white farmers. White schools and agents stubbornly refused to share knowledge with black agents and purposely kept them outside the information loop. The FES's national structure invited political intrusion at every level, and the American Farm Bureau Federation, the Farmers Union, and other farmer organizations used extension to spread their message to rural people. It was the Farm Bureau, however, that became most closely allied with the Extension Service.

The extension idea of disseminating information on better farming techniques fit easily into the plans of corporations such as International Harvester that had salesmen selling products in rural areas, bankers who furnished credit, and chambers of commerce that promoted business. Bradford Knapp, Seaman Knapp's son, headed the FES in 1918 and faced requests from aggressive businesses that wanted the names and addresses of extension agents in order to send them brochures and information on their line of goods. Knapp feared that a failure to cooperate would undermine agents while cooperating too closely would impugn the integrity of government workers. He decided that names and addresses were public information, thus moving the Extension Service into a closer relationship with corporations.[2] By the Great Depression, the Extension Service had matured into a powerful national bureaucracy with strong political and corporate allies. As New Deal agricultural programs proliferated and challenged extension's supremacy, the FES countered by placing agents as ex officio members on local, state, and federal agricultural committees and, in the case of the Farm Security Administration, watching its congressional supporters annihilate the FSA during World War II.

Although African American land-grant schools hosted the Negro Extension Service, they shared little of the FES largesse. If the Negro Extension Service had been separate but equal, in 1941, there would have been 1,000 black agents instead of 549 in sixteen southern states. Looked at another way, there were 1,303 white farm operators per white extension agent and 2,781 black farm operators per black agent. In 1941, the total Extension Service budget was $14.8 million for whites and $988,000 for African Americans.[3] Black schools and extension workers carved out zones of autonomy but were beholden to white funding and priorities. Black farmers had few resources to subsidize extension agents, so counties with black agents were scarce and agents were underpaid. As U.S. Commission on Civil Rights interviews in 1964 revealed, separate but equal was a myth, a sick joke. Black extension workers had cramped offices, hand-me-down furniture, scarce telephones, and few typewriters or mimeograph machines. Still, extension and home-demonstration agents made an enormous contribution to the lives of black farm families. They traveled over rough country roads to call on farmers, demonstrated the latest cultivation practices, held meetings, dipped livestock, castrated pigs, promoted home improvements, advised on terracing the land, organized 4-H and home-demonstration clubs, wrote letters, handed out pamphlets, and organized displays at county fairs. White administrators often denied blacks access to specialized classes on dairy farming, livestock, and other specialties they deemed unsuitable for African Americans.

Black agents received less pay than white agents for the same work and endured numbing discrimination, but at the same time, extension jobs offered respectability and the opportunity to serve rural people. Recent scholarship on rural women furnishes examples of both black and white women who found immense personal satisfaction in home-demonstration work. In a culture that allowed few opportunities for women to work outside the home, travel, organize clubs, or exert authority, home demonstration offered women rewarding professional lives. "This has been the most enriched and rewarding type of work that I have ever engaged in," Juanita Y. DeVaughn wrote in 1962 when she resigned her African American home-demonstration position in Alabama to spend more time with her family. "The work has linked me very closely with down to earth women, men, girls and boys in the rural farm and non farm areas as well as the urban."[4] In many cases, white agents worked comfortably with African Americans, but because black agents answered to white county agents, tension, spite, and even jealousy sometimes stymied black initiatives. Whites tolerated only those black agents whose advice did not challenge white supremacy or disrupt farm labor.

Extension agents shied away from civil rights workers as if they feared being infected. In Mississippi, during the summer of 1964, COFO's Michael Piore invited representatives from the Madison County FHA and the SCS, white county agent Buddy Smith, and black agent R. M. Machie to attend a farmers meeting and explain agricultural programs. After initially showing some interest, Smith explained that the county board of supervisors had instructed him not to attend because such meetings were "unwanted and unnecessary." When Piore complained to state extension leader William Marlin Bost about the reluctant county agent, Bost lamely replied that county agents were extremely busy. Bost, a Mississippi native and World War II veteran, staunchly defended white control of the state's extension program. After earning a B.S. at Mississippi State University in 1949, he worked in extension and continued his education, earning an M.S. and Ph.D. and in 1962 becoming director of the Mississippi Cooperative Extension Service. After receiving Bost's evasive reply, Piore wrote to FES deputy administrator Lloyd H. Davis that he was "under the impression that it was the specific duty of the County Agent to arrange meetings at which the farm programs could be explained." Buddy Smith, he continued, wrote a column about farming in the local newspaper and was well known among whites in the community. His refusal to meet with black farmers, Piore concluded, "is a deliberate attempt to suppress the development of Negro farmers and deny them even *equal* information about and access to the agricultural programs in the County."[5]

Robert Church, born in Clarke County, Georgia, in 1909, personified the aspirations and frustrations that marked a career in black extension work. Church completed high school and college at Hampton Institute and began teaching in Wilkes County, Georgia, in the fall of 1934, but at Christmas break, he packed his clothes and vowed never to return. "They weren't paying me and working me to death teaching, coaching basketball, coaching football, all that," he complained. When the holiday was ending, his mother washed all his clothes, packed them, and, despite his protests, said, "Oh, you're going back." He had purchased a Chevrolet with a rumble seat and boasted that he "was the hottest thing in town," so he needed work. His next job was in Jenkins County, where, he said, "I was a principal, vocational agriculture teacher, coach." He also worked with the black county agent and visited farmers, and when the county agent died, Church took over the extension position while continuing to teach agriculture.[6]

In addition to teaching better farming methods, Church encouraged industrious sharecroppers to buy land and become independent. After hearing Church's encouraging words, one sharecropper told his wife he wanted to own a farm. She was less than enthusiastic and maliciously told their landlord, "Mr. Church going to get us a farm." The landlord put out the word that Church would soon be buried in a local swamp. Admitting that he was "a little disturbed," Church began carrying a pistol, vowing to protect himself. After all, he reasoned, he was only doing his job. One year, he was awarded a $25 per month raise, but the white supervisor spitefully omitted it from the budget. After ten years, Church moved to Peach County, where he became county agent.[7]

Much of his work in Peach County, the home of Fort Valley State College, was in what was known as "ham and eggs shows," which allowed African Americans to display their goods, but he also worked with 4-H clubs and farmers. Otis S. O'Neal started the ham and egg shows in 1914 to organize young people and give them outlets for their farm work. O'Neal had a remarkable career as teacher, county agent, and Fort Valley State College professor, and Church continued the ham and egg work. In the 1950s, he promoted a cooperative, and peanut farmers banded together to purchase a thresher. "I sent the first black boy to the national 4-H convention in Chicago," he boasted. Church had hoped to study veterinary medicine and was admitted to the University of Michigan but lacked the finances to attend. "I became very efficient in castrating pigs, hogs, cows, inoculating for cholera back in those days," he recalled. "I had my syringes and my equipment for inoculating hogs, cows." In 1958, he earned an M.A. degree from Tuskegee Institute.[8]

Georgia had segregated 4-H camps for blacks and whites, and after the Civil Rights Act of 1964, whites were slow to comply with integration orders. After agreeing to wait a year before sending black children to Rock Eagle, the white camp, Church was determined to go there the next year. The black camp at Dublin, Church judged, "wasn't up to standard." He wrote a letter to the University of Georgia extension administrators in Athens requesting that his 4-H kids go to Rock Eagle. "Before that letter got to Athens," he joked, "I said they must have met it on the way. I had an answer, Rock Eagle is filled." At a meeting in Americus, the white supervisor intimated that black agents were not cooperating with the 4-H program by pushing for Rock Eagle. Disgruntled by the supervisor's not-so-subtle intimidation, Church led a group of agents who confronted the supervisor. The next morning, he recalled, word had spread about his insubordination. "I knew then I was a bad boy," he remembered. "And from then on, I caught hell." At every meeting after that, the white supervisor encouraged Church to retire. When the supervising county agent retired, Church was passed over for a white man who lacked an M.A. degree, and Church ran the county agency until the man received his degree.[9]

In November 1965, William Seabron complained bitterly to Robert J. Pitchell, FES deputy administrator, about overt discrimination in the Georgia Extension Service. He raised questions about tests that were being given to extension personnel and asked what purpose they served. He also suspected that the freedom-of-choice policy for 4-H camps put undue pressure on black children and their families who wanted to choose Rock Eagle. Seabron suggested that a policy be established for assigning children to camps. Black agents continued to serve only black farmers, and the hopes for goodwill among county black and white agents had not materialized. African American professionals, he complained, were often not even assigned work "and sit merely wasting time." At some point, he feared, "they may be asked to report on their non-existent activities." Later in the month, he told Pitchell of a visit to Dougherty County, Georgia, where black farmers complained that they had been without a county agent for eight or nine years. The recently hired home-demonstration agent, he learned, "is jammed into a corner of the office and carries out no useful work." He furnished other examples of discrimination and asked Pitchell to look into the situation.[10]

Integrating 4-H camps preoccupied white extension administrators, and by the late 1960s, their tentative steps to integrate often failed, perhaps on purpose. W. M. Bost, Mississippi's extension head, admitted a lack of success and complained in June 1969 to Lloyd Davis, "I will not attempt to justify this except to say that integrated camping has been our most difficult hurdle." Bost promised to continue his efforts with "conservation clinics and special interest camps," insisting that closing camps would be "disastrous for 4-H Club work in our state." Yet in October he announced to Davis that he had canceled the spring 1969 camping program and, in January 1970, explained to his extension staff that due to civil rights regulations that could not be met, the entire camping program for 1970 was canceled with no future plans.[11] Robert Church had touched a sensitive nerve when he insisted on integrating Rock Eagle.

Black agents did not dwell on their subordinate positions in the Extension Service, Robert Church explained, because "that was the policy. That was tradition." Instead, they used common sense. "And, of course, a lot of us were buying homes, raising families, raising our children, educating our children." Church had a distinguished career, and although white administrators criticized him, African Americans showered him with impressive awards. "I achieved and I received the History Makers Award," he modestly observed. "I received the Distinguished Service Award from Fort Valley State, and I received the Hall of Fame Award from Tuskegee and on and on, various awards from churches, county commissioners, and so forth." He looked back on his career with pride. He also understood that the civil rights movement had not created prosperity or equal rights. Eventually, he recalled, some black farmers won seats on the ASCS committee, "but they had a lot of Uncle Toms on there."[12]

Obviously, African American extension agents varied enormously in their aptitude for addressing the needs of black farmers on the one hand and preserving white goodwill on the other. When Richard M. Shapiro and Donald S. Safford interviewed South Carolina African American extension personnel in April 1964, they found segregation and discrimination deeply embedded. All extension offices were segregated, and many black agents lacked access to slide projectors or tape recorders, were denied information on some programs, were excluded from participating in radio broadcasts, and watched new equipment they had ordered end up in white offices while they got hand-me-downs. Before 1955, the interviewers learned, the assistant in agricultural extension at South Carolina State College in Orangeburg had direct supervision of county agents, but since then, black agents answered to white county agents. In some cases, white agents demanded menial work from black agents. Although there was no written rule about participation in civil rights activities, black agents understood the unwritten prohibition. Black agents seemed wary of Shapiro and Safford and, according to one source, were sometimes untruthful. "He gets the most praise when he does the least," one observer said, adding, "The only thing a Negro agent does is what he thinks the white people want him to do." Both black and white agents sometimes claimed untruthfully that black agents served on various committees, hoping to impress commission interviewers. Harold McNeill, who supervised African American extension agents in western North Carolina, and Minnie Brown, assistant state home economics agent, reported much the same attitude.[13]

Many black agents endured prejudice and maintained their pride and resourcefulness. In his search for family roots in Mississippi, author W. Ralph Eubanks pondered how to explain his father's work as an extension agent to his children. "I should tell them about how my father had the title 'Negro County Agent,' was paid a fraction of the salary of his white counterpart, and worked in a tiny cinder-block building with a tin roof and no bathroom, in spite of being a college-educated professional." Eubanks's father bought a farm and built a house but had problems getting sufficient funding from the FHA to complete it. Ultimately, he became the assistant county supervisor for the FHA and was the boss, friend, and quail-hunting buddy of the man who had denied him the funding. Five years after Eubanks's father died, his family received back pay for the discrimination he had endured in the Extension Service.[14]

While Ralph Eubanks's father lived to witness benefits from the civil rights movement, the Extension Service remained obstinate and endured as a formidable segregated fortress that fiercely resisted implementing civil rights laws or serving its black clientele. Intractable and devious, the FES and white land-grant colleges smiled agreeably while feigning integration and demeaning black workers. In October 1965, administrator Lloyd H. Davis shamelessly claimed that the FES was in the forefront of civil rights compliance. William Seabron pointedly reminded him of "dissatisfaction among Negro State Extension employees" and "open complaints of discrimination." After 1964, African American extension agents existed in a second-class twilight zone, and leaders were folded into offices in white land-grant universities, in many cases with no meaningful work.[15]

In the mid-1960s, the Georgia Commission on Civil Rights advisory committee discovered that white administrators at the University of Georgia assigned black agents to work only with low-income farmers but provided no guidelines. Despite the FES's assurances of integration, many black agents continued to work out of separate and unequal offices. They also lacked secretarial help, were still called by their first names, and received scant information to distribute to black farmers. Most 4-H clubs and camps remained segregated, and in 1965, no blacks were among 249 Georgia youth who received awards. Black 4-H agents did not learn of crucial meetings, announcements, or projects. Georgia's extension magazine, *Cloverleaf*, was sent only to whites and over six years had carried only one article on black members.[16]

Georgia director of extension Linton W. Eberhardt Jr. failed to convince the advisory committee that the program did not continue discrimination and segregation. Fifty-year-old Eberhardt became head of Georgia's Extension Service in May 1963 after serving as a county agent and 4-H leader. Reared on a Jackson County farm, he held a B.A. in forestry from the University of Georgia and, before his promotion, had been associate director of the Georgia Extension Service. Despite the Civil Rights Act of 1964, which prescribed the integration of black and white administrative structures, no African American had been given primary responsibility and a "commensurate title." Black men and women in many cases had longer service and more advanced degrees than whites yet earned less. White associate county agents, for example, earned $1,130 more per year than blacks. In an egregious attempt to limit black opportunities, Eberhardt ruled that black assistant agents must enroll in a graduate program either at the University of Georgia or at a comparable land-grant college. Attending Fort Valley State College or Savannah State College would not count. He even declared that black extension workers with higher degrees in agriculture from such schools as the University of Minnesota, Iowa State University, and Michigan State University were unqualified for county agent positions "because of the lack of agricultural technology and lack of training in agricultural technology." The state committee interpreted Eberhardt's educational requirements as a transparent plan to demote and demean black extension workers. Eberhardt himself had no advanced degree.[17]

Despite their control over programs and information, white extension administrators were extremely wary of the changing attitudes among African Americans. At the meeting at Fort Valley State College in March 1964 discussed in chapter 2, white agents misinformed black agents when they warned that participating in civil rights activities was a violation of the Hatch Act. White Georgia Extension Service officials expressed "immediate hostility and defensiveness" when blacks challenged discrimination. White extension administrators spoke condescendingly to the black audience, and misleading, irrelevant, defensive, and insulting words and actions suggested that they held African Americans in contempt. Still, to appear interested in black farmers, extension reports exaggerated contacts with African Americans. In July 1965, William Seabron questioned a statistical report issued by Nyle Brady, the USDA's director of science and education, claiming that the Extension Service contacted 312,000 nonwhite southern farmers. Seabron pointed out that census statistics showed that at most 200,000 nonwhite farmers remained in the South. "How can figures like these be used to evaluate the actual extent of participation of Negroes in Extension programs?," he questioned.[18]

Seabron attended the national convention of black extension and home-demonstration agents on July 26, 1965, admitting that this would doubtless be the last convention of the National County Agricultural

Agents Association and the National Home Demonstration Agents Association. Most agents agreed that "the dissolution of these associations
was a desirable step forward," Seabron learned, but many members
wanted to keep some organization to address the needs of black farmers.
There was universal skepticism about the integration of black and white
staffs, especially since black agents were being placed under whites in
every county in the country. The new FES organization stripped black
supervisory personnel of their responsibilities, and Seabron discussed
this problem with Lloyd Davis and requested that he create some criteria "that we can defend and support as fair and equitable."[19] Between the
lines of Seabron's memo ran a subtext that the FES intended to destroy
black organizations, ostracize black agents within white organizations,
and prohibit blacks from having a voice in policy making.

In October 1965, Durham attorney C. O. Pearson, general counsel for
North Carolina's state conference of NAACP branches, complained to
Secretary Freeman about the FES reorganization. After reviewing the
FES's historical discrimination, Pearson observed that in North Carolina all policy decisions continued to come through white administrators
from North Carolina State College. "No Negro has ever been appointed
County Chairman or District Chairman in the history of the organization," he observed, "yet, there are qualified Negroes in the Extension Service." Former black administrators of 4-H programs were reassigned,
and whites took on supervision of black and white activities. The fate of
the African American extension staff at North Carolina A&T College in
Greensboro was uncertain. Pearson insisted that, when integration came
in North Carolina, it allow blacks to preserve their administrative duties.[20]

In October 1965, two retired black extension agents met with Thomas R.
Hughes and other USDA officials in Washington. "I was quite impressed
with their sincerity," Hughes confided to Seabron, "and I believe we owe
an obligation to run these complaints down in view of the fact they took
their own time and money to come to Washington to present them to us."
There was no magic bullet, Hughes admitted, but he wanted to address a
long list of discriminatory practices. The two agents complained that the
Georgia Extension Service had no black representatives on the advisory
committee for compliance, that no blacks were serving as county chairmen, that titles had been changed to the detriment of black agents, that
most blacks were working only with low-income families, that black state
staff members' titles and supervisory responsibilities had been reduced,
that black agents were forbidden to communicate with county sponsor-

ing boards, and that all civil rights actions, as well as other issues, were handled by whites.[21] The Extension Service in Georgia and throughout the South nullified the Civil Rights Act and disregarded Freeman's memo demanding compliance. The two black agents returned to Georgia realizing that they had no champion in Washington.

W. W. Law, president of the Savannah branch of the NAACP, complained to Freeman in November 1965 that Earlene G. Malone, who had worked as a secretary in the African American extension office in Liberty County since 1958, had been dismissed. She was not allowed to apply for openings in the white office either before or after integration. "There are no Negroes serving as secretaries or clerks in any county office anywhere in Georgia, to my knowledge," Law wrote. He observed that extension work "is still as segregated as it ever was and continues to discriminate against Negroes." He asked for an investigation and insisted that Malone be reinstated. Seabron replied that Malone was hired and paid by the county and thus was not an extension employee, but he did refer her case to the inspector general. He also admitted receiving other complaints from Georgia about discrimination.[22]

Annye H. Braxton complained directly to William Seabron about being fired on December 31, 1965, from her secretarial job in Linden, Alabama. She had agreed to move to Linden when the Demopolis African American extension office was combined with the white Linden office in July 1965. "I don't know if you are familiar with the situation in our area," she wrote to Seabron, "but it is an area dedicated to White Supremacy." On July 7, three days before she was scheduled to testify before the Alabama Civil Rights Advisory Committee about discrimination at the Bryan Whitfield Hospital, county extension chairman Frank Jones visited her office and asked what she was going to say and if she thought testifying was a good idea. "I told him yes, because I would tell only the truth about the hospital as I knew it to be," she had replied. As a private citizen, she argued, "I felt I was with in [sic] my rights to appear." In September, she enrolled her daughter in first grade at the all-white Demopolis elementary school. "For this," she continued, "the whites in Marengo County cut off our credit, and my husband was fired from his job." Her firing, she insisted, was "total racial discrimination." At the end of the year, both she and her husband would be unemployed, and she asked Seabron to look into the matter. Fred R. Robertson, director of the Alabama Extension Service, dismissed the complaint, justifying his decision to terminate Braxton "solely as an administrative matter." The commission's William

Taylor forwarded what he considered her "well founded" case to the Justice Department in December 1965, but two years later, he had received no response.[23]

Pauline F. Smith had much the same experience in Oktibbeha County, the home of Mississippi State University, and in April 1966, she appealed to Secretary Freeman to intervene. She had worked four years in the extension office with no problems until August 1965, when she enrolled her daughter in a white school. Her supervisor asked why she had not consulted him, and later she was informed that she lived outside the district and must send her daughter to the black school. She was later reprimanded for eating in an all-white café, even though she explained that she was "in the kitchen and . . . there were no whites in that part." Pressure continued on her to resign. When she applied for maternity leave to begin on February 28, she was terminated. "I could go on and on relating other things that are happening that I would rather not put on paper," she added.[24]

An inspector general investigation substantiated Smith's complaints and found that her supervisor's "discussions with Mrs. Smith were not in accord with the intent of the Civil Rights Act of 1964." Mississippi extension administrators insisted that maternity leave was "an earned privilege" that allowed discretion depending on "circumstances involved in the individual case." On July 29, 1966, Mississippi extension director W. M. Bost ruled that Pauline Smith "was denied maternity leave because of doubt concerning her marital status and her failure to provide evidence that she was married." Bost admitted that he knew the marriage had taken place, but the marriage certification had not been submitted to him. Her attempt to enroll her child in a formerly all-white school "had no bearing on her termination of employment," he insisted. He dismissed the OIG report that found her supervisor's conversation regarding her child's enrollment in a white school in violation of the Civil Rights Act of 1964. Bost itched to reprimand Smith for even filing a complaint, irrationally declaring that "her action was a serious abuse of a legal right." In a similar case dealing with Esther W. Reed's discrimination complaint on July 29, Bost menacingly wrote, "The penalty for making false charges under oath is removal from office, fine and/or imprisonment," even though some of Reed's charges had been substantiated by an inspector general report. Bost was less than subtle with his threats and intimidation.[25]

Inadvertently, Bost's fulminations revealed several layers of prejudice. One wonders, for example, how many white employees had to supply proof of marriage to their supervisor, had to consult with their supervisor about where they sent their children to school, or were reprimanded by their supervisor for their restaurant choice. Bost vented his exasperation in a letter to Lloyd Davis. "So long as the misdirected word of a disgruntled, poorly performing Negro employee is to be accepted at face value as a fact in preference to the word of the Extension Director," he bristled, "FES and State Extension Services will continue to do 'cartwheels' at their beck and call." He suggested that Esther Reed "should be dismissed for non-performance." Bost, as did most whites, recoiled at black aggressiveness. At an executive board meeting in January 1965, he warned that it was a "critical year" and blamed civil rights workers for demanding change. "We have our own way of life in Mississippi which has been disturbed." In 1981, after thirty years with the Mississippi Extension Service and numerous honors, Bost retired, and the Bost Extension Building at Mississippi State University was named in his honor.[26]

While Earlene G. Malone, Annye H. Braxton, and Pauline F. Smith ran afoul of white prejudice, Sadye Wier showed unusual mettle when moved to white offices in Columbus, Mississippi. She was given two weeks' notice of the move, and after discarding detritus from eleven years, she recalled, "I moved down on second avenue in Columbus with the whites." She knew her job and "wasn't going to be pushed around" or "dictated to" in her work with African Americans. The white woman in charge, she remembered, "kind of felt like she knew everything," but she depended on Wier for demonstrations. "Once or twice they knocked my black 4-H club girls out of a number of prizes they were going to get because she had some white person they wanted to give the prizes to," she said, but she was accustomed to such treatment in segregation days.[27]

Wier worked solely with African Americans, and she wanted it that way. She told a white female colleague, "You haven't been Black one day, and you don't know how these people live and I'm trying to give them a better way of life." Wier reminded the woman that her white clients had money and better living conditions. She also made a scathing observation about African American men, saying they were "lazy and they weren't too ambitious." Women, she insisted, "had to take the major role and see that they got where they were supposed to be as far as buying a home and educating the children and keeping clothes and things." Men, she complained, "were out there in the street with alcohol and doing something else that didn't or wasn't helping the family at home." Boys grew up with few responsibilities, she judged, "and then they turn around they haven't done anything but run out and get on alcohol or drugs."[28]

Ample evidence had accumulated revealing how the FES intended to handle integration. Reduced to its essence, its intention was to bury African American administrators in white land-grant schools, strip them of their administrative duties, place black county agents in inferior positions, cripple black 4-H initiatives, and assign blacks to work only with low-income black farmers. Despite the flood of memos circulating about FES interest in being fair to black workers, evidence reveals a plan to annihilate the remnants of the black program. No FES official ever suggested that the administration of a state program be transferred to a black land-grant school or that the white agent organizations be integrated into the black groups or even that highly qualified black administrators assume decision-making positions. State extension leaders dared opposition to their decisions, and, sadly, African American challenges found no support in Washington. The federal civil rights apparatus increasingly opened gaps for discriminatory treatment. The commission's Richard Shapiro analyzed this process and, in September 1965, insisted that without outside pressure, USDA extension administrators in Washington would never challenge state staffing changes. Even as the commission gathered evidence, southern land-grant schools implemented their discriminatory blueprint.[29]

Black land-grant schools faced a crisis as they lost extension work and denied charges that their graduates were not prepared to compete for federal positions. Land-grant colleges enrolled some 1 million students, but only about 43,000 were in black schools, some 4 percent. The secretary of agriculture's Citizens Advisory Committee on Civil Rights discussed the issue on December 3, 1965. R. P. Daniel, president of Virginia State College in Petersburg, explained that black land-grant colleges varied enormously and that originally "these colleges were not envisioned to be Land Grant Colleges" and "were not assumed to be significant in the Land Grant field." They primarily educated agriculture teachers and received no research funds. Virginia State had upgraded over the past decade, no doubt to amend the separate-but-equal gap. L. C. Dowdy, president of North Carolina A&T College in Greensboro, looked to the future of black land-grant education. He explained that these schools had been "existing on a starving diet all along" and the solution was not "to get fat overnight." Rather, he envisioned a patient climb to excellence. In 1960, he explained, A&T had added "Ag-Business and Ag-Science" and was placing some graduates in good jobs and sending others to Ph.D. programs. "We have about 37 white students in our institution," he told the committee. After a long discussion about funding and curriculum, Dowdy declared, "I didn't come to this meeting looking for a suggestion to phase out our program." With better faculty salaries and research funds, he judged, "I know our colleges will be better."[30] Because of discrimination, African American institutions of higher learning were constrained by miserly state funding, parsimonious foundation support, minute endowments, low salaries, weak infrastructures, and grave accreditation issues. Of the 105 predominately black colleges in the South, only 70 were accredited and at most 2 offered Ph.D. programs. In the South at large, only 8.3 percent of college students were African American.[31]

When the secretary's Citizens Advisory Committee met in February 1966, it heard from William Seabron about his trip to Mississippi and the evictions and complaints of former sharecroppers. "The people there were living in such poverty and squalor," he told the committee, "that it would be difficult to visualize it unless you see it firsthand." The committee thoroughly reviewed the case of Delores G. Morse, a home agent in Mecklenburg County, Virginia, whose claim of FES discrimination had prompted an inspector general investigation. Lester P. Condon, the USDA's inspector general, had examined Morse's complaint and found her allegations correct. Earlier in the meeting, he had conceded that his office had a backlog of some 800 cases. Condon grew up in Mt. Vernon, New York; took his law degree at Georgetown University; and worked for the FBI before accepting the inspector general position in July 1962. Delores Morse, who had always received a high efficiency rating, should have been classified at a higher grade, and FES administrator Lloyd Davis reported that she had been promoted on February 1, 1966. As he discussed Morse's complaint, Davis digressed into the complex cooperative relationship between the USDA and land-grant colleges and insisted that state and county FES employees "work under State laws that provide a State personnel system." The Smith Lever Act, of course, authorized the withholding of federal funds and other punitive actions if a state did not accept federal mandates. Davis joined the chorus of complaints about a lack of qualified African Americans. "In the past FES had lower standards for Negroes than whites," he claimed, ignoring the fact that many black agents had graduate degrees from excellent northern universities and years of experience. "We need Federal pressure," frustrated advisory committee member Father Louis J. Twomey insisted, but his voice was lost amid the complexities of the discussion.[32]

Lloyd Davis was less than forthcoming about resolving the Delores Morse case or revealing the continuing discrimination in Virginia's extension program. In March 1966, Seabron acknowledged that Morse had experienced discrimination during her twenty-year tenure and concluded,

"She has borne the economic penalty for all of that time." In July 1966, Morse complained to Seabron, "I am not satisfied with my salary." Despite a raise on May 1, she was still not at the top of her grade's salary range. Seabron urged Davis to support her request and suggested to Morse that her raise would come in August. But in August, she learned that because of her May raise, she was not eligible for another raise for eighteen months. Seabron weakly replied in October that he was still working on her case.[33] The slow and inadequate response to Morse's complaint seemed a calculated affront, another slight that revealed the USDA's malevolent discrimination.

In mid-June 1966, John W. Slusser from Seabron's office visited North Carolina and wrote a rich and detailed report on the Extension Service. Not all offices were integrated, nor was there much work across the color line, he found. Only 25 of the state's 2,400 4-H clubs were integrated. In Chowan County, he learned that the office furniture and supplies were "equally poor for Negroes and whites" but added, "There is neither functional nor physical integration of staff here." The state had six summer camps, one of which had been for blacks. Slusser visited the Betsy–Jeff Penn camp during a session that was supposed to be attended by 90 whites and 52 blacks, but the whites withdrew. An earlier session with 6 blacks and 145 whites worked well. Camp director Charles Page had a staff of ten, including one black counselor. "All staff attends church on Sundays in a group," Slusser wrote. "They get some stares from the citizens, but they are a swinging young group and make no apologies." Slusser observed the staff at an evening program working with black children. "I am convinced they could not have given any campers better staff participation than these children got. Color makes no difference whatever to these staff members. I was vastly impressed." Half of the black children's parents were sharecroppers. Slusser reasoned that if the white children signed up for the camp session had been allowed to decide for themselves whether to attend, things at the camp would have been fine. While North Carolina's efforts were promising, at the national level, tokenism sufficed. In September 1966, Secretary Freeman, attempting to demonstrate 4-H progress, requested that President Johnson meet six 4-H Club members who were on a national tour. "One of the delegates is a fine Negro girl," he assured the president.[34]

African American extension agents faced discrimination not only by white county administrators but also by the white National Association of County Agricultural Agents, which refused to share membership and power with black agents. Writing from Edenton, North Carolina, in July

1966, black extension agent Fletcher L. Lassiter complained to the Commission on Civil Rights' William Payne that blacks had been forced to abandon their county-agent organization because it was judged a remnant of segregation. In 1964 and 1965, when the NACAA invited black agents to its convention as guests, not members, they refused to attend. Since the NACAA influenced policy that affected black extension workers, Lassiter insisted that blacks participate on equal terms with whites. It made sense to Lassiter that leaders in the segregated organization should be leaders in the integrated one. In July 1966, the commission's Walter B. Lewis expressed dissatisfaction with the FES claim that it lacked jurisdiction over the NACAA's policies. Lewis, a Vicksburg, Mississippi, native who graduated from Tougaloo College, was deputy director of the Washington Urban League before becoming head of the commission's federal programs division. In 1968, he would head the civil rights office at the Department of Housing and Urban Development. Lewis learned that at its August 1966 meeting, the North Carolina NACAA branch did not even consider Lassiter's request for membership and again sent "a lily-white delegation" to the NACAA convention. It would be ideal if the state NACAA organization could work out its problems, Lassiter informed Seabron in September 1966, but "I am sure that you will agree that almost nowhere has this approach worked." Lassiter called on the federal government to end segregation and stressed that he was weary of hearing sympathy not backed by action. "I am asking for an equal opportunity to contribute to the ultimate objectives of the program rather than to be given what someone else decides that I should be satisfied with."[35]

Lassiter's dilemma personified how some civil rights initiatives worked to the detriment of African Americans, but it was only one example of sweeping discrimination embedded in the Extension Service. White 4-H members participated in a wide variety of activities, such as tours, conventions, and stock-raising and judging contests, while African American youngsters worked on a narrower spectrum of projects. While black agents were reluctant to state outright that civil rights activities would work against a farmer seeking an FHA loan, they understood that their own participation in civil rights activities would at least be frowned upon and at most cost them their jobs. Extension work, then, was divided into two hostile camps: one white, well financed, and adequately housed, and the other black, financially starved, and demeaned.[36]

Throughout the spring and summer of 1965, the commission, William Seabron's office, and Lloyd Davis met, corresponded, and argued about compliance rules. Seabron only saw the draft FES compliance form on

November 22, 1965, and said he "felt immediately that it was unaccept-able, and drafted a form, for discussion purposes, which we felt would do the job better." On December 1, Seabron discussed his draft with three FES staffers, but the next day, with no input from Seabron, they decided on a watered-down version. "We were disturbed to see that 3 critical items were substantially altered and 10 others omitted completely," he complained. The FES also missed the deadline for submitting its report on civil rights progress.[37] The Extension Service choked compliance for-mulation and dawdled on civil rights reports, and Seabron, lacking the authority to coerce cooperation, suffered another disappointment.

The Extension Service entered into a drawn-out process to delay equal opportunity employment regulations. In January 1966, Seabron com-plained that the FES had dragged its feet on submitting reports and had sabotaged a proposed rules agreement. On July 29, all USDA agencies involved with civil rights enforcement and FES representatives finally signed a draft on equal opportunity employment regulations that was the result of excruciating negotiations. Even as this hopeful scenario played out in Washington, Extension Service head Lloyd Davis was circu-lating the draft to state extension leaders, who insisted on forty changes that would weaken if not eviscerate enforcement. Seabron carefully ex-amined whether federal, state, or county governments were responsible for enforcing the employment regulations. In October, he complained that all land-grant university presidents had agreed that extension per-sonnel were employees of the university. African American extension agents insisted that they were federal employees because they, as were white agents, were part of the federal civil service retirement system, held civil service appointments, had franking privileges, and in many coun-ties worked in federal office buildings. If states retained control over civil rights in the FES, Seabron feared, "the Department and the Secretary will be subjected, in my opinion, to the worst civil rights publicity accorded any Federal Department to date." Freeman's dilemma, according to Sea-bron, came down to the choice between offending powerful land-grant university presidents or offending vulnerable African Americans.[38]

State extension leaders, meanwhile, portrayed equal employment guidelines as a source of endless litigation. "Complaints may be filed by the representative of or an organization for the aggrieved individual (SNICK [sic], COFO, NAACP, CORE, Delta Ministry, Sharecropper Fund, etc.)," Mississippi extension head W. M. Bost presumed. In Bost's estima-tion, adhering to federal employment guidelines raised a host of issues that included penalties for violations, endless reviews of civil rights initiatives, focus on complaints rather than "reports on constructive programs," suits for previous racist policies, and, most important, sweeping away the power of land-grant universities to manage their own affairs. As he would in the future, Bost stressed the tripartite composition of the Extension Service, insisting that county staffs should have control over hiring. "I do not believe that State Extension Services should submit to these proposed regulations," Bost advised Mississippi State University president William L. Giles, "and recommend that should they be implemented by FES, we should ignore them and carry our case to the courts." As Bost made his case for nullifying federal guidelines, he portrayed the FES in Washington as a "policing agency." He had no problem accepting equal opportunity "as a policy," he clarified, but wanted to ensure that state and county entities carried out, or failed to carry out, the policy.[39]

Lacking equal employment guidelines, the Washington FES office could only audit noncompliance with Title VI civil rights laws. In December 1967, Lloyd Davis sent Bost the inspector general's audit of Mississippi's Extension Service and asked for corrective action, exactly the kind of intrusion Bost had feared. Fifteen counties continued segregation in 4-H work, office space, home demonstration, and access to programs. After listing five pages of shortcomings, Davis concluded his letter by commending Bost "for the splendid progress you have made in your State."[40]

As 1966 came to an end, the secretary's Citizens Advisory Committee on Civil Rights reported on its hearings and addressed nine areas of discrimination. "The unemployment, underemployment, and displacement taking place in southern rural areas deriving from accelerating technological change, the new cotton program, and the pressures arising from the civil rights revolution," the report began, "are by no means meeting with an appropriate response at any level of government." The committee targeted the Extension Service for focusing on "the more prosperous and successful farmers" while neglecting "a large percentage of those who are most in need of help." The committee urged Secretary Freeman to use his influence with the president, his cabinet, and other powerful players. Local control over extension personnel meant continuing discrimination, the report argued, and the committee recommended placing all appointments under federal guidelines. Agriculture as a way of life "is dwindling rapidly," it cautioned, pointing out that in 1950 6.7 million people worked in agriculture but in 1966 only 3.25 million remained. Many displaced farmers moved north, the committee noted, underlining a major concern that urban concentrations of African Americans were flash points of violence. The resolutions addressed the uneven distribution of food to poor people and the existence of hunger despite bulging warehouses. In states where African American farmers constituted at least 15 percent of farm operators, the report suggested that "the Secretary appoint at least one Negro farmer to the State ASC Committee." Despite federal programs, it observed, "the small Negro farmer is descending to a distressful serf-dom."[41]

The delay in setting equal opportunity guidelines put Seabron in an uneasy situation. In July 1966, African Methodist Episcopal minister Daniel J. Zeigler reminded Seabron that the case of Union, South Caro-lina, assistant county agent M. B. Jackson seemed no nearer to resolution than it was two months earlier or, according to his count, seventy-seven days before that when he first complained. Seabron explained that regu-lations still were not in place. In January 1967, Zeigler upbraided Sea-bron again not only for failing to press the case but also for sending it back to the Extension Service for resolution. Zeigler wrote that he "should prefer to eat locusts and drink branch water" than suffer Seabron's "hu-miliation" of sending Jackson back to "the abuses of southern tyrants." The case, he continued, "is just another indication that America is going backwards instead of forward despite the chief-executive's nice language which for all practical purposes contains only promises." Seabron wrote a memo to Thomas Hughes explaining that he was attempting to find "some remedy" for such discrimination.[42] Zeigler's words upset Seabron because they hit so near the mark of Extension Service perfidy that he seemed helpless to redress.

No matter what good intentions were paraded in official memos, agency administrators had neither the will nor the inclination to push states, land-grant schools, or county committees to abide by equal op-portunity guidelines. The Extension Service, Seabron observed, was orga-nized in such a way "as to lodge effective control of the entire program in the County Board or Police Jury," that is, the white elite. "Individual per-sonnel of the Extension Service are hired, assigned duties, given raises, transferred, and terminated," he continued, "as the direct result of inter-est shown by the County Boards." The resulting snarl, he concluded, was the product of complex memos of understanding generated by local, state, and federal components of USDA policy. The only way to imple-ment equal opportunity, he suggested, was for the federal government either to fund or administer all Extension Service budgets or to revise the memos of understanding.[43] Seabron, along with everyone from black farmers to civil rights organizations, knew aggressive federal enforcement was the only way to crack discrimination. Inexplicably, the USDA seemed exempt from White House pressure to rigidly enforce civil rights laws.

Whites continued to ignore the inequities that grew out of the segregated 4-H clubs. The U.S. Commission on Civil Rights' Walter Lewis discovered in 1966 that of the 119 4-H fellowships awarded since 1931, none had gone to an African American. In October 1967, Seabron reminded FES head Lloyd Davis that once again no African Americans had been selected for fellowships. Seabron asked Davis to provide him "with complete information relative to the candidates, their qualifications and race, the judges and their race, and the manner of selection."[44] Davis stonewalled on this and other requests, in particular his pledge to take 4-H programs out of schools that did not comply with civil rights policies. Davis lamely explained that it would be unfair to withdraw a program only to have a school come into compliance shortly thereafter. Seabron adamantly disagreed. Davis evidently thought it better to allow widespread noncompliance than to enforce the law. In April 1967, a frustrated Seabron wrote to Secretary Freeman, "Fifteen months after the target date, sixteen counties in five states still maintain separate offices for white and non-white Extension Service clientele."[45]

Like ASCS executives, Extension Service administrators were masters at turning negative OIG reports to their advantage. FES civil rights liaison Luke M. Schruben interpreted inspector general investigations and forwarded letters for Seabron's signature. Seabron often found discrepancies between OIG findings and Schruben's drafts. An OIG investigation in Dougherty County, Georgia, for example, found that black 4-H members "were denied the same services provided white 4-H Club boys" and that county staff members doctored compliance reports. Yet Schruben's letter concluded that "none of the facts affords us with persuasive support that violation of Title VI has occurred as charged in this case." Seabron refused to sign and raised questions about many of Schruben's distortions.[46] Such malign indifference to facts and pernicious misconstruing of discriminatory behavior indicated not only a breakdown in departmental ethics but also a willful nullification of the Civil Rights Act.

Tom L. Lambert from the OIG investigated Richland Parish, Louisiana, in November 1967 and found egregious civil rights violations. "If a 'nigra' comes to see me with his hat in his hand, I will help him," the county agent stated. Lambert indicated that this explanation meant "the Negro has to relinquish his pride and self-respect in order to obtain the service that is supported by Federal funds." The county agent retaliated against

G. D. Merrick of Rayville for writing to Seabron about discrimination in the extension office. Somehow the agent obtained copies of Merrick's letter and showed it to local white farmers, who "discontinued veterinary business with him." Lambert, with careful understatement, suggested that "the attitude of the County Agent is such that it may be a barrier to Civil Rights implementation in the CSES [Cooperative State Extension Service] program and services in the parish." Seabron discovered that in 1965 Louisiana extension director John A. Cox claimed that with the exception of housing, "everything was either in good shape or moving adequately toward a situation of civil rights compliance." The assurances in 1965, Seabron judged, were misleading. "We were all misinformed." In fact, nothing in Richland Parish had changed in over two years except that black and white offices had been consolidated. Seabron argued that the county agent was not fit to serve in the federal government and should be fired. He knew exactly what the FES was doing and shared his thoughts with Inspector General Lester Condon. "I feel, Les, that your office and mine have been treated poorly in this matter," he confided. "What can be done to assure us that we, both of us, are not being led down the primrose path in other places and cases?"[47] This was as close as Seabron ever came to acknowledging the Extension Service's duplicity.

Often discrimination was hidden beneath surface compliance. On January 5, 1967, Calvin L. Beale, a statistician with the USDA's economic research service on holiday in Texas, visited the Fayette County extension office hoping to get insight into the county's outmigration. The most recent telephone directory, he discovered, listed an extension agent and the Negro Extension Service. He found that the extension office was a modern facility with "small private offices for each of the four staff members." He asked the secretary if he could speak with a county agent. She advised him that "they are *both* out at the moment" but said she expected them back shortly. Beale noticed a black staff member, and the secretary acknowledged that he was an agent. "It apparently never entered the mind of the secretary that a white man coming into the office asking to talk with an agent," Beale observed, "would be interested in talking with the Negro staff member." Mr. Randolph, Beale discovered, was in his late fifties, had lived in the area all his life, and was proud of his work with both whites and blacks. He deplored the lack of extension work among "the Mexican population" and the decline of interest among African Americans after whites took over much of the work that he used to do. The white agents controlled the county advisory committee, which consulted Randolph only as an afterthought. The district agent sent the advisory committee's plan back for more input by African Americans, but the revisions did not result in blacks being appointed to committees or subcommittees. Everyone in the office was friendly and helpful, Beale observed, but, as evinced by the secretary's disregard for the black agent and white agents' disregard for Mexicans and by the freezing out of black input in the advisory committee, the vestiges of segregation were unmistakable.[48]

In July 1967, Ruth W. Harvey, chair of the education committee of Georgia's Dublin and Laurens Counties NAACP, suggested that since the passage of the 1964 Civil Rights Act, whites had increased discrimination and cleverly discovered ways to nullify each new piece of legislation. Harvey wrote of the "death knell" and "those awful death blows which are being dealt to Negro workers by their white counterparts." For years, she recalled, extension in Georgia had been a "Guiding Light" to rural people, but it was "fastly fading into folklore and folksong." Since 1964, whites had consolidated power and "have all but expelled the Negro agents." Paradoxically, Harvey pointed out, extension agents and black farmers "are really suffering more from the so-called Civil Rights Act than their white counterparts." She analyzed the way whites had isolated blacks and denied them a voice in policy and listed eleven unfair practices and five examples of discrimination. Protests had not changed the situation, and she suggested in her letter to Freeman that a lawsuit might soon be initiated. Two months later, C. L. Tapley, who identified himself as chairman of the Georgia NAACP committee on complaints, sent a long list of violations to the U.S. Commission on Civil Rights and explained that he had reported them to state extension leaders and to Lloyd Davis in Washington. "They seemingly are together in their organized program of seeing to it that fewer and fewer Negroes receive the benefits of the service and that Negroes are gradually phased out of employment in the service." Tapley, like Harvey, suggested legal action.[49]

In 1964, Tuskegee Institute requested that Auburn University recognize its School of Agriculture as a substation of Auburn's agricultural experiment station system and requested a $335,000 grant. The proposal languished in the Auburn system for a year before the dean of agriculture and experiment stations, E. V. Smith, reminded President Harry M. Philpott that "the University will have to come to grips" with the request. Smith had done preliminary research and discovered that the Texas and Virginia experiment station systems had small branches at Prairie View A&M University and Virginia State College, respectively, both black land-grant schools. Smith also discussed the issue with white southern administrators, but it was unclear whether he was searching for a way to implement or avoid such a relationship with Tuskegee. Since Alabama A&M College was the official Alabama African American land-grant school, he was unsure whether Tuskegee would qualify or whether politics would necessitate two African American substations. "In all fairness," Smith observed, "Tuskegee Institute does appear to have a larger and better trained staff than any of the Negro Land-Grant Colleges." As the issue moved slowly through the Auburn bureaucracy, Tuskegee dean L. A. Potts appealed to Thomas Hughes in Freeman's office. "For three-fourths of a century Negroes have with a few exceptions been denied an opportunity to make a contribution to society through the state agricultural experiment stations in the twelve southern states," he reminded Hughes. In his tardy reply in February 1966, President Philpott explained that the ten Alabama substations were tightly controlled by Auburn and had no permanent scientists in residence. Even if funds were available, he continued, "it would still represent quite a radical departure in our operation" and, he hinted gravely, provoke similar requests from other institutions.[50]

African American land-grant universities became centers of intrigue as they lost valuable extension programs and watched federal funding move to larger white schools. In Georgia, a complex scenario unfolded as Fort Valley State College, evidently cooperating with the University of Georgia board of trustees and other stakeholders, lured the state's Negro Extension Service away from Savannah State College. Fort Valley president Cornelius V. Troup and Cozy L. Ellison, chairman of the Division of Agriculture, apparently convinced the USDA and the state legislature to approve the transfer. Troup followed the noted educator Horace Mann Bond as president of Fort Valley and, in addition to successful fundraising, added master's degrees in education, home economics, and agriculture to the curriculum.

Troup and Ellison's plan surprised both University of Georgia president O. C. Aderhold and extension director L. W. Eberhardt Jr. "I did know, of course, that Fort Valley had been designated by resolution of the Legislature as the negro [sic] Land-Grant institution," Aderhold admitted in May 1963, "but I was not aware, or I had forgotten, that the Board had authorized the moving of this personnel when facilities were available at Fort Valley."[51] Eberhardt secured a copy of Ellison's appeal to the USDA and bristled that the letter "goes over the heads of everyone in the State in attempting to bring pressure from Washington." Ellison's letter, Eberhardt insinuated, manifested a "lack of knowledge of the purpose of the Cooperative Extension Service." In April 1964, Aderhold wrote that he was "astounded" that the university board of regents had "passed a resolution directing that we move our Negro Extension workers from Savannah State College to Fort Valley." He had expected the board to postpone action. The black agents, he had learned, did not favor the move. Aderhold released a statement suggesting that he opposed the move to Fort Valley and recommended that if the agents were to be moved, it should be to the University of Georgia. Evidently, Fort Valley was preparing for the move by shifting its course emphasis from agriculture to research and science.[52]

Only a thin evidential trail for this intrigue remains, and no doubt much of the negotiating took place over the telephone or in person. Still, a *Wall Street Journal* story by Joe Western indicated that the move from Savannah State to Fort Valley was a complex scheme by the University of Georgia's board of regents to avoid integration. "On June 10, over Agriculture Department protests, Georgia's Board of Regents moved a six-man Negro Extension headquarters staff from one Negro college to another, instead of integrating it with the University of Georgia's all-white extension staff," Western reported on September 14, 1964. "To Mr. Freeman's advisers, the move looks like a direct challenge," the article continued. Aderhold and Georgia white extension leaders, of course, saw the move as an end run by the board of regents, with the complicity of the state legislature and Fort Valley's President Troup. Although the move probably violated the recently passed Civil Rights Act, Secretary Freeman and his staff were at a loss how to respond, the reporter learned, and they suspected that they would be unlikely to cut off funds. "We don't believe the new law gives us power to cut these off," a USDA lawyer offered. "Besides, we don't think it can be shown that any benefits have been denied any farmer for racial reasons."[53]

The extension move to Fort Valley lasted only a year, and in March 1965, Eberhardt transferred four African American agents to the extension annex building on the Athens campus and sent two 4-H specialists to the 4-H Club Center in Dublin. "This," he concluded on March 8, "would move all of our personnel from the Fort Valley State College Campus." The move was scheduled for July 1, 1965. Eberhardt evidently traveled to Washington in May to explore policy on the logistics of integrated offices and learned that black extension staff must be "housed with white staff members of similar responsibility." As other extension efforts would show, placing black and white agents in proximity did not accomplish integration. Fort Valley's Cozy Ellison struggled to maintain agricultural education at Fort Valley and, in August 1967, reminded its new president,

# THE NEGRO FARMER

PUBLISHED IN THE INTEREST OF FARM AND HOME DEMONSTRATION AND 4-H CLUB WORK THROUGH THE NEGRO
DIVISION OF THE AUBURN UNIVERSITY EXTENSION SERVICE, AT TUSKEGEE INSTITUTE

VOLUME 24          TUSKEGEE INSTITUTE, ALABAMA, OCTOBER, 1962          NUMBER 10

## Pringle's New Home
Mr. and Mrs. Ransom Pringle of Burkville, Ala. sit on the lawn in front of their new home discussing crop prospects with

## Success Comes Double To Family

*William Bailey Hill (left) talking with Mr. and Mrs. Ransom Pringle. From
The Negro Farmer, October 1962. Courtesy of National Agricultural Library.*

W. W. E. Blanchet, that many of the state's vocational agriculture teachers had graduated from the school and that enrollment in agricultural courses was increasing.[54] In 1979, the Agricultural Mechanics Building at Fort Valley was named in Ellison's honor.

In Alabama, William Bailey Hill headed the Negro Extension Service at Tuskegee Institute. He grew up on a farm in Bibb County, Alabama, and his schoolteacher parents encouraged his education. After attending local schools, Hill entered Tuskegee Institute's high school and then took several teaching jobs before returning to Tuskegee for college. After earning a degree in agricultural education, he taught in Arkansas before returning to Alabama in 1934 and joining the Extension Service a year later as an agent in Marengo County. A decade later, he moved to Tuskegee Institute as a district extension agent. In 1948, Hill earned an M.S. degree at Cornell University, and in 1949, he became state leader for black extension work. Later he earned a Ph.D. in cooperative extension from the University of Wisconsin. Hill supervised seventy-two county agents and home-demonstration agents, four district agents and district home-demonstration agents, a state 4-H Club agent, an assistant editor, and five clerical workers.[55]

As black units of the Extension Service were incorporated into the white structure, agents were consigned to serve only "limited-resource" families, those with $3,000 or less annual income and less than $10,000 in gross income from farming. Bailey Hill, who in 1965 moved from Tuskegee to Auburn University, spoke of his assignment in August 1967. While black agents worked only with black farmers, Hill observed that 87 percent of the 92,530 farm operators in the 1964 census sold less than the prescribed $10,000 and that 39 percent had yearly incomes of less than $3,000, thus falling into the limited-resource category. Hill did not break this category down by color, but he insisted that extension, especially demonstration work, was reaching out to such farmers. Despite discriminatory practices, Hill spoke of progress, of farmers learning to grow cucumbers and other crops to find a niche to survive. Hill did not mention black farmers whose incomes exceeded the limited-resource guidelines or indicate where they might find useful information. Such speeches were a far cry from Hill's earlier work at Tuskegee administering the Negro Extension Service.[56]

Extension Service discrimination operated in the context of continuing civil rights activities throughout the South. Compliance officer John Slusser informed Seabron in February 1968 that Simpson Clark, an assistant county agent in Morgan County, Georgia (which was 47.6 percent African American), was being dismissed. Clark, thirty-one, was married with two children, held a master's degree in animal science, and had worked in extension for eight years. Morgan County authorities claimed that they could no longer afford the county's $1,200 portion of his salary. At the time Clark was being terminated, the white county agent was retiring, and African American farmers wanted Clark as his replacement. Clark was not only well educated but also active in a six-county nonprofit program to help low-income farmers, a program attacked by the white elite. In adjoining Walton County, blacks were, in Slusser's words, "demonstrating, lying down in the front of buses and picketing" because three black teachers were fired. Slusser warned Seabron of the volatile situation and that the provocative actions of the Morgan County authorities could lead to trouble. This example of the Extension Service's callous disregard for equal opportunity caught the eye of Secretary Freeman, who met with Georgia extension head L. W. Eberhardt Jr. "I was impressed and delighted by what you're doing with Extension in Georgia," Freeman gushed before expressing his concern for conditions in Morgan and Walton Counties. He recommended that Eberhardt "take a strong step forward" and hire Simpson Clark as county agent. Eberhardt twisted Secretary Freeman's praise into sanction for further discriminatory actions, and there is no record that Freeman profited from Eberhardt's lesson on southern charm and duplicity. In August 1968, Seabron's office reported that in Georgia no African Americans attended the annual Home Show, communication school, forestry camp, or leadership conference and that there were no black officers in district or state 4-H clubs, no expense-paid trips for blacks (whites had 150), and no college scholarships for blacks (whites had 15).[57]

In Warren County, North Carolina, Ernest A. Turner, chairman of the county NAACP chapter, reported in August 1968 that the extension office was under one roof but still segregated with whites on one side and blacks on the other. The white county agent had recently hired an assistant, overlooking the black agent, who had far more experience. "Of course," Turner added, "his new agent was white and Mr. Cooper is black." Turner tabulated the county office staffs of the ASCS, SCS, FHA, and Extension Service and came up with twenty-one whites and three blacks. African Americans composed 65 percent of Warren County's population. An audit of Virginia's extension program in 1967 revealed continuing discrimination in pay and training, inadequate compliance reviews, and a long list of discriminatory practices.[58]

White land-grant schools fought equal employment guidelines and worked through the National Association of State Universities and Land-Grant Colleges (NASULGC), a powerful lobby organization. Mississippi State University president William L. Giles complained of the proposed guidelines to the USDA's George L. Mehren on July 14, 1967. As currently written, he whined, land-grant universities were to abide by "certain prescribed methods" that had a "'policing' ring" and would incur unreasonable personnel expenses. Russell I. Thackrey, executive director of the NASULGC, commiserated with Giles on July 18, arguing that the regulations "in effect by-passed the university completely." Thackrey had worked behind the scenes to weaken the regulations and assured Giles that "they are at least a vast improvement over the original draft version." A week later, Giles boasted to Thackrey of "the wonderful progress" in Mississippi in equal opportunity and claimed that state and local people "have worked so diligently and in such good faith" that constant federal pressure, or, as he put it, "constant abrasions," was unnecessary.[59]

Seabron judged that a September 1967 FES equal opportunity draft contained fatal problems and that it "should be abandoned" and await a ruling by the attorney general. "The matter of EEO in the Extension

Services is the most persistent of all complaints of racial discrimination within the Department of Agriculture," he pointed out to Assistant Secretary George Mehren on October 20. Only a ruling from an impartial Justice Department, he reasoned, could settle the matter and end the bickering about who was and who was not covered by equal employment regulations. "The drafted proposal requires the employee to seek as his judge, in the first instance, the very person whom he accuses of racial discrimination," Seabron observed. This route led to the complaint graveyard. "I can only conclude that the management of the Federal Extension Service is unable or unwilling to compel the necessary changes," Seabron warned Secretary Freeman in December 1967. FES intransigence outraged the Justice Department, the Office of Economic Opportunity, the U.S. Commission on Civil Rights, and the Civil Service Commission. Lawsuits were "imminent" in Texas, Georgia, Virginia, and North Carolina. "No one," Seabron concluded, "can defend the Department against the wave of honest criticism which is about to hit." In March 1968, Seabron again urged Secretary Freeman to issue equal employment guidelines for the Extension Service.[60]

Yet the Extension Service sidestepped complaints and continued to stall. Freeman had hesitated for nearly two years to issue regulations binding the Extension Service to equal opportunity and had allowed extension personnel to tie the issue into knots. After Justice Department assurances, in January 1968 Seabron at last pushed ahead with the rule that was published in the *Federal Register* in May 1968.[61] In a forceful March 1968 memorandum, Seabron had reminded the secretary that "racial discrimination in hiring, assignment, promotion and compensation of employees" continued throughout the South. The Extension Service would have no structure to handle the flood of equal employment opportunity complaints until Freeman approved procedures. "Deservedly or not," Seabron added with understatement, "USDA's equal opportunity image is not good." The employment regulations were legal under the Smith-Lever Act, Seabron stressed, and the Justice Department had assured him that they could be defended in court. He added, "It is morally right that we not stand by and let discrimination continue when we have it in our power to institute measures to put an end to it." Unless the USDA provided "moderate civil rights leadership," Seabron warned, African Americans could embrace "the concept of black power, which basically preaches that democracy is impotent insofar as solutions to the problems of black people are concerned."[62]

In May, Secretary Freeman sent copies of the regulations to the presidents of the white land-grant schools, which set off Mississippi State University president William L. Giles. He armored himself in a state- and even county-rights argument tailored to the university and insisted that Freeman's meddling in employment issues that were better left to the Extension Service was "without precedent this side of Moscow itself." Giles faulted Freeman for focusing on enforcement of equal opportunity with hints of, in his mind, "the goad and the fetter." In Giles's estimation, there was a "field day already being enjoyed by the flotsam of the legal profession, the inciters to complaint, the latter-day carpetbaggers, in residence and *in absentia*, who aggravate civil strife in the name and at the expense of civil rights." Giles and the university extension staff had been in touch with other schools, and he expected the pressure to ease on equal opportunity. After praising Giles's letter as "a masterpiece," Mississippi congressman Jamie Whitten assured him that he was working to undermine the regulations. "I have been trying to do my part to stir up public opinion," he confided to Giles, "feeling that would be the best way to make the court and others respond in the right direction." When Clifford M. Hardin became secretary of agriculture in 1969, Giles sensed sympathy for noncompliance, and his tone changed drastically. In response to Secretary Hardin's pleas for cooperation among black land-grant schools, Giles assured him that a group of representatives from Alcorn A&M College and Mississippi State "meets regularly to discuss problems and possibilities for helping each other."[63] Such a thin approach to cooperation seemed to placate the new secretary.

In 1968, Congressman Whitten inserted a statement in a House of Representatives conference report denying the secretary's authority to meddle in extension business. "You may be forced to make a political decision," a staff member warned Secretary Freeman, "regarding the desirability of opposing Whitten versus the advice of the Department of Justice that you have such authority." The statement implied that the USDA could not enforce equal opportunity by cutting FES funds.[64] John C. Lynn, legislative director of the Farm Bureau Federation, thoughtfully sent Senator Herman Talmadge a copy of the regulations, balefully adding, "I believe it is the beginning of the end of the Federal-State Cooperative Extension Service." After a commentary period, the equal employment opportunity regulations were finalized in August 1968, and Freeman danced around Whitten hoping to escape retribution.[65] Equal opportunity in extension hiring threatened the near monopoly that whites enjoyed in well-paying and prestigious federal jobs.

As integration orders came down, white women objected strenuously to the prospect of black home-demonstration agents visiting their homes and sharing equal status in space where black women had formerly been servants. Mrs. A. L. Carson of Smyrna, Georgia, on the other hand, directed her ire at the Georgia Extension Service. "It is high time Georgians quit playing the 'game' of waiting to act justly towards their fellows until they are forced into such action by the big bad Feds," she chided, "so we can holler 'I didn't want to—they made me.'" "We have never fooled anyone except ourselves," she concluded, "and the day may come when even we are not fooled by such misdirection."[66]

As the 1968 election approached, Seabron became more adamant about civil rights enforcement. As poor as the USDA record in civil rights appeared at the time, a Republican administration elected in part with a "southern strategy" might reverse even the modest gains of the mid-1960s. In September 1968, Seabron impatiently scolded Extension Service administrator Lloyd Davis: "It is imperative that you either act to terminate discrimination and segregation immediately or, alternatively, prepare and present to the U.S. Commission on Civil Rights that material which will convincingly demonstrate that discrimination and segregation do not exist in the Extension Services. To fail to do either is to be unfair to the Secretary who seeks the successful discharge of his responsibilities."[67] As usual, Davis answered with inaction.

State extension leaders duly distributed numerous civil rights directives from Washington, but the remnants of segregation festered. There is no indication that black extension workers who were transferred to white land-grant schools had a voice in any civil rights planning. Compliance reviews piled up, but no recommendations were forthcoming. Indeed, the paper storm helped hide continuing discrimination for distribution of memorandums had no impact on equal enforcement of civil rights laws. Despite the neutral compliance language, there was a subtext that Auburn University would protect as much of segregation's legacy as possible. The U.S. Commission on Civil Rights understood Seabron's frustrations and offered a telling analysis. "Not only is Mr. Seabron limited by the arbitrary response of agency administrators, he is limited, in practice, by the interposition of other staff advisors to the Secretary." The commission pointed to Seabron's admission that he had to deliver an important document directly to Freeman "because if he sent it through the Department's mail system, 'somebody else usually decides if the Secretary should see it.'" Isolated among white bureaucrats, Seabron pushed for implementation, but as the commission astutely observed, "interposition" prevented effectiveness.[68]

With Richard M. Nixon installed as president of the United States and Clifford Hardin as secretary of agriculture, extension administrators aggressively attacked affirmative action employment rules for land-grant universities. Joseph M. Robertson and Ned D. Bayley submitted a new employment plan, explaining to Secretary Hardin on February 5, 1969, that it had "limited provisions for *'affirmative action'* for equal opportunity as compared to programs developed for regular Federal employees." Other federal programs required that "qualified minority personnel are to be 'sought out' by deliberate, aggressive, positive action in recruiting, assigning, upgrading and training personnel" and mandated that minority personnel have "full access" to training and advancement. Robertson and Bayley assured the secretary that they supported affirmative action efforts but "we do believe that it would not be appropriate to *require* them as part of the EEO regulations with the threat of withholding funds or taking other action if they are not included." Stripping the power to withhold funds was, of course, a transparent plan to protect discriminatory hiring in land-grant universities and county offices and to safeguard white privilege. Robertson and Bayley admitted that civil rights groups would likely criticize their plan, and, they added, "Bill Seabron . . . strongly disagrees with our position." Indeed, Seabron sent a memorandum to Freeman the next day reviewing the procedure for enforcing equal opportunity complaints.[69]

Hoping to undermine earlier decisions, Lloyd Davis suggested that Secretary Hardin review statements by Jamie Whitten, consult his congressional conference report, and request a fresh opinion on equal employment requirements in the Extension Service from the attorney general. Davis was especially interested in preserving the right of county offices to control their own hiring. "The issues raised in this letter should be resolved before State EEO plans are signed by you," he advised the secretary. Davis sent his memorandum through Ned Bayley, and the two men formed a troika with Joseph Robertson that undermined equal employment rules for Extension Service personnel.[70]

Telephone calls became the preferred communication among FES executives even as compliance forms created a mountain of paper. The higher the pile of paper, the less action on civil rights. When Albert B. Britton Jr., chair of the Mississippi advisory committee to the U.S. Commission on Civil Rights, circulated a report on the Mississippi Extension Service's dismal equal opportunity record, director W. M. Bost claimed the data was "outdated" and boasted of notable progress. The report observed that "rural minority groups have not benefitted and are still today the victims of both overt and inherent discrimination and lack of equal opportunity in extension programs." Camps, 4-H clubs, and home economics clubs were segregated if they existed at all. At Mississippi State University, tokenism resulted in only 6 black professional staff among 90, only 3 of the 63 nonprofessionals were African American, and there were no black clerks. None of the 6 black professionals worked in the major subject areas such as engineering, agronomy, nutrition, family life, and rural sociology. Salary differentials continued, and highly qualified African American supervisors who had been integrated into Mississippi State "were not assigned work commensurate with their previous experience." Two-thirds were assigned resource-development work, code, the report stressed, for "work with Negroes." The same discriminatory treatment occurred in county offices.[71]

It took Bost ten pages to explain to Davis the charges in the advisory committee's report, going to great lengths to convey very little. As if the argument were not worn to shreds, Bost, sometimes failing to capitalize "Negro," labored over the unique tripartite county, state, and federal extension organization, the voluntary nature of clubs, and the accelerated incrementalism achieved by his office. After reading Bost's rejoinder, William Giles complimented him "on the restraint showed in your reply to the absurd and untrue allegations made in the paper." While Bost and Giles were no doubt satisfied that a proper response had been filed, black Holmes County extension agent Charlie F. Wade sued for being passed over by a less-qualified white applicant, and the case would ultimately expose the depths of continuing discrimination in the Mississippi Extension Service.[72]

USDA leadership further isolated William Seabron as it minimized civil rights efforts. Joseph Robertson, assistant secretary for administration, on March 12, 1969, prepared a briefing paper on department civil rights initiatives for Secretary Hardin. In the section dealing with problems raised by the U.S. Commission on Civil Rights, Robertson observed that the commission had not evaluated Seabron's performance, an issue that had never before been mentioned. "The Assistant to the Secretary had adequate authority and responsibility in the area of civil rights," Robertson shamelessly exaggerated. He then used this inflation of Seabron's power to suggest cynically that creating a more powerful equal opportunity office "is unlikely to remedy the underlying causes for slow progress." Keeping Seabron's office weak and isolated and targeted for failure offered a convenient scapegoat for USDA civil rights failings and, since Seabron was African American, a convenient personification of black incompetence.[73]

FES leaders had done everything in their power, including flagrantly violating civil rights laws, to prevent equal opportunity regulations from applying to state and county extension workers. They had negotiated in bad faith, delayed implementation of rules, accepted falsified compliance reports, winked at egregious discrimination, and set up William Seabron, first by denying him authority to carry out civil rights edicts and then by blaming him for not carrying them out. Secretary Freeman, like so many liberals of his day, embraced civil rights and even issued supporting memorandums, but he had no stomach for confronting southern politicians or even his own deceitful staff. Faced with overwhelming evidence that the USDA was a deeply flawed institution, Freeman allowed prejudice to run its course. For many African American farmers, Freeman's fecklessness doomed their rural life.

The Citizens Advisory Committee on Civil Rights appointed by Secretary Freeman made its final report to Secretary Hardin on March 14, 1969. The committee criticized the FES for failing to take an active role in enforcing civil rights. It deplored the contradiction between subsidizing research that aided agribusiness and ignoring the concerns of poor farmers. "It is, therefore, plain that the Department suffers from operational schizophrenia in dealing with rural commercial resources as against concern for and with rural people," it judged. The committee gave low marks to the FES, ASCS, and FHA. Months before the last meeting, advisory committee member Elmer Ellis encouraged Hardin to keep Seabron on as head of the new committee, praising his "excellent judgment." Ellis also argued that Seabron "has the confidence of the Civil Rights leaders and deserves it."[74]

Robert C. Edwards, president of Clemson University, was the sole holdover when in September 1970 Joseph Robertson reported to Secretary Hardin that his advisory committee on civil rights had been selected. When Clemson had hosted the University of Maryland football team in November 1963, Maryland's Darryl Hill became the first African American to play in "Death Valley," Clemson's stadium. As he warmed up before the game, his mother, Palestine Hill, arrived at the gate but was not allowed to enter. Hill left practice and found his mother and, perhaps providentially, President Edwards and his wife, who invited Dr. Hill (she had a Ph.D.) to see the game from his box. Darryl Hill caught ten passes in the game and set a Maryland record. Edwards invited Hill to spend the night at his home and dropped her off at the train station the next morning.[75]

Edwards had Senator Strom Thurmond's backing for Hardin's committee. "All of these people have received appropriate clearance," Robertson assured the secretary. Indeed, nearly every member had been recommended by a U.S. senator or congressman or by the Republican National Committee. There were six African Americans, three Hispanics, and seven white members among the thirteen men and three women. Overton Rexford Johnson, one of the African American members, directed the School of Agriculture at Virginia State College, a black land-grant school, and he had the endorsement of the Republican National Committee and Governor A. Linwood Holton Jr., Virginia's first Republican governor since the Reconstruction era. Robertson volunteered to serve as chairman of the committee, a position that Seabron formerly held. In May, Seabron had met with Secretary Hardin and encouraged him to continue the committee. Obviously, Robertson carefully constructed the committee and pushed Seabron aside. Hardin's committee lacked the aggressive edge of Freeman's, and it held only four meetings during the next two years, which were "mainly orientation in nature." Evidently an entirely new committee was approved in October 1972 and had met twice by June 1974.[76]

The burgeoning compliance paperwork and complaint files clogged enforcement machinery and further masked continuing discrimination. School lunches, food stamps, private clubs, and other such issues gained more importance than the everyday problems of farmers or the mechanics of ASCS, FES, or FHA discrimination. Local offices were battlegrounds, but local black farmers lacked support to confront continuing discrimination. A November 1969 audit of the North Carolina Extension Service showed not only noncompliance but also that the extension staff "clearly misrepresented their status of compliance in the 10 counties audited by OIG." Inspector general investigations showing discrimination were routinely edited from agency reports. In 1970, a Mississippi State University administrator admitted to claiming that twenty-three African Americans were on the staff when there were only nine and belatedly clarified to subordinates that "an Affirmative Action Plan is result-oriented." Reports on what "you say you are doing are all well and good," he went on, but the Office for Civil Rights wanted accurate numbers.[77] Despite claims that the ASCS encouraged black farmers to participate in elections, in 1969, only two were elected among the 4,100 ASCS committee members who disbursed $3.5 billion to farmers, and community committee representation fell from 418 to 380 among the 30,000 elected. "Until the U.S. Department of Agriculture moves forcefully to end patterns enforced by decades of dominance by Southern white supremacists," the Southern Regional Council charged in December 1969, "black Southern farms will be denied equal protection under the law." The USDA had created a machine that produced mountains of compliance paperwork, but that paperwork, ironically, defeated compliance initiatives. Hardin responded to criticism by issuing a memo demanding that all USDA agencies hold training courses on civil rights.[78]

When Reverend Theodore H. Hesburgh, chairman of the U.S. Commission on Civil Rights, requested a report from the USDA on implementation of civil rights directives, Secretary Hardin admitted that progress "has not been as rapid as I had hoped" and itemized his agenda. Among his projected tools were "a new performance rating and a career objective form," "a career review index," a "computerized personnel skills data base," "statistical reports which will measure performance," and, of course, his carefully vetted Citizens Advisory Committee. In addition, Hardin would make efforts to fill appointed FHA and ASCS slots with African Americans. Other than generating enormous stacks of paper, it was not obvious how his agenda would promote equal opportunity, but it suggested the extent to which the USDA papered over discrimination. No doubt it was such sophisticated instruments that in 1972 allowed 23,000 USDA on-site compliance reviews on over 213,000 Title VI recipients to show complete compliance. A commission report suggested with careful understatement that the perfect compliance scores raised "serious doubt about the quality" of USDA reviews.[79]

When Earl L. Butz became secretary of agriculture, commission chair Theodore M. Hesburgh sent him a four-page letter praising his "firm position" on *Meet the Press* on December 12, 1971, and stressing areas that needed attention. Despite training programs, reviews, and statistical studies, Hesburgh observed, the USDA "has one of the poorest equal employment opportunity records of any major agency of the Federal Government." Few African Americans held positions in the decision-making chain of command in the department. Discrimination in USDA programs, he stressed, "constitutes a clear violation of Federal law and policy," and he pointed to the Extension Service and the FHA in particular. "Examples of our concerns could be multiplied," he added. Hesburgh admitted that the USDA "has shown an increasing awareness" of discrimination but lamented that it had not taken action to address the problem.[80] Indeed, the more the USDA studied the problem, the worse it became.

Moving extension from black land-grant schools created chaos for the loss of budget and staff in the midst of the civil rights movement elicited aggressive countermeasures, such as Tuskegee's request for an experiment substation. By 1970, the USDA showed interest in establishing closer contact with and providing better funding for the black land-grant schools, admitting that its former policy "has ranged from virtually nothing to very little." By this time, white USDA officials actually met with black staff, and in 1970, the USDA placed personnel on all black land-grant campuses after a tortured run through the 1862/1890 land-grant school, Smith-Lever, Smith-Hughes gauntlet. "The 1890 colleges are interested in their own welfare," one report observed, and "the 1862 institutions, quite frankly, fear opening a 'Pandora's box' of scattering of funds to all sort of institutions."[81] There was little justice in this attitude.

Then in 1972, Jim Hightower's book, *Hard Tomatoes, Hard Times: The Failure of the Land Grant College Complex,* pulled back the curtain that had obscured the cozy relationship among agribusiness, land-grant universities, and Congress. The book offered not only a brief history of the USDA and land-grant schools but also a biting and penetrating analysis of how agribusiness shaped land-grant research agendas to the exclusion of average farmers. Congressional hearings on land-grant budgets became a convocation of true believers awash in the spirit of technology. "Overwhelmingly," Hightower wrote, "agricultural research continues to be committed to the technological and managerial needs of the largest-scale producers and of agribusiness corporations, and it continues to omit those most in need of research assistance." Land-grant schools had fostered a rural revolution, he wrote, one that continued to push farmers off the land. A synopsis of the book remained in the papers of Harry Philpott, president of Auburn University. B. F. Smith, the executive vice president of the Delta Council, proclaimed to Mississippi State University president William Giles that Hightower's book "sounds like something that would be initiated by the 'Socialist Scholars Conference'" and likened such "left-wing groups" to "a group of jackals or hyenas surrounding and harassing their prey." On the other hand, at a congressional hearing, Richard David Morrison, president of Alabama A&M University, testified that it was 1972 before the black schools received USDA funding for research. "For too long," he fretted, "the 1890 land-grant colleges have been forced to deal with 'Rotten Tomatoes and Hard Times.'" Several organizations, including the NSF and the Georgia Council on Human Rights, requested a court order to stop the $750 million appropriation to land-grant universities and experiment stations, arguing that the program focused on agribusiness and not small farmers or environmental issues. Despite congressional hearings and the glare of publicity on Hightower's book, the USDA absorbed his critique and rolled on.[82] It was, after all, USDA policy to support agribusiness.

In 1976, USDA secretary Earl Butz told a joke that the *Christian Science Monitor* termed "an obscene characterization of black Americans." Butz's unprintable racist jocularity suggested a USDA culture that tolerated humor at the expense of minorities and that perpetuated stereotypes and undermined equal opportunity reforms. The controversy awakened the press to other incidents in Butz's tenure. *Los Angeles Times* reporter Grayson Mitchell discovered that in November 1973 Butz had held two secret meetings with Extension Service leaders from seven states—Arkansas, Georgia, Illinois, Kansas, Louisiana, Maryland, and Texas—to discuss how to preserve federal funding despite evidence of equal opportunity violations in hiring. At the time, there were 8,418 minorities (9.9 percent) among the 85,088 full-time USDA personnel, a poor showing in a federal bureaucracy that averaged 15.9 percent. The Civil Rights Commission in 1975 found "blatant and widespread" violations of equal rights laws as well as "the continuing complicity of the USDA secretary and other high-level USDA officials," adding that the USDA seemed more concerned with "protecting noncomplying recipients than those people whom the law seeks to protect." Violation of Title VI equal rights laws required the USDA to halt funding, but instead, Butz ensured funding by a backroom deal.[83]

As Butz was cleaning out his office after resigning, he denied Mitchell's report, and acting USDA secretary John A. Knebel released the astounding statement that during Butz's tenure, "the department has made great strides, perhaps strides more [*sic*] in the last two years than were made in the first 100 years of the department[,] toward getting a proper (racial) balance." When the Emergency Land Fund's Joseph F. Brooks criticized the FHA under Butz for discrimination in granting loans and hiring minorities, an FHA administrator countered by stating that the USDA took "great pride in not discriminating against any individual on the basis of sex, religion, race, marital status—you name it." Brooks scoffed and reported that many black farmers had succinctly concluded, "Those white folks don't want us to have land."[84]

Secretary Clifford Hardin and then Earl Butz, both of whose educations and careers had been spent in land-grant colleges, sanitized civil rights initiatives during their tenures at the USDA. Hardin had been a student of Earl Butz at Purdue University, and like his mentor, who doubled his salary by sitting on various agribusiness boards, Hardin moved easily among corporations and deans. It was easy to ignore discrimination when the USDA focused on larger farmers and businesses. In addition to creating Hardin's appropriately cleared advisory committee, USDA bureaucrats invented machinery that mishandled or mangled complaints. Black and white employees in civil rights and equal opportunity offices turned on each other, complaining of discrimination and reverse discrimination. A cursory examination of Nixon's USDA might suggest that civil rights initiatives continued, but African Americans found themselves more marginalized than ever. A series of court cases chipped away at discrimination, and then the Alabama Cooperative Extension Service pushed Willie Strain too far.[85]

*The racial discrimination in this case has so permeated the*
*employment practices and services distribution of the defendants*
*that this Court finds it necessary to enter a detailed and specific decree*
*which will not only prohibit discrimination but which will also*
*prescribe procedures designed to prevent discrimination in the future*
*and to correct the effects of past discrimination.*
—*Judge Frank M. Johnson, in* Strain v. Philpott

# 7

# THE CASE OF WILLIE STRAIN

Reverend K. L. Buford, Alabama field director for the NAACP, wrote to the state's extension director Fred Robertson questioning a statement reported in the *Birmingham News* on January 16, 1968. Robertson had claimed that there was no discrimination in the state's Extension Service. Buford filled three and a half pages with incidents of bias that had come to his attention. As a Tuskegee resident, Buford focused on the former African American extension administrative staff that had moved to Auburn University in 1965. He recalled Willie Strain with some fondness because when Buford built a "concrete finishing parlor for hogs," Strain visited his farm and took photographs that Buford said were published in *The Negro Farmer*. "Mr. Strain and his paper are conspicuous by their absence," Buford sadly observed. "The Negro community is presently being denied the service he was rendering." Buford had lost track of Strain, but, he ventured, "he evidently is neatly tucked away in a corner somewhere."[1] He put his finger squarely on what black farmers lost to the integration of the Extension Service when Willie Strain's creativity, energy, and vision were shunned and he was indeed "tucked away" and given nothing to do at Auburn University.

As Willie Strain's career in the Negro Extension Service made clear, even under white control there were significant opportunities for agents to work with rural men, women, and children. By 1965, both Willie Strain and Bertha Jones, who headed 4-H work for girls, had risen through the ranks and were among the top extension administrators at Tuskegee In-

*Bertha M. Jones, 2007.*
*Photograph by Pete Daniel.*

stitute. Then they both endured the unintended consequences of civil rights initiatives for their remarkable careers in the segregated Negro Extension Service wilted when they were transferred to Auburn University. Their experiences illustrate the fragile career paths of black extension workers, the weight of racism that proscribed their work, and the unyielding discrimination that permeated the Alabama Cooperative Extension Service (ACES).

Bertha Jones's career demonstrated the intelligence, grit, and determination that guided her to the top of Alabama 4-H work. She grew up in Huntsville, Alabama, and worked her way through nearby Alabama A&M College waiting tables and then rising to become student manager of the canteen. Huntsville at that time, she recalled, "had a lot of rural to it," and the street by her house was unpaved. The "stove room" for cooking and dining sat a short distance from her house to prevent kitchen fires from spreading. Jones majored in home economics and graduated in 1945. Her first job was in Luverne, Alabama, where she energetically attended 4-H Club and home-demonstration meetings. "We taught better family living, and we taught clothing, we taught food preservation, we taught the whole gamut of things that you do in the house." She worked both with sharecroppers and with farm owners in Crenshaw County and stressed the importance of record keeping to parents and the value of education to children. She recommended that her best students attend Tuskegee Institute and would make unauthorized trips to a nearby city to buy material for their uniforms. Then, if the student's parents lacked sewing skills, "I would sit up at night and make their uniforms." She stressed that her students "had to look the part, and you had to do the part."[2]

Jones's Crenshaw County office was partitioned off inside a restaurant. "So we had to smell all of that grease all day," she recalled. She stayed up late learning to hunt and peck on an old typewriter so she could send material to Tuskegee Institute for reproduction. "We didn't have any reproduction machines, duplicating machines," she said. Jones worked there for ten years and then took a sabbatical and completed an M.A. degree at Penn State University. She became African American state leader of 4-H and moved to Tuskegee in 1956.[3]

Jones took extraordinary interest in giving her students special opportunities. African American 4-H members were not allowed to attend the National 4-H Club Congress in Chicago but could attend the National 4-H Club Conference in Washington. Each year, Jones took four girls and Thomas R. Agnew, her male counterpart, took four boys, and they drove from Tuskegee to Washington. "That was before we could go in restaurants to eat," she explained, "so we had to pack a lunch so that we would have something to eat on the first day of travel." They would stop at an African American college in North Carolina and spend the night in a dormitory, completing the trip the next day. Jones worried about having an automobile accident and, in the early 1960s, secured funding for taking a train to Washington.[4] Her remarkable career stalled in 1965 when she was moved to Auburn University. "They changed our titles when we went up there," she bitterly recalled. "They sent us a letter, and they told me, see, I had been a state leader of 4-H, but all of a sudden I was a state 4-H specialist. They didn't tell me what I was to specialize in." Isolated with no assignment or direction, Jones's creative supervisory work with the 4-H program came to an unfortunate and unnecessary end.[5]

Willie L. Strain grew up near Priceville in Morgan County, Alabama, where his family farmed near the banks of the Tennessee River. Strain attended the Morgan County Training School in Hartselle and graduated in 1949. He then enrolled at Tuskegee Institute on a work-study program, joined the ROTC, and graduated in 1954. After returning from service in the air force, Strain took a position with the Extension Service in Butler County. He remembered that "the conditions that we were working under were very deplorable." In the air force, "you had secretaries and you had all up-to-date equipment and everything." When he got to Butler County,

*Willie L. Strain, 2007.*
*Photograph by Pete Daniel.*

"all we had was an old broken down desk, old typewriters that weren't worth having, and no secretaries. You had to do all of your secretarial work yourself." His office, he complained, "was located in the area where they had the road service folks for the county," while the white extension office was in the courthouse. The conditions in Coosa County, the location of his next assignment, were no better, and his office was on the floor above a shoe shop. His office in Tuscaloosa was near a service station and poolroom.[6]

In June 1959, Negro District Agent G. W. Taylor congratulated Strain "for the very fine job you are doing in Tuscaloosa County" and mentioned his achievements in several areas. "To me," Taylor continued, "it is reassuring to observe a conscientious worker like you in operation, one who, it seems to me, is interested in helping people improve themselves both socially and economically." Three years later, an administrator at Selma University (founded in 1878 as the Alabama Baptist Normal and Theological School) thanked Strain for teaching an extension course on campus. "It is certainly out of the ordinary," A. P. Torrence wrote, "for individuals to work as diligently as you have worked in the interest of others."[7]

In addition to performing his regular duties in Tuscaloosa, Strain wrote articles for the newspaper and, in cooperation with a white FHA staffer, produced a radio program. He recorded his programs on a tape recorder, and his radio broadcasts ran from 1957 to 1962, when he moved to Tuskegee Institute to take a position as assistant editor of the monthly newspaper, *The Negro Farmer*. "I never did find out who the editor was," he joked, knowing it was an example of withholding full titles from African Americans. He worked under William Bailey Hill, the state leader for African American work, as did two district agents for men and two for women and a 4-H leader for boys, Thomas Agnew, and one for girls, Bertha Jones. Strain's territory covered the thirty-seven counties that had black extension agents. "It was my responsibility to train these folks how to write for newspapers and how to do radio programs," he explained. He also gathered material for his radio program and for *The Negro Farmer*. "I was responsible for getting that paper out, going and visiting counties, doing my photography work myself, doing my editing and writing myself, laying out the paper, and doing everything," he continued. He then took the paper to Auburn University for printing. "Everything that we did had to go through Auburn."[8]

To obtain basic supplies, Strain was forced to appeal to the white extension editor at Auburn. In July 1962, he asked Robert R. Chesnutt for a 35mm camera to photograph 4-H functions, create slides for demonstrations, and do other projects. Some black agents with county radio programs, he had discovered, had "tape recorders of poor quality" that detracted from the programs. Better equipment, Strain suggested, "would help them represent the Extension Service in a more dignified manner."[9]

*The Negro Farmer* appealed primarily to rural families and carried stories about successful farm families, including photographs of people, houses, equipment, and stock as well as women's kitchens and appliances. Several articles about successful black dairy farmers gave the lie to the claim that African Americans could not be dairy farmers because they lacked the industry to work seven days a week. In the June 1963 issue, John Richardson, owner of a 162-acre dairy farm, told Strain, "Buying a farm is the most important thing I have done in my life." Strain thoroughly covered 4-H work and featured photographs of 4-H contest winners and an occasional column by Bertha Jones. Significantly, the newspaper was inclusive in its focus on men, women, and children and insistent on high standards of achievement. The "Things You'll Want to Know" column offered helpful hints on planting, storing, and safety, as well as other general information. Conferences at Tuskegee Institute and Alabama A&M College received thorough coverage. In January and February 1965, the paper ran "Points about the Economic Opportunity Act of 1964." There was nothing, however, on the civil rights movement or ASCS elections.[10]

The June 1965 issue carried a front-page story by Willie Strain, "'The

# "The Negro Farmer" Closes Shop After 25 Years Of Publication

By W. L. Strain

Assistant Extension Editor
Auburn University
Extension Service

This issue of The Negro Farmer closes the pages of a paper that has come into your home since April 1940.

For one-fourth of a century we have attempted to connect the lines of communication with the rural, urban and city the farmers and it will give recognition to those who are doing so well that their story will be an inspiration to others."

The above educators and the 4 different editors of The Negro Farmer — C. E. Trout, Joseph Bradford, Lorenious McDonald and I—saw the necessity of trying to keep the rural people informed with the latest news in agriculture, home economics, youth work from the many vast channels of communication? Have the people moved so far up that they are subscribing to the numerous state or local newspapers? Are these papers putting all the information before these groups in a manner that they will understand and interpret it? Do they read and interpret the information furnished them on the latest in agriculture, home economics, and youth areas?

*Willie Strain's last column in* The Negro Farmer, *June 1965.*
*Courtesy of National Agricultural Library.*

Negro Farmer' Closes Shop after 25 Years of Publication." The newspaper began in 1940 under Thomas M. Campbell, the legendary figure who had largely been responsible for inaugurating black extension work. Its goal, Strain explained, had been to inform both rural and urban African Americans about higher standards of living and farming. "Since the beginning of this paper," he insisted, "all efforts have been geared toward communicating with the small, disadvantaged and low-income groups." *The Negro Farmer* had a nationwide circulation of 16,000, primarily distributed by extension agents. Strain doubted that black farmers no longer needed such a newspaper.[11]

When the Extension Service was ordered to integrate, Bailey Hill and Willie Strain went to Auburn University on July 15, 1965. "That was the end of the Negro Division of Extension," Strain sadly judged. Three of the staff who moved to Auburn had doctorates, two from the University of Wisconsin and one from Cornell, but, Strain explained, they were all "made subordinate to whites." Strain learned that some of the white Auburn staff, lacking such credentials, left immediately for graduate school. When the Extension Service staff left Tuskegee, its president, Luther Hilton Foster Jr., insisted that white extension staff be sent to Tuskegee just as black staff had been sent to Auburn. When this suggestion was refused, Tuskegee lost its connection to extension. "So they destroyed everything that was at Tuskegee," Strain recalled, "all of the records and everything."[12]

The U.S. Commission on Civil Rights quickly grasped the significance of the Extension Service's discriminatory integration plans. In late October, William Seabron and his top staff joined commission representatives in a meeting with Robert Pitchell, a newly appointed deputy at the FES who had just returned from a trip to the South. Pitchell ignored the demotion of black agents in the South and failed to grasp that a substantial number of these men and women had been but no longer were supervisors. Instead of facing such urgent issues, he suggested collecting data and conducting surveys. Seabron's staff found the Extension Service's compliance forms "abysmal." When Pitchell asked commission staff for recommendations, William Payne declared that a tough statement from FES head Lloyd Davis would be a good first step. John Slusser from Seabron's staff revealed that some of the best black extension agents and administrators were leaving because the FES was not committed to equal opportunity. "This meeting," Payne wrote later, "was frankly one of the most depressing experiences I have ever been through."[13] At the time, Seabron was desperately trying to win Secretary Freeman's support "against the agency administrators who are dragging their feet." Seabron planned to show Freeman the flawed Kentucky compliance response, which failed to measure up even to the weak FES standards, and strongly recommend that he not approve it. "Slusser indicated that Seabron hopes to use this example of duplicity to show Freeman that the Administrators cannot be trusted to carry out their own instructions," Payne reported.[14] There is no record that Seabron's plan succeeded.

The ACES's flawed integration design resonated not only with that of other state extension programs but also with the merger of black and white public schools. Private schools cropped up throughout the South, giving some students an alternative to attending integrated classes. But in many school systems, white administrators shut down African American schools, some recently constructed and superior to the white facilities, and integrated black students into the white system, shedding black principals and teachers. By 1970, an estimated 31,504 black teachers lost their positions even as numerous white teachers were being hired. In his study of Hyde County, North Carolina, *Along Freedom Road*, David S. Cecelski reported that the number of black principals in the state declined from 620 to 170 between 1963 and 1970, and only 3 remained in 1973. In North Carolina, 3,051 black teachers lost their positions, and the remaining teachers and administrators were often assigned lower-status jobs than they had held in the segregated system. Cecelski found that when African Americans merged into the white system, they left behind more than schoolhouses, for "their names, mascots, mottos, holidays, and traditions were sacrificed with them." Against incredible odds and Ku Klux Klan intimidation, Hyde County blacks, including a host of exceptionally brave and intelligent students, successfully fought to save their schools and made integration a two-way process. This was unusual. "Neither the federal courts nor federal agencies felt obligated to intervene against local school leaders," Cecelski concluded.[15] Equal opportunity did not apply to southern school systems and USDA county offices.

The ACES expected approval of its integration plan, which shut down the Negro Extension Service and boasted of full compliance with civil rights laws. Other Tuskegee extension personnel joined Bailey Hill and Willie Strain at Auburn, while some were assigned elsewhere. Hill had been administrator of the Negro Extension Service but at Auburn had no administrative responsibilities, and Strain had edited *The Negro Farmer* but had no editorial duties at Auburn, as if the information he published was irrelevant. Auburn had been forced to integrate bodies but persisted in discriminating against African Americans and totally controlling the Extension Service.[16]

Until he moved to Auburn, Strain had steadily moved up in the Extension Service hierarchy and expected his career to flourish. Because the Negro Extension Service had some autonomy, talented and ambitious blacks like Hill, Strain, and Jones could rise to responsible positions. With integration, white extension administrators ignored the talent and experience of black extension workers. White administrators demeaned and shunned highly trained African Americans, especially those with advanced degrees and administrative experience, brushing aside an important perspective on rural life. When blacks arrived at the Auburn extension offices, they were given no duties, reflecting the bias that informed segregation and white condescension toward black extension personnel. Robert Chesnutt's title, meanwhile, changed from extension editor to agricultural editor and chairman of the extension information division.[17]

The Auburn staff regarded Strain as a "radical," he supposed, because he was aggressive and innovative. Bailey Hill, he ventured, "was a kind of docile type of individual." Strain's frankness did not go down well. "When I came into the office in the morning, they would be standing around in the lobby near several offices," he recalled, and "if they saw me coming in, everybody would dash to their offices." Neither Strain nor Bertha Jones were assigned any work, but both were grudgingly given office space. After being shunned each morning, Strain went to his office for ten minutes and then spent the remainder of his workday in the library. After six months of this routine, he decided in January 1966 to take leave and enter the University of Wisconsin to pursue a master's degree in journalism and a Ph.D. to add to his master's in education. He completed the work for the M.A. and began doctoral work, but after experiencing health problems and academic difficulties, he returned to Auburn and continued his old routine of reading in the library. Bertha Jones was likewise restless in her undefined position. She began writing pamphlets on health and nutrition, but she was prevented access to state 4-H work, which had been her specialty.[18]

In early June 1969, Strain attended a Sunday afternoon extension meeting with a visiting delegation from Australia. After department head Robert Chesnutt introduced the speaker, he sat down beside Strain and suffered a fatal heart attack. Although Strain had more experience and education than others on the staff, the department selected John L. Parrott to replace Chesnutt. "So when they appointed him, that was when I filed the complaint to the U.S. Department of Agriculture," Strain recalled, "and then they insisted that they had to give me a hearing and everything."[19]

Strain's letter to Secretary of Agriculture Clifford Hardin on July 11, 1969, went to FES administrator Lloyd Davis, who forwarded it to Harry Philpott, president of Auburn University, who handed it off to ACES director Fred Robertson, the official who Reverend K. L. Buford had upbraided for claiming there was no discrimination in the Extension Service. The Auburn FES staff diligently investigated Strain's University of Wisconsin leave and focused on why he did not complete the Ph.D. In addition to health problems, they learned, Strain had difficulties with statistics and other courses for which his Tuskegee education had not prepared him. His Wisconsin adviser suggested that he pursue an M.S. in communications instead of a Ph.D. Since Strain had requested leave to work on his Ph.D., Auburn authorities seized on that issue, suddenly paying attention to the shunned staff member. By attacking Strain, the ACES obviously hoped to shift the focus away from the complaint and impugn Strain's integrity before a sympathetic university forum. Strain postponed the hearing set for August 21, 1969, hired an attorney, and on May 26, 1970, brought suit against the FES, changing the jurisdiction from the ACES to the federal courts.[20] He had decided that discrimination throughout the ACES system transcended his case. Strain's lawyers decided that the Extension Service's discrimination deserved more than a hearing before an Auburn ACES committee and filed a class-action suit.

Realizing that the Auburn ACES staff was attempting to conflate his discrimination complaint with his difficulties at the University of Wisconsin, Strain upped the ante dramatically with his class-action suit. The case named numerous black extension and home-demonstration agents, black women who had been denied admission to white home-demonstration clubs, black farmers who received inferior service from extension agents, and black children denied membership in white 4-H clubs. The list of infractions provided a revelation of ACES's false compliance. Blacks earned lower salaries and were denied in-service training and promotion, and of the 70 specialists on the staff, only 5 were black. All 67 county extension chairmen were white males, and only 13 of the 109 clerical staff in Alabama's 67 counties were black. Some county offices were segregated, and 32 counties had no black agents at all, while black agents in other counties had higher workloads and dealt solely with blacks. Nearly all of Alabama's home-demonstration clubs were segregated, but in counties lacking black extension programs, there were no black clubs, meaning that black children in those counties were denied participation in 4-H clubs since most clubs remained segregated. Black children were excluded from being judges in 4-H contests, and no black 4-H member had attended the National 4-H Club Congress in Chicago and very few went to the National 4-H Club Conference in Washington. The ACES paper mountain purportedly showing compliance burned to the ground. The *Strain* case, like other assaults on white institutional power, attracted little media attention in Alabama.[21]

Auburn's attorney, Thomas D. Samford III, entered a point-by-point rebuttal that predictably admitted no discrimination. Even as the Washington FES office was rewriting equal opportunity guidelines to justify discrimination, Strain's case exposed bias that infected the entire state of Alabama extension apparatus.[22] On July 31, 1970, the Justice Department intervened as a plaintiff, and the case refocused on Auburn University and the ACES. It was no longer *Strain v. Hardin*, against the secretary of agriculture, but *Strain v. Philpott*, against the president of Auburn University. Whatever the Justice Department's motivation, its intervention removed both the secretary of agriculture and the head of the FES from the suit. Interestingly, the Justice Department brief supported the legal basis for holding county governing boards liable for discrimination even as FES administrators in Washington were fighting such guidelines.[23]

The depositions, preliminary hearings, and records generated from both sides were substantial. Five briefs were submitted, one from the Justice Department and several from Auburn's attorneys and from county governments. As Strain recalled, after Judge Frank M. Johnson evaluated the material, "he said he didn't see any need of having a trial because the segregation was so blatant that he didn't need a trial to write an opinion and an order." Strain praised Judge Johnson's order and guidelines. By this time, Johnson, chief judge of the U.S. Middle District Court in Alabama, had earned a reputation for his bold civil rights decisions on school integration, voting rights, and gender discrimination. He did not consider his rulings those of an activist judge but rather he saw them as carrying out the court's constitutional duty to provide relief to those suffering particular deprivations.[24]

On September 1, 1971, Judge Johnson derailed the ACES's train of noncompliance and provided a concise analysis of the ACES structure. The state director of extension was responsible to Auburn University's president and board of trustees, and a white man had always held the position. Under the segregated system, the African American state leader reported to the white state director, who made all personnel decisions and supervised every level of extension work. Despite the discriminatory structure, Johnson pointed out, African Americans "did hold positions of supervision and had the responsibility of administration within their branch." It was this structurally separate and unequal African American organization that on July 1, 1965, merged with the white Auburn ACES structure. Bailey Hill, the former state leader for black extension work, was made an assistant to the director, but the other 5 assistants, 4 associate district extension chairmen, and all 67 county extension chairmen were white. The two exceptions were Carolyn Brown Williams, a black woman who in 1968 became associate county extension chair in Macon County, the location of Tuskegee Institute, and Evelyn Blackmon, who headed Greene County demonstration work. Johnson inferred that stripping two other highly qualified African Americans of broad supervisory duties led in 1965 to their resignations.[25]

Johnson diagrammed the Extension Service's dual structure prior to 1965 and its residual segregated skeleton after the merger, discussing hierarchy, title changes, employee assignments, discrimination, and the failure to allow African Americans the opportunity to participate as equals in the Extension Service. All black extension and demonstration staff, Johnson observed, were moved to a second-tier extension farm-agent and extension home-agent category "regardless of their tenure, ability, job effectiveness or educational background," clearly a demotion, while white agents "were either not affected or received increased responsibilities." The ACES no doubt hoped that moving talented black staff to Auburn and giving them offices with no duties would mask continuing

discrimination. "Thus, the stipulated and uncontroverted facts affirmatively reflect, and this Court specifically finds," Johnson ruled, "that the
purported merger of the ACES' dual systems in 1965 has thoroughly and
effectively perpetuated and to a substantial degree aggravated the racial
discrimination which existed previously." He also found that the newly
instituted practice of allowing county commissioners three choices for
extension positions, basically permitting them to exclude black candidates, "was designed to achieve and has resulted in racially discriminatory hiring on the county level." The ACES director had claimed that
blacks lacked proper training to become county chairmen and that only
Auburn University provided such training. "Negroes were denied positions of responsibility within the ACES because they lack an educational
background which was refused them because of their race," Johnson declared. The ACES also withheld from African Americans information on
job openings, giving whites an advantage. Black agents were locked into
programs aimed exclusively at poor farmers and were excluded from promoting beef cattle and dairy farming. In all of the ACES, there were only
eighteen black clerical personnel, all working in counties where there
were black extension agents. "Not a single Negro secretary has been assigned to work in an all-white county office in the years since the merger,"
Johnson observed. "This practice is clearly a result of racial discrimination." The sorry record and duplicity prompted Johnson to go beyond
simply finding against the ACES. "The racial discrimination in this case
has so permeated the employment practices and services distribution of
the defendants," he ruled, "that this Court finds it necessary to enter a detailed and specific decree which will not only prohibit discrimination but
which will also prescribe procedures designed to prevent discrimination
in the future and to correct the effects of past discrimination."[26]

Judge Frank Johnson's decision dismembered the ACES rationale for
its purportedly integrated structure. It attacked white presumptions and
exposed qualifiers that perpetuated discrimination and obstructionist
strategies that typified extension programs throughout the South. Johnson directed the ACES to give former African American county agents
and county home-demonstration agents "first priority for consideration
for all future promotions to County Extension Chairman and Associate
County Extension Chairman positions respectively." If for some reason
the ACES assigned a white person to such a position, it had to justify the
appointment with ample documentation. Willie Strain, Thomas Agnew,
and Bertha Jones "shall be given first priority for consideration for all
future promotions in the subject area in which they have had training and

experience whether prior or subsequent to the merger of 1965." Strain's lawyer had unsuccessfully demanded that Strain immediately replace John Parrott as chairman of the extension information division. Johnson issued instructions that black former district agents Cleo Walker and Ruth Rivers be assigned to "positions with supervisory responsibility which are commensurate with their experience and educational background." He decreed that the ACES should abandon submitting three names to county governing boards, cease consulting with county boards about the acceptance of African American applicants, ignore the rule of requiring technical education at Auburn University, eliminate segregation in 4-H clubs, and submit a plan to achieve his directives within ninety days.[27]

The ACES, of course, did not yield to the decree gracefully. After all, it was tightly linked to the powerful Farm Bureau Federation and its conservative political agenda as well as to the National Association of State Universities and Land-Grant Colleges and would not abandon discrimination without a fight. In the 1920s, the ACES supported the formation of the Farm Bureau to oppose the Farmers Union, which supported small farmers. County agents even collected dues for Farm Bureau membership, and with extension agents in every county, the bureau heavily influenced ACES policy. The ACES was unaccustomed to being challenged, and it struck back at Johnson's order. Its final plan submitted to the court on September 29, 1971, relied on a points system for promotion that stressed educational attainment, scholarly achievement, tenure, and subjective traits such as leadership, ability to work with people, drive, and initiative. "The present proposal of ACES to deemphasize tenure as a selection criterion would have the effect of freezing the incumbent Negro employees out of the supervisory positions of county extension chairman and associate county extension chairman," the court decreed. "This Court cannot tolerate the downgrading of tenure as a selection standard." The points for tenure were raised from 15 to 50. An elaborate schedule established pay equalization for black extension workers. In December 1971, the Department of Justice issued guidelines for implementing the *Strain* decision, and the U.S. Commission on Civil Rights staff met with the USDA's Ned D. Bayley to discuss implementing the decision. In January 1973, Reverend K. L. Buford, who fondly remembered Willie Strain's reporting for *The Negro Farmer*, urged the NAACP to weigh in on continuing discrimination in the Alabama Extension Service. "If implementation is thwarted, or revisions circumvented by the ACES," Buford warned, "it is reasonable to assume that other states will follow suit."[28]

The ACES also implemented directives to promote certain staff. Ruth

Rivers, for example, accepted a job as state supervisor for the expanded food and nutrition education program, and Cleo Walker was promoted to associate district extension chair. Alex Brown became a specialist in 4-H visuals, and Mariah Brymer became associate county extension chair in Coosa County. Thomas Agnew had retired in May 1971 and did not benefit from the *Strain* decision. When a position as district extension chairman opened in August 1972, the ACES did not offer it to Willie Strain but opened it to all applicants, arguing that the decree did not require promoting Strain automatically to this position. Ultimately, the court agreed with the ACES, and the position went to someone else. After substantial confusion over a proper assignment, Bertha Jones accepted the position of state leader of the urban 4-H program on January 15, 1973. In May 1973, Willie Strain considered what he called a "reassignment of duties" as radio-TV editor in the information division but did not accept it. When John Parrott retired, Strain became head of the Extension Service's communications department and worked there until retirement.[29] The *Strain* decision and its implementation went a long way in erasing white privilege in Alabama's Extension Service.

When Willie Strain reflected on his case's significance, he stressed that its ramifications went far beyond correcting racial bias. "Now you have white women that are district agents, you have black women," he noted. "The suit of *Strain* helped black males, white women, black women, and white men that the system didn't like, but they had the qualifications." Strain had made that point at a meeting, and Gaines Smith, who later became acting ACES director, countered that the case resulted in promoting, as Strain recalled Smith's words, "a lot of blacks that weren't qualified to be in positions." Strain replied, "Yes, maybe it did, but there were very few, but it made it possible for a lot of white males to be in positions that they weren't qualified for, and you are one of them."[30]

Strain believed that his competitive spirit came primarily from his mother; his father died when he was young. His mother finished only third grade and worked "in folks' kitchens." She insisted that her son get an education and "that whatever you do, you do right to folks, you treat people right, and you do what you can to help people, because that's the only reason that God put you on earth is to help one another." He also took a lesson from his high school football coach, whose philosophy, Strain recalled, was that "in order to be a good football player, you had to be agile, hostile, and mobile." He boasted, "I have never been afraid of no one."[31]

The *Strain* case did not put an end to discrimination complaints, but actually, as Strain observed, it encouraged minorities and women to con-test their treatment. From Monroe County, Georgia, Carolyn R. Newton appealed to Secretary of Agriculture Earl Butz to investigate her com-plex case. She had worked for the·Extension Service for seven years and, on November 1, 1972, learned that she and a few other women had been appointed to serve as county extension chairs. In December, she discov-ered that the appointment had been postponed after a meeting of county commissioners, state extension officials, and local citizens. On Janu-ary 10, 1973, she told the district agent that she was pregnant and was due in June, and a week later, the appointment of county chair was put off until July. In February, Newton discussed filing discrimination charges with extension officials in Athens. Suddenly, the Monroe County agent was transferred, and the county's extension program was terminated. Newton listed six examples of discrimination, including unannounced meetings, salary inequities, and illegal transfer of the county agent. "For these reasons," Newton ended her letter, "I have decided to terminate now."[32] Obviously, the Monroe County commissioners, state extension leaders, and local white elite did not want a woman heading extension work, but Carolyn Newton did not go quietly.

As compliance paperwork multiplied after *Strain*, the president of Vir-ginia Tech, after assuring Secretary Earl Butz in December 1972 of the university's "strong commitment" to equal opportunity, estimated that the extra staff needed to prepare "unnecessary paper work" cost $151,000. Virginia Tech was nearing completion of a university-wide plan ap-proved by the Justice Department, T. Marshall Hahn Jr. explained, and similar material generated for the USDA was unnecessary. Ned Bayley, the USDA's director of science and education, considered Hahn's point but argued that, given court decisions involving civil rights, land-grant schools should subscribe to USDA guidelines. Virginia Tech duly submit-ted a plan on June 27, 1973, that proved unsatisfactory and needed sub-stantial revision. The USDA's requirement of five copies of the plan within thirty days certainly did not amuse Hahn.[33] White land-grant schools, of course, were accustomed to having their way, and being regulated by outside forces, even a usually sympathetic USDA, did not go down well. Hahn correctly judged the illusory nature of paperwork, while Bayley re-lied on paper to cover discrimination.

The Alabama Extension Service epitomized the duplicity throughout the southern land-grant university network. In North Carolina, *Baze-more v. Friday*, another challenge to Extension Service discrimination, dragged on for more than a decade, complicated by the question of the role of county commissioners in hiring extension agents and the question of whether pre-1964 salary inequities were relevant after Extension Service integration. In August 1965, when North Carolina merged the black and white branches of the Extension Service, not all salary disparities were settled, and an action was brought in 1971. The district court, ruling that the petitioners failed to supply proof of discrimination, decided that the Extension Service had operated in a nondiscriminatory manner since it integrated, and the court of appeals affirmed the decision. The case was also complicated by the Equal Employment Opportunity Act of 1972, which applied to public employees as well as those in the private sector. The Supreme Court rejected the Extension Service's argument and ruled that the fact "that the Extension Service discriminated with respect to salaries *prior* to the time it was covered by Title VII does not excuse perpetuating that discrimination *after* the Extension Service became covered by Title VII. To hold otherwise would have the effect of exempting from liability those employers who were historically the greatest offenders of the rights of blacks." The district court and court of appeals decisions had rested on what both considered a flawed regression analysis of the petitioners, but this argument, according to the Supreme Court, "was erroneous in important respects."[34]

The Mississippi Cooperative Extension Service (MCES) used the same tactics as Alabama in evading the implementation of equal opportunity, including demeaning job nomenclature, no assignments of blacks to supervisory positions, and continuing salary differentials. The case, *Charles Wade v. Mississippi Cooperative Extension Service*, takes its name from Charlie F. Wade, a black extension agent in Holmes County who, in 1969, was passed over when the white agent retired. Without announcing the vacancy, the MCES hired James Nelms, who had a bachelor's degree and only two years experience. Wade had a bachelor's degree and some work toward a master's and thirteen years experience. Although the MCES claimed that Nelms received a higher evaluation than Wade, "more credible evidence," the court observed, pointed to "the belief that Holmes County was not ready to accept a black county agent."[35]

Laverne Y. Lindsay, another plaintiff in the case, held a master's degree and had worked in all phases of extension activities, while Jocelyn W. Frizzell, a white woman, had a bachelor's degree and had worked primarily with youth activities. The MCES justified Frizzell's appointment as extension home economist by claiming she worked better with county residents, but the court held that Lindsay was not promoted "because of racial considerations." The larger picture of MCES integration, the court observed, showed that African Americans "were summarily demoted in job title, authority, and responsibility." It was also significant, the court observed, "that during the merger not a single black was offered the position of district agent, county agent or extension home economist." The court demanded "that qualified blacks in substantial numbers at all levels of MCES employment be employed throughout Mississippi and that plans to accomplish that goal be formed without delay and implemented within a reasonable time." The court also observed that since the case originated, the secretary of agriculture and federal extension officials had first denied but then admitted "discriminatory practices," a warning that MCES officials were under pressure to implement equal opportunity.[36]

Despite the *Strain* decision and other rulings favoring equal opportunity, the Extension Service continued to delay and obfuscate audits and whine about paperwork. In January 1973, the OIG scheduled audits, but extension head Edwin L. Kirby asked to delay them until states completed affirmative action plans, although there was no direct connection. Instead of rescheduling the audits, in a September 1973 meeting, extension administrators complained that the audits "asked for too much information and did not give the States long enough to prepare answers," and they were postponed. Meanwhile, the Extension Service continued to avoid dealing fairly with minorities.[37]

When the Arkansas audit report arrived, it made misleading conclusions in every category. Black staff worked primarily with black clients and in only eighteen of the state's seventy-seven county offices. African Americans working in state and county offices had gained one promotion in nine years, and no black held an extension secretary position. "It should be noted that there are no black state supervisors, no black District Agents and no black County Chairmen." As in Alabama, the black state professional staff disappeared into the white structure with no supervisory duties. The audit report lacked sufficient information on vacancies and promotions. Only 6 of the 98 professional hires and 9 of the 122 clerical hires since July 1970 were African American, and the report again lacked sufficient explanatory documentation. Salary information seemed to indicate that blacks were approaching equality, but, again, the data seemed fuzzy. Homemaker clubs (1.8 percent integrated), 4-H clubs (20 of 1,128 integrated), and many camping sessions were segregated. The analysis revealed that the Arkansas Cooperative Extension Service had drawn conclusions on integration based on dubious data. After all the whining and complaining about data demands, the final product showed that a straightforward request for information had been transformed into a cover-up of continuing discrimination. The Justice Department analyzed the audit and found "material violations of Title VI and USDA's equal employment regulations," yet no action was taken.[38]

In 1974, the U.S. Commission on Civil Rights compiled a massive study of civil rights enforcement and chronicled in detail Extension Service obstinacy, evasions, and shortcomings. It also focused on the USDA's failure to implement enforcement of sexual discrimination laws. According to the 1970 census, women composed 5 percent of the 1.5 million farmers in the country, and they were becoming more assertive. In the Extension Service, only 28 women worked in the GS-11 to GS-18 levels compared to 81 men, and no women were in the GS-16 to GS-18 levels. Several offices had Title VI civil rights responsibilities: the assistant secretary for administration, the OEO, the civil rights compliance staff of the FES, the Office of General Counsel, and the Offices of Investigation and Audit. Secretary Clifford Hardin had vested the assistant secretary with directing civil rights initiatives, but supervision of compliance reporting splintered among different offices. Although Extension Service units submitted data for compliance reports, the commission discovered that it was often fabricated. Extension policy to a large extent excluded the 13 percent Hispanic population eligible for services. Audits revealed that in one state minority employees received $2,000 a year less than whites, and one-third of audits showed noncompliance with the law. To enforce compliance, the USDA could terminate funding, but in a typical action, FES administrator Edwin Kirby overrode an OEO finding that a Tennessee county's actions were inadequate and ruled it in compliance. Apparently, the USDA had never terminated funding of a USDA program or sent a case to the Department of Justice. Instead of pushing for compliance, the USDA invented delays, added paperwork, and refused to enforce civil rights laws, and by 1975, many compliance cases had been in limbo, some for fourteen months.[39]

Despite court rulings, local discriminatory customs persisted, and county agents often simply ignored federal guidelines. On September 21, 1974, St. Helena Parish, Louisiana, agent Grafton Cooper invited 4-H Club members to his home to discuss their beef projects, but he failed to invite African American Paxton Gordon, who also had a beef project. Assistant district agent Buck Greene, demonstrating daft reasoning, explained "that such action was contrary to policy, but could be carried out if and when an agent felt it necessary."[40] In the spring of 1975, the Extension

Service in seven states—Arkansas, Georgia, Illinois, Kansas, Louisiana, Maryland, and Texas—was in noncompliance with equal opportunity guidelines. The USDA hierarchy spun in circles attempting to reconcile shortcomings in recruitment, hiring, promotion, and salary equalization while avoiding taking punitive action and holding off the Justice Department. The acting head of the USDA Office of Equal Opportunity listed five questions for consideration, the first being "Where are we?" He mentioned remedies aimed at noncomplying states but did not specify what they might be.[41]

In 1983, the United States filed suit charging that Alabama operated a segregated higher educational system, and a series of decisions and appeals continued under *Knight v. Alabama* until May 21, 2007, when the U.S. Court of Appeals for the Eleventh Circuit upheld the October 2006 decision of the U.S. District Court for the Northern District of Alabama. Willie Strain had joined the appeal, charging that vestiges of segregation remained that would persist without further action. The case addressed the gap between funding for African American and white institutions of higher learning, in particular inferior facilities, discrimination in faculty hiring, and program duplication. Years of segregation and discrimination had created a dismal disparity between the black and white colleges and universities, and, of course, the white power structure had little interest in ironing out the differences. In 1995, the district court found continuing segregation and issued a remedial decree. Among other remedies, the district court combined the land-grant programs of Alabama A&M University and Auburn University. The court of appeals admission that vestiges of segregation remained did not satisfy Willie Strain. In his mind, the *Knight* case pushed aside reforms initiated by the *Strain* decision. "Now," he concluded in 2007, "they have just about destroyed all of the guidelines and have almost gone back to where they were in 1965 before the court order of *Strain v. Philpott*."[42]

The experiences of Willie Strain and Bertha Jones and many African American extension agents represent the wreckage created not by civil rights legislation but by white land-grant administrators and FES employees who nullified the laws. Without Strain's class-action suit and Judge Johnson's mandate to rectify discrimination, the ACES would have perpetuated its discriminatory policies, and the same was true of the Extension Service in other southern states. The extension culture at Auburn made it impossible for whites to accept blacks as equals or even to use the substantial talents of the African Americans who arrived there in

1965. Despite Strain's experience as an extension agent, editor, radio producer, photographer, and student of rural life, he could not recall ever being asked his opinion by a white Auburn staff member. Even as integration was enforced in the ACES headquarters, whites presumed they knew best.

*How smoothly one becomes, not a cheat, exactly, not really a liar,*
*just a man who'll say anything for pay.*
—*Sloan Wilson,* The Man in the Gray Flannel Suit

# CREDITWORTHY

The need for better rural housing was apparent to anyone who traveled southern country roads. Many farmers' homes were poorly constructed and ill maintained, and as late as the 1960s, a substantial number lacked running water, 30 percent depended on outdoor privies, and nearly all lacked central heating. Sharecropper houses in the South were notoriously shabby, and farmers who moved frequently paid no more attention to upkeep than did the stingy planters who owned the sagging structures. Instead of focusing on housing for poor people, the FHA increasingly made loans for developing rural water supplies and in 1962 established loans for swimming pools, picnic areas, ball fields, tennis courts, and golf courses, projects that served primarily white middle-class communities. Investigations by the NSF and the NAACP uncovered an array of discriminatory practices at the FHA, and in 1964, the U.S. Commission on Civil Rights focused on the agency as it reviewed discrimination in the USDA. In 1968, commission general counsel Howard A. Glickstein provided a biting analysis of FHA programs and predicted that they would "result in replacing the existing population with a new one of tourists, trees and 'educated, able young people.'" These projects, he pointed out, "ignore the needs and interests of the least educated, the most disadvantaged, the poorest and most discriminated against populace of the locality." The FHA had strayed far from its New Deal origins, Glickstein chided, when "rural renaissance" was regarded "as virtuous—no matter what the money is spent for."[1]

In 1946, the Farmers Home Administration succeeded the Farm Security Administration as the primary USDA lending agency for economically strapped farmers. Now remembered mostly for its documentary photographs, the FSA and its predecessor, the Resettlement Administration, focused on helping sharecroppers and tenants, an agenda not always popular with an increasingly conservative Congress or pleasing to rapidly coalescing agribusiness components. While the tens of thousands of FSA photographs now reside safely in the Library of Congress's Prints and Photographs Division, Congress strangled the agency. Its replacement, the FHA, widened its reach far beyond poor farmers, and its loan programs expanded exponentially until by the 1960s it not only funded home construction and improvement but also provided operating loans to farmers, support for cooperatives, and money for rural-development projects. Farmers followed seasonal routines dictated by their crops, and in the spring, nearly all borrowed money to buy seeds, fertilizer, and pesticides, expecting to pay off the loans in the autumn when they sold their crops. FHA personnel supervised loans by establishing budgets, providing advice on agricultural practices, and addressing financial questions. Such supervision by biased administrators could be turned against farmers.

To obtain FHA loans, farmers endured bureaucratic red tape, first submitting to an exhaustive and subjective, if not intrusive, eligibility test: were they creditworthy, were they good citizens, did they have farming experience, did they need farm income, were they of good character, were they good managers, what was their credit history, and was their operation a family farm? A farmer who passed the eligibility test could apply for a loan and, if approved, would be asked to sign three crucial but seldom fully explained documents: a promissory note indicating the interest rate, installments, and the FHA's right to accelerate the loan if the farmer defaulted; a security agreement listing the property backing the loan; and a farm-and-home document that outlined an annual financial plan.[2] The FHA had been created as the lender of last resort, and it continued to disburse credit but increasingly to the less needy. Indeed, few federal agencies failed so completely to address their intended constituency. With complicity from the Washington FHA headquarters, state and county offices twisted programs intended for struggling farmers into subsidies for established farmers and ultimately into segregated community projects. In the 1960s, marginal farmers left the land by the millions, and many were stranded in stagnant pools of low-paying jobs or were unemployed. The FHA and other USDA agencies purposely sacrificed small farmers for their dream of large operations wrapped in capital-intensive science and technology.

Given the numerous loan categories and sheaves of forms, uneducated farmers needed assistance from FHA offices. First, of course, they needed to know what kinds of loans were available, but with FHA offices largely ignoring minority farmers and poor whites, loans increasingly went to literate farmers who were within the circle of USDA information distribution. The rural class structure operated to the benefit of wealthy farmers with representation on agriculture committees, and committee members favored family and friends with benefits, so the conventional wisdom held.

In some cases, it seemed that the more promising a black farmer was, the less chance he had of getting an FHA loan. In 1952, Joe Henry Thomas requested a loan from the Lee County, Alabama, FHA office. In his mid-fifties, Thomas and his wife Beatrice had an eleven-year-old son and two daughters, twelve and eighteen. He owned an eighty-five-acre farm that he bought doing public work, rented other land, and tended nine acres of cotton. Beatrice Thomas worked at the Pepperrell Manufacturing Company in Opelika. In 1964, Thomas sold cattle and vegetables valued at $260, cotton worth $1,440, and $480 of wood, plus Beatrice earned $2,400. He owned a 1954 automobile, a 1950 pickup, a mule, three cows, and twenty-five laying hens. Thomas dreamed of building a modern house on his own farm, and to consolidate his debts and finance the house, he needed an FHA loan. The Lee County FHA supervisor turned down his loan request "without coming out to my farm to see what I had for portion of the security," he wrote to the NSF's Fay Bennett.[3]

While the Lee County FHA supervisor simply ignored Joe Henry Thomas's loan request, other FHA bureaucrats became as arbitrary, capricious, and insulting to African American farmers as the meanest planters and supply merchants. In the spring of 1961, Marshallville, Georgia, FHA supervisor Carl Grant urged Fred Amica to take out a $2,280 operating loan that came due on November 1; Amica customarily borrowed much less. He had a good crop year and, before paying the FHA loan, paid off part of his tractor loan. In the next months, Grant "obtained a judgment against Mr. Amica on the unpaid loan," U.S. Commission on Civil Rights interviewers Richard M. Shapiro and Donald S. Safford learned in their 1964 interview. In May 1962, Grant and a Tri-County Tractor Company employee visited Amica's farm and, as Grant stood by and cursed, removed Amica's farm equipment. Amica had paid two-thirds on his tractor and had money on hand to pay off half of what remained, but the company

repossessed the tractor. When the sale of Amica's machinery did not satisfy his FHA debt, Grant garnished Amica's cotton crop. "Grant indicated
that he was going in the hospital to have a head operation," Amica confided to Shapiro and Safford, "and that he had received authority from
Atlanta to 'clear Amica out.'" Grant had also spread the word among local
businesses that Amica had not paid off his FHA loan, thus damaging his
credit in the community. Pressuring Amica to assume more debt and,
as the crop-lien terminology put it, "cleaning him out" resonated with
the treatment of many sharecroppers and indebted small farm owners.
Grant's strategy of encouraging Amica to take a larger loan, refusing to
set extended terms for repayment, auctioning off machinery, having the
tractor repossessed, and cleaning Amica out had evolved into a common
FHA strategy. In Amica's case, as in many others, securing an FHA loan
led to his ruin.[4] Carl Grant may have been a rogue supervisor in need of
a head operation, but his agenda was all too common.

Civil rights laws often collided with local customs encrusted with bar-
nacles of the discriminatory past. South Carolina attorney Ira Kaye was
tortured over conditions among black farmers and outraged that USDA
agencies remained aloof, distancing themselves from their discriminatory policies. In a February 1962 letter to Fay Bennett, Kaye expressed his
hope that the cycle of black dependence on white administrators would
end. At times, he mused, he would almost favor "separate but equal"
structures so that blacks could gain administrative experience. "Most of
the farm owners that I have interviewed," Kaye predicted in December
1962, "are going to lose their farms either this year or next year." They
were so indebted that even a good harvest would not help. Still, in February 1963, he asked Bennett if there was some way to find $15,000 to help
a group of farmers to continue for another year. "To most of them, giving
up right now would mean selling out their small homes and farms at sac-
rifice prices or losing them with the hammer of a mortgage foreclosure
sale."[5]

Kaye described to Bennett the segregated FHA offices in Sumter
County and stressed that the main office was for whites only. "They had
a very dinky, dingy office for Negro applicants which was no less than
a national disgrace," he wrote in January 1963. Kaye had complained of
the shabby office all the way up to the FHA director in Washington, who
lamely promised to look into it. Office space could be corrected, but so
long as FHA committees were all white, Kaye stated, "the imbalance of
funds will continue to exist." After weathering several bad years, farmers
were facing drastic allotment cuts in tobacco and cotton and, even if they

had a good crop year, would be unable to make their mortgage payments or deal with other debts. "Frankly, I am beginning to think there is no possible solution since no one in authority is really interested in the problems of the small farmer," Kaye dejectedly wrote. In March 1963, Kaye reported that nearly all African American farmers were at risk. Landowners and renters alike lacked operating funds to begin the crop year. "All are victims of past neglect by Farmers Home Administration and," he exasperatedly judged, "the people in charge of those offices blame them for their poverty." Despite evidence to the contrary, Fay Bennett assured Kaye that things seemed to be improving. "I think we are coming closer to getting across to the Agriculture Department your concern and ours that something be done," she wrote early in 1964.[6]

As Kaye suggested, black farmers were at the mercy of white FHA administrators practiced in discrimination. Leonidas S. James worked for both the NSF and the South Carolina Council on Human Relations investigating FHA loans in South Carolina. A graduate of Morgan State College, James held an M.A. from Columbia University and was far along on his Ph.D. at the University of Pennsylvania. He had worked as a psychologist with the Veterans Administration for twenty years and had been a guidance counselor at South Carolina A&M College in Orangeburg. James reviewed several loan cases that illuminated not only FHA discrimination but also the sagacity and talent of black farmers. Muldrew Burgess had farmed near Gable for thirty years, James discovered, and for sixteen of those years, had rented 100 acres, on which, with the help of his two teenage sons, the fifty-three-year-old farmer grew tobacco, soybeans, and cotton. In 1962, the FHA supervisor deceived Burgess by assuring him that his operating loan was approved and directing him to a merchant, then telling the merchant the loan was approved, and finally, after Burgess had purchased fertilizer from the merchant, notifying him that the committee did not approve the loan after all. The merchant, obviously in on the scheme, then billed Burgess twice the amount of the original purchase. Despite the duplicity, Burgess went about his farming, and his tobacco crop grossed $1,000, his soybeans $1,500, and his twenty-one bales of cotton $3,100. In 1963, the industrious Burgess applied for a loan to purchase 165 acres in Sumter County and a tractor, but the FHA committee, perhaps judging him too capable, denied him the loan. James appealed to FHA state director Edwin P. Rogers to "give this case careful consideration."[7]

James also urged the state FHA director to review the case of Willis P. Canty, who farmed with his wife and three children near Pinewood. Refused an FHA loan in 1962, Canty borrowed from a local source and paid his expenses for the crop and some on his debts. Still, he had a $7,500 mortgage on his farm and owed $1,785 on farm machinery. In the spring of 1963, Canty applied to the FHA for a $1,500 operating loan and $9,000 to cover other expenses. The FHA turned him down without explanation. He received a postcard from the Federal Land Bank that loans were available for cattle, but when he presented the card at the bank, the manager told him cattle loans were only for whites. The man who held the mortgage on Canty's farm offered to buy it for $100 per acre and then let him sharecrop. The 112-acre farm was worth twice as much as he owed on it.[8]

Mamie Deschamps, widowed for nine years, ably farmed her 109 acres near Pinewood with her four sons, aged 27, 19, 16, and 13. With her husband, she had borrowed $4,800 in 1943 to buy the farm and a new house. Shortly before their final payment, the house burned, and FHA insurance paid off the note. The family then moved to a dilapidated house, and Mamie Deschamps mortgaged the farm. In 1963, her seventeen bales of cotton brought $2,805, her .47-acre tobacco crop sold poorly, and the drought took her soybeans. She ended up $2,500 in the hole. The FHA denied her loan application. A neighboring white farmer of some substance offered her $100 an acre for her farm; this farmer, James learned, had received a $70,000 FHA loan in 1962.[9]

Forty-nine-year-old John H. Wheeler owned a 150-acre farm, had borrowed earlier from the FHA, and had paid off his farm in ten years. Shrinking acreage allotments left Wheeler with less income. After the FHA refused him a loan, he borrowed $5,600 from a private party to renovate his home and repair a barn. "He has a well built frame home just large enough for him and his family," James observed. The Wheelers had a fourteen-year-old daughter. In the spring of 1963, Wheeler applied for a $6,620 FHA loan to cover operating expenses and consolidate his mortgage, but the FHA committee without any explanation denied him the loan. The man holding the mortgage offered to buy Wheeler's farm for $50 an acre or else he would foreclose, generously promising Wheeler he could stay on as a sharecropper. Wheeler's tobacco crop earned $1,750, but a white neighbor's hogs got into his soybean field before harvest, consuming the crop.[10] Being turned down for an FHA loan inevitably seemed to lead to an offer to purchase the farm, the owner promising to keep the former farm owner on as a sharecropper.

The case of thirty-seven-year-old Lawton Byrd, who farmed near Lamar, illustrated the vulnerability of even the most successful farmer. Byrd, his wife, and their two daughters lived on a 214-acre farm. He had ample machinery for his 44 acres of cotton, 40 acres of soybeans, and 6.8 acres of tobacco. In 1963, his corn crop yielded 122 bushels per acre, a state record. Farming at that level, Byrd had debts, all of which he kept up with until 1960, when crop failure caused him to miss a payment. Although he paid some of the debt during the next two years, he was threatened with foreclose unless he paid off the entire $8,000. Byrd applied for a $40,000 FHA loan to consolidate all of his debts, but the agency told him he was ineligible.[11] These cases epitomized how supervisors and committees handled loans for African American farmers. FHA loans were intended for just the kinds of farmers that James investigated. Besides displaying overt discrimination, FHA denials had an iniquitous edge, a smirking delight at breaking successful black farmers so they could be bought out. As Ira Kaye judged, there was no appeal to USDA perfidy. The record does not reveal the outcome of L. S. James's appeals.

As discussed in chapter 2, in the spring of 1964, the commission interviewed FHA bureaucrats at the federal, state, and county level as well as farmers. What bureaucrats reported as progress only dramatized the sense of privilege and lack of understanding that plagued the USDA. E. T. Fatheree, the Mississippi FHA director, for example, boasted in August 1964 that the state had three African American assistant county supervisors, one in Jackson, another in Lexington, and a trainee in Greenville, plus a black staff assistant. All were federal GS-7 level employees, he explained to commission investigators, but even if blacks were better qualified than whites for a promotion to county supervisor, Fatheree cautioned that advancement would depend on "the personality of the individual and possible community reaction." At the height of Freedom Summer, the Mississippi FHA had made several token changes in its segregated operations, but no African American had penetrated decision-making committees or staff positions.[12]

Fatheree oversaw the working conditions of African American assistants. John S. Currie had served as FHA assistant county supervisor in Hinds County since graduating in 1938 from Tuskegee Institute, where he majored in vocational agriculture. He explained to the commission's Thomas T. Williams in May 1964 that he handled applications from black farmers in Hinds and Rankin Counties. In Jackson, his unidentified, dingy office was across the hall from the white FHA office where he could use the office equipment and supplies, but he had no telephone and had to walk two blocks to the federal building to use a restroom. "No one told me I couldn't use the rest room," he explained, "but it is one of those things you already know and accept." His clients preferred to take out operating loans rather than home or machinery loans, and landowners were reluctant to mortgage their property for fear of losing it. Blacks continued to plant cotton, but most lacked the capital to mechanize. Those who wanted to go into raising beef cattle, Currie explained, often lacked sufficient land. In addition, many were old and set in their ways, Currie added, and most did not know about FHA loans until he told them.[13]

In Georgia, the state FHA director managed ten area supervisors who oversaw county supervisors, some responsible for more than one county. W. C. Thigpen, for example, handled Brooks, Thomas, Lowndes, Lanier, and Echols Counties. County supervisors submitted nominees for committee membership for the approval of the state FHA director. No blacks had ever been nominated. The committees determined applicant eligibility, loan amounts, and foreclosures, but the county supervisor, embodying immense discretionary power, could override committee decisions. When a loan was denied, the applicant received an explanatory letter that recommended a meeting with the committee. If still dissatisfied, the applicant could appeal to the state FHA director, but Thigpen could not recall such a case, nor had he ever heard of the director overturning a county committee decision. He stated that he did not like to foreclose on housing loans and would usually extend the loan period instead.[14]

Marian Yankauer and the commission staff had bored into the FHA structure and located layers of discrimination. Only eighty blacks were involved with the FHA in the South, and only two, Robert L. Hurst in South Carolina and Cozy L. Ellison in Georgia, served on state advisory committees. There were a scattering of alternate county committeemen, some twenty assistant county supervisors working solely with black farmers, and blacks in statewide positions in ten states, again serving only black farmers. Loans to black farmers were often for emergencies, living expenses, and seeds and fertilizer. Unlike the pattern with white loans, "very little FHA money goes to Negroes for the capital investments, such as livestock, equipment, farm improvement, farm ownership, and housing, which would lead to economic improvements." Nor did black farmers receive technical assistance on a par with whites. The FHA was initially created to aid low-income farmers, but loans had crept up to $60,000 for farms and continued escalating. "It is partly this natural bureaucratic tendency to serve the successful that accounts for the poor service to Negroes," the analysis concluded. The commission also made recommendations to improve hiring practices and provide better services to black farmers.[15]

County USDA offices and white farmers monitored African American farmers closely for civil rights activity. A. B. Hamerter, who farmed near Eufaula, Alabama, owned 80 acres of land and rented 69 additional acres. In 1965, he borrowed $2,600 from the FHA to plant 28 acres to peanuts with an expected 20-ton yield, and he hoped for 8 bales from his 15 acres of cotton. Near harvest time, he contracted with a white man to dig his peanuts, but when the man learned that Hamerter had been active in the civil rights movement, he "removed his machinery from Mr. Hamerter's farm, leaving the crop unpicked." The FHA office refused his loan request for funds to hire help to dig the peanuts. Ultimately, he lost 12 tons of his crop. "Mr. Hamerter feels he is ruined because of his inability to get minimal financing through the FHA at a critical time," William Seabron reported. Hamerter requested an OIG investigation of the FHA office in Barber County, Alabama.[16]

Despite the power of landlords, banks, and merchants to punish black farmers, the civil rights movement provoked militancy. Still, African American farmers who participated in the movement faced huge risks. In July 1965, tenant and mechanical-cotton-picker driver Willie Joe White and his wife signed up their three children to attend the white Haynesville High School in rural Alabama. His landlord, Julian Bryant, sent word for him to come to his office, but on the way, White saw Bryant sitting in his truck. After briefly talking about White's work, Bryant allowed that he had heard that White's wife had signed up his children to attend the white school and demanded that she remove the children's names. "The hell you preach," White responded, and the two men parted with the question unsettled and no doubt with Bryant scratching his head at his tractor driver's aggressive reply.[17]

In Lowndes County, Alabama, black farmers paid a price for their civil rights activism. In the fall of 1964, Cato Lee borrowed $4,000 from L. R. Higgler to purchase building material. In November 1965, when Lee asked FHA supervisor James A. Garrett for a $2,500 housing loan, he was sent to get recommendations from three white people. Since Lee had been involved in civil rights work and had enrolled his children in the white Haynesville school, he realized it would be difficult to find three white people in Lowndes County who would vouch for him. Meanwhile, Higgler demanded repayment. On December 28, Lee and his wife signed an FHA loan agreement, but a week later, Garrett demanded Lee's deeds. Seven months later, Garrett still had the deeds, but Lee had not received a loan. In a similar case, Threddie Stewart, a tenant, was evicted in the fall of 1965 after his landlord discovered his involvement in civil rights activity. He had worked on the Mary James farm since 1946 and had even added a room to the house to accommodate his five children. He applied for an operating and house loan in 1966, but the FHA denied his application. "I have 12 cows, 7 acres of corn, an acre of okra and an acre of cucumbers to help me make a living," he explained. Lowndes County farmer Eugene Peoples asked the FHA for a loan to finance the balance he owed on an International tractor. FHA supervisor James Garrett promised to visit Peoples's farm but never showed up, and he also promised that he would notify Peoples of the committee's action. "The last week in December of 1965 the tractor company came out to pick up the tractor," Peoples related, "and I had not heard anything from Mr. Garrett yet." He added that the FHA office staff did not use courtesy titles.[18] Clearly Garrett was using FHA loans as a punitive tool against farmers who were active in the civil rights movement.

The white network operated with malevolent efficiency to control black farmers, carefully monitoring and punishing any contact with civil rights activists. Bobby Storeef rented farmland in Sumter County, Georgia, and over the years, he obtained small operating loans from the FHA. When he applied for an operating loan of $3,500 to farm 233 acres in 1966, he ran into trouble. "He alleges that he is being denied this loan, not because of any outstanding balance, but because he has agreed to rent and farm some land owned by the Barnum family, who are very active in civil rights activities in Americus," Civil Rights Commission attorney Robert A. Cook learned. Relishing his power, the FHA supervisor told him, "In fact, I may not let you farm at all." By renting the Barnum land, Storeef also experienced "other harassment from creditors in Americus."[19]

As the Hamerter and Storeef cases made clear, at the county level, African Americans were at the mercy of FHA supervisors, committees, and office staffs. Greene County, Alabama, FHA supervisor John D. Pattillo and his office staff personified coarse, insulting, dismissive, and unhelpful biased bureaucrats. Annoyed at Pattillo's habitual delays and misinformation, thirty black farmers applied for FHA economic opportunity loans in the summer of 1965 and cataloged their complaints to state FHA director Robert C. Bamberg. Pattillo had announced that there was no such thing as an economic opportunity loan, but after using "coarse language" and chiding blacks that such a loan was "something they heard about in the jungle," he qualified his answer. When poorly educated black applicants asked for staff help, they were often dismissed, told to come back later, or instructed to take the forms home and fill them out. "You will not get the loan until next year, even if you qualify," a presumptuous secretary announced to one applicant. The office staff misinformed blacks, telling them they were ineligible for loans if they owed money, which inspired Reverend Percy McShan to petition the Civil Rights Commission. "This becomes an evil tool against Negroes because the system of farming and white domination has always kept the Negro in debt," McShan reported. He suggested that if Pattillo's actions were "due to racial prejudice," he should be dismissed. Being illiterate, not owning radios or TVs, and ignored by white extension and FHA staffs, many black farmers did not know about FHA programs. When asked in a public meeting in Demopolis if black farmers received the same information as white farmers, Pattillo replied, "Yes, sir," to a chorus of black voices shouting, "No, no." With little access to FHA loans or even useful information, Greene County black farmers were at a distinct disadvantage.[20]

In 1964, when word came down from the FHA to find black alternate committeemen, John Pattillo and state FHA director Julian Brown contacted Harry Means, a successful Greene County black farmer who had attended Tuskegee Institute and who grew cotton, hay, and corn and raised cattle on his 600-acre farm. Brown had explained that a black alternate would serve "in a training and advisory capacity and he will act in the place of a regular member when one is absent." Pattillo described the position and the FHA loan process to a skeptical Means. When Pattillo told him that only farmers with at least fifteen acres of cotton could qualify for loans, Means replied, "I thought about this and concluded that over 60% of the Negroes wouldn't be able to qualify if this standard was used."[21] Obviously, Pattillo and Brown only wanted Means as a token to fill a quota and did not intend to allow him a voice in policy.

Harry Means's experience resonated with that of Adam White, who served as an alternate FHA committeeman in Jefferson Davis County, Mississippi. White recalled that two executives from the county FHA committee visited his eighty-two-acre farm to tell him "that they had been watching me and found out that I was an honest person" and ask him to serve as alternate committeeman. "I received no training for the appointment," he stated. He attended the twice-a-month meetings, discussed the applications from both black and white farmers, and voted, but he could not sign official papers, nor did his name appear on the official list of committee members.[22]

Because county agencies controlled hiring and firing and federal agencies invariably supported local decisions, investigations and remedial action often brought retribution. After Mildred E. Meadows testified in a racial discrimination case against Greene County FHA supervisor John

Pattillo in 1966, she was called to Montgomery to meet Ella Mae Berdahl, the national FHA home economics supervisor. "Miss Berdahl was very abusive of Miss Meadows and told her she was appalled at the testimony she had given in the Pattillo case," William Seabron learned. Meadows asked the inspector general staffer she had talked with earlier to investigate since he had promised that if her testimony caused repercussions he should be notified. As happened so often, the Meadows case dragged on for months, and the OIG evidently failed to follow up. Finally, in January 1968, Seabron notified Meadows that "none of the parties concerned knowingly intended to harass or intimidate you," contradicting what he had written earlier. He observed that things had improved in Greene County and expressed confidence that Meadows's work there was appreciated. For one thing, John Pattillo had been transferred out of the county in 1966.[23]

Black FHA employees in southern states worked out of segregated offices, served only African American farmers, were barred from county FHA committee meetings, and were warned to avoid civil rights issues. State program staff assistant Joshua A. Lloyd found no respect in the white Louisiana FHA hierarchy. After receiving a B.S. in agriculture and industry from Southern University in 1932, Lloyd worked in four USDA agencies before taking a GS-7 position with the FHA in 1951. He was not promoted to GS-9 until 1964, but the promotion brought no increased responsibility. He still served only black farmers from his phoneless and secretaryless office at Southern University. In northern Louisiana, he had "just not been accepted," and white FHA office workers there kept him waiting and addressed him by his first name. When told about the alternate-committee-member scheme, he had urged administrators to select intelligent leaders. Instead, he complained, one alternate was "afraid to death" and several others lacked intelligence. Obviously, he observed, they were selected to demonstrate black incompetence. The powerful FHA state program chief, Lloyd revealed, hated blacks and insisted that hiring them "won't work."[24]

Communities were especially adept at deviously undermining federal civil rights laws, as Reverend Jim Bryant discovered in the summer of 1966. Bryant and a group of Perry County, Alabama, black farmers visited Washington to complain about ASCS discrimination. After Bryant's return from Washington, field reporter William A. Tippins noted that "he has been treated fairly and decently by everyone, even though courtesy titles have not been accorded him." The FHA, uncooperative in the past, approved an operating loan. While USDA officials cooperated, other local institutions stepped in to punish Bryant. Soon after his return, the bank foreclosed the mortgage on his 110-acre farm, which had an 8-acre cotton allotment. Bryant continued to rent land on thirds from black landowner Mary E. Jackson, who had a 21-acre cotton allotment. That fall, he hired a crop duster from the Magnolia Aviation Company to spray for boll weevils eleven times at $2.00 per acre—with no effect. White farmers who hired the same duster got results with only two applications. When the historian Lu Ann Jones interviewed Henry Woodard near Tunica, Mississippi, in 1987, he recalled much the same experience. A neighbor told Woodard that, yes, they sprayed, but there was no poison in the tank. Crop dusters also harassed civil rights meetings, as demonstrated by a duster who sprayed cotton poison on people gathered at a church in Eutaw, Alabama, in July 1965.[25]

While ASCS officials smiled with feigned cooperation, the ASCS measurer determined that the acreage in two of Bryant's fields, which had matched his allotment in previous years, was now too large, meaning not only that he had overplanted and would need to destroy part of his crop but also that his yield calculations would be reduced. When Bryant questioned the ASCS official who measured the land, he smartly replied, "You have fertilized so much, the land has grown." The ASCS meanwhile made undocumented changes to his yield figures and reduced acreage allotments to his disadvantage. Investigator William Tippins studied Bryant's gin receipts and ASCS accounting notes and found deductions "for an unknown purpose" each year. "The foreclosure on his farm, the enlargement of cotton acreage (which adversely affects his projected yield), the obvious inequity of the initial projected yield, and his allegation with respect to ineffective poisoning of his cotton," field agent Tippins suggested, "should be sufficient grounds for a more thorough investigation." Although the Washington ASCS office directed the county ASCS committee to treat Bryant fairly, his visit to Washington not only brought down the wrath of white businessmen who conspired to ruin him but also generated more widespread hostility among county USDA staffers.[26]

While the Extension Service isolated blacks, denied them opportunity, and gave lip service to integration, the FHA perfected tokenism. In the two years since the Civil Rights Act of 1964, the Georgia FHA had hired 1 professional and 1 clerical African American at the state level (out of 34) and 8 professional and 4 clerical African Americans at the county level (out of 172). Gradually blacks appeared on county FHA committees and received more rural housing and economic opportunity loans. Still, some applicants complained that they were denied loans despite being good risks. As organizations such as the NSF spread information about FHA loans, blacks applied, were turned down, often were insulted, and lost confidence in both the FHA and the NSF. Those who got loans often were not supervised, or supervised poorly, so many were unable to repay and were in worse financial shape than before the loan. Other farmers went into debt for tractors, could not pay them off, got bad credit ratings, and could no longer borrow from the FHA. The FHA sometimes lured black farmers into debt, failed to supervise them, and then foreclosed and, at other times, refused to loan to them at all. Either way, blacks were put at risk and were not given information or advice or treated with courtesy. Federal funds remained tightly controlled by the white elite, which seemed determined to eliminate black farmers and certainly prevent their serving on committees or working in FHA offices. Farmer Will Bacon summed up his battle with the FHA in Tchula, Mississippi. The office staff would humor him for months, he wrote in May 1970, "playing a waiting game, saying it aint no money available," all the time demeaning him. "The way I see it they don't want to lone poor negroes no money for nothing."[27]

In July 1966, William P. Mitchell, executive secretary of the Tuskegee civic association, analyzed how practices in his area of Alabama violated civil rights regulations. In the FHA office, black employees assisted only black applicants. In Bullock County, where most borrowers were black, the entire office staff was white. In no offices were blacks addressed with titles of respect, and many whites mispronounced "Negro." When appraising applications by blacks, the staff paid inordinate attention to white merchants' opinions of applicants' "reputation or honesty, debt payment, etc." At a November 1970 NAACP hearing, Thomas Reed, an African American with an M.A. degree, reprimanded whites and suggested "that if you gentlemen cannot pronounce the word Negro that you refer to us as black."[28]

William Seabron rejected investigations that he deemed inadequate or that came to a conclusion not warranted by the evidence. On December 5, 1966, he disagreed with FHA administrator Howard Bertsch's decision that there was no discrimination in the Wilcox County, Georgia, FHA office. The county supervisor, Seabron charged, "misinterpreted program regulations to the detriment of clientele"; did not communicate with African Americans, who were 77.9 percent of the rural population; failed to start a self-help housing program; and "conducted some FHA business in such a manner that reasonable men could say that discrimination appears to be present." Seabron also detected discrimination in the granting of FHA operating, housing, ownership, and economic opportunity loans and saw a conflict of interest when a county committeeman leased his farm to a client with an FHA housing loan. "I would very much like to have your comments on this entire investigation report and what significance it has," he wrote Bertsch.[29] Bertsch tolerated the county supervisor's discrimination and incompetence as well as the comfortable arrangement of one of the committeemen with his tenant. Nothing seemed outrageous enough to warrant action.

On April 29, 1968, Alabama FHA director Robert Bamberg testified before the U.S. Commission on Civil Rights in Montgomery, rambling and intermixing a fixation on FHA rules with homely advice to the commissioners. Bamberg had spent his career working for the Extension Service and the ASCS, serving as the Alabama commissioner of agriculture, and finally serving as state FHA director. When asked why the average loan to black farmers was less than the average loan to whites, Bamberg rudely replied, "In many cases our nigger population has small acreage." Commission attorney Howard A. Glickstein then asked Bamberg whether it was the policy to loan more to prosperous white ($4,200) than to poor black farmers ($1,300), when the figure for blacks was "approximately the value of the goods advanced by furnishing merchants every planting season." Since the traditional credit system had changed, Glickstein questioned "whether the FHA isn't duplicating or replacing the activities of the furnishing merchants." He then asked if Bamberg had tenants, and Bamberg replied, "Somewhere between 18 and 25." Bamberg charged them 6 percent interest until September 1 and settled up on December 1. Glickstein observed that the FHA charged 3 to 5 percent and slyly suggested that Bamberg's tenants might do better with FHA financing. Bamberg denied that FHA loans resembled those of furnishing merchants. He then insisted that FHA loans were closely supervised. Otherwise, he claimed, playing on the stereotype that blacks are irresponsible, "they would go out and buy a pickup truck and that would be the end of it." Bamberg observed that many farmers were leaving for urban areas, and when Glickstein asked if he was likely to leave, Bamberg said that Washington was frightening to him. "Among all those mausoleum buildings up there, I have a depressing feeling," he testified. The more Bamberg insisted on the complexity of factors considered in deciding whether to grant an FHA loan—the size of the farm, the interest rate, the allotment, the potential client—the more it was apparent that county supervisors had enormous discretion in disbursing federal funds. When Bamberg took the opportunity to argue for higher parity prices for farmers, commission staff director William Taylor replied, "You almost got me shedding a tear for the plight of the large commercial farmer."[30]

After the hearing, Taylor warned Secretary Orville Freeman of Bamberg's conflicts of interest and his paternalist attitude toward African Americans. In addition to holding the state FHA position, Bamberg owned a 4,200-acre plantation in Perry County. His statements in the hearing, Taylor wrote, "indicate he believes no serious effort should be made to assist Negro farmers in poverty," and his "personal animosity" toward black aspirations "is based in part on his experience as a landlord over his Negro tenants." Taylor found it indefensible that Bamberg, who headed a state office charged with providing farmers affordable FHA credit, charged his own tenants higher interest. Bamberg's testimony, Taylor concluded, "raises a question of whether he can fairly administer FHA programs which, increasingly, are directed toward assisting poor farmers—the largest proportion of whom are Negro in Alabama." Alabama's 1967 FHA budget was $37 million, so Bamberg supervised an enormous enterprise, as did directors of the ASCS's $104.7 million budget and the Extension Service's $2.6 million budget.[31]

A complaint by Bob Mants from Lowndes County, Alabama, in April 1966 raised the issue of FHA committeemen who, like Robert Bamberg, supplied credit to black farmers at far higher interest rates than FHA loans. Mants also complained of delays and discrimination in administering FHA loans, and while William Seabron admitted that the Selma FHA office was guilty of "unreasonable delays in processing applications," he attributed the delays to inadequate funding. He was more pointed with FHA head Howard Bertsch about another case. "This investigation record is clear that County Committeemen Roy O. Loftin and Henry H. Gates denied, in March 1966, the application of Mr. Elzie Lee McGill on the basis of Mr. McGill's race and civil rights activity." Seabron recommended a warning to Loftin and Gates.[32]

Although poorer farmers were starving for operating loans and other credit, the FHA pushed loans that would transform rural land into golf courses, shooting ranges, and other tourist attractions. Some FHA loans went to construct buildings used solely by whites. Kenneth L. Dean, executive secretary of the Mississippi Council on Human Relations, advised Seabron in February 1967 that segregated social and recreational institutions aided by federal funds "will not fade in a season" and would "perpetuate segregation—and thereby hate—for years and generations to come." It disturbed him that "the Federal Government has one hand busy trying to tear apart segregated schools, yet is quite busy with the other hand constructing segregated social and recreational institutions." Seabron advised FHA head Howard Bertsch that he considered Dean's observations "terribly correct." The situation disturbed Seabron "as deeply as any other I can think of." Such loans also upset poor white people. A Georgia farm woman wrote to Senator Herman Talmadge in 1969 condemning loans to country club developers and large farmers. Observing that her veteran husband could not get an FHA loan, she wrote, "I will have to remodel my thinking about justice."[33] Three years later, a Government Accountability Office report requested by African American Missouri congressman William L. Clay showed that throughout the country most swimming pools and golf courses financed by FHA funds were segregated. The 682 recreational projects served 130,000 white people and 14,643 blacks, but only 21 clubs had as many as 5 black members. Nineteen of those 21 were all black. There could hardly be a clearer example of the federal government's hand in continuing segregation and discrimination.[34]

The FHA became ever more adept at shaping statistics to cloak discrimination. After 1966, the number of loans to black farmers increased in Alabama, and in 1967, African Americans received 35 percent of the operating loans but only 16 percent of the total funds. As evidence made clear, some of these loans were made to put black farmers at risk by loaning enough to plant but not to finish the crop year. The arbitrary decisions of many white administrators often ignored the human costs of the decisions, as if their petty and discriminatory rulings had no implications for people's lives. In their minds, purging black farmers became an agenda item in maintaining white supremacy, and they eagerly pushed blacks off their land, unconcerned about their hopes and aspirations. Many whites held that blacks were unintelligent, unskilled, and childish. Strident speeches, confrontations, demonstrations, and sympathetic TV coverage of the civil rights movement confused and angered whites, for in their thinking blacks had no right to challenge white supremacy. In spite of their Christian beliefs, many whites found it difficult to readjust their prejudices to support brotherhood. In a larger sense, agrigovernment bias toward black farmers dovetailed with its larger policy of erasing all small farming operations and ushering in the golden age of rural science and technology.

Black landowners found it increasingly difficult to endure the baneful odor exuding from many local agricultural offices in the South, especially when officials insisted on counterproductive farming practices. George McElrath owned 400 acres in Decatur County, Tennessee, and needed

legal help to sort out his FHA loans. He was the only black farmer in the
county, and McElrath's lawyers argued that local whites had conspired to
destroy his farming operation. USDA officials in Nashville had insisted
that McElrath take seven actions that went against his better judgment
and common sense, including growing ten acres of cotton when he lacked
the personnel to grow cotton and leasing ten acres of his corn land to the
government when he needed it to grow corn to feed his livestock. After
citing the USDA suggestions, the law firm called for a "thorough investigation" of Decatur County agriculture committees, which had a "history
of racial discrimination and ascendancy of white Citizens' Councils and
Ku Klux Klan organizations." The law firm insisted that the person who
did the investigation be free of bias.[35]

Inspector general investigations, responses from the Extension Service or FHA, and the noncooperation of local offices often clouded rather
than clarified matters. William Seabron reviewed a file on the Lee County,
Georgia, FHA office and agreed that there was no discrimination in the
case. "I am disturbed, though, by the fact that while Negroes are 62.7% of
the rural population, they have only 31% of the loans for only 24% of the
dollars," he wrote to a deputy FHA administrator. "I do not know whether
these figures demonstrate lack of communication between County Supervisor Fox and the Negro community, racial discrimination, economic differences, or something else." He might have rephrased his sentence to
include "or all of the above."[36]

Other reports from the field substantiated the reports of biased federal personnel. A Tennessee investigator reported in September 1967 that
in Dyer County poorer farmers did not receive FHA loan information.
The county welfare coordinator predicted that African American farmers
would all leave farming in "the next two or three years." His reasoning
was couched in the hackneyed rhetoric of white supremacy: "The reason for this is that the Negro is not consistent; he works three days and
not for the next month; there is not enough consistent labor." In Tipton
County, the investigator discovered that blacks were denied loans "for
almost any minor excuse." In Fayette County, a black FHA office worker
got along with everyone but one white woman, who constantly criticized
her work. When the black woman made a mistake and offered to correct it, the white woman snipped, "You're too dumb to straighten out the
mistakes you have made and if you stay here 20 years you will still be too
dumb to learn this work."[37]

W. L. Rouse, who farmed in Burke County, Georgia, completed high
school and, after returning from World War II, used the GI Bill to study

agriculture for two years. He lived with his wife and eleven children in what NSF agent William S. McElhannon described as "a four-room shack with no modern facilities." In 1961 and 1962, Rouse secured an FHA operating loan and paid it off, but 1963 and 1964 were drought years and threw him $4,000 in debt. The FHA supervisor arrived at his farm as Rouse was butchering hogs and took his farm machinery, livestock, and even the hog carcasses, claiming he still owed $3,000. The supervisor made no attempt to arrange repayment terms. Without machinery or credit, Rouse had no way to pay off the loan. McElhannon's August 24, 1968, report upset Fay Bennett. "It makes absolutely no sense whatever to put a small farmer out of business," she argued, "when what he needs is help to stay in business." Bennett wrote to Howard Bertsch asking him to investigate the case. Rouse was one of few African Americans who managed to take advantage of the GI Bill. The historian Adrienne Petty argued that GI Bill education benefits went primarily to whites and veterans who had high school educations. As with USDA programs, the benefits accrued to middle-class whites. The GI Bill loan program, Petty concluded, "was administered on the local level, allowing local committees to steer aid to more prosperous farmers."[38]

Because African Americans under segregation seldom complained to local authorities but rather attempted to find a seam in the wall of discrimination, they were reluctant even after *Brown* and the Civil Rights Acts of the mid-1960s to challenge the white elite. Equal opportunity specialist Ferguise Mayronne interviewed several farmers in the Stantonsburg, North Carolina, community in June 1968. Moses Forbes was assured of an FHA operating loan, but on May 9, after planting season, he was notified that funds were exhausted. In 1943, Lillian M. Edwards purchased a forty-six-acre farm through a bank and had never missed a payment. Her house had deteriorated, though, and she sought an FHA home loan but was told to seek other sources of funding. In January 1968, she consulted the new county FHA supervisor, who also directed her to a private agency. She felt she was "getting the run-around" and refused to go back. She had never heard of "diversion payments or price-support payments on crops," nor had the ASCS contacted her about other programs. Greene County had experienced a high turnover of FHA supervisors, and black applicants received conflicting information.[39]

By the time the Nixon administration entered office in 1969, USDA programs had resisted changes that would have led to equal treatment of black farmers. "The problems are enormous," M. John Bundy and Allen D. Evans argued in a law review article in 1970, "and made even more difficult to attack because the federal government has helped to create and perpetuate them." Bundy and Evans analyzed FES, ASCS, and FHA programs and suggested legal remedies for discrimination. There was no magic bullet, they realized, and ending discrimination in USDA programs would "require a lengthy and arduous campaign of lawsuits, lobbying, publicity, education and political organization." The bureaucracy, meanwhile, absorbed challenges by manipulating compliance reports and feigning cooperation.[40]

Francis B. Stevens, a white Mississippian, worked for the Lawyer's Committee for Civil Rights Under Law in Mississippi and had been instrumental in setting up the North Bolivar County Farm Cooperative, which handled vegetables primarily for the rural poor. The co-op also attempted to aid poor black farmers and supported David Dulaney, who also rented forty acres, in purchasing a forty-acre farm with an FHA home loan. Dulaney experienced three crop failures and, as Stevens explained to Secretary of Agriculture Clifford Hardin, became "hopelessly behind in his FHA loan payments—both home and operating loans." After a careful evaluation, the co-op offered to lease Dulaney's land, an action that would provide funds for his home payments and ultimately get him back to farming. The FHA denied Dulaney permission to rent his farm unless he signed a second deed of trust to cover what he owed on the FHA operating loan. Stevens pointed out that the FHA customarily "waives its right to recover operating loans in hardship cases" and speculated that Dulaney was denied because of his ties to the co-op. He enclosed a letter from the FHA that failed to use courtesy titles and explained the FHA's handling of the case as typical of its "demeaning treatment." The FHA's discrimination transcended this case involving a black man, Stevens argued, for Dulaney "might very well have been an Appalachian white man, a Mexican American of the South West or a migrant farm worker." Stevens located FHA bias in its attitude toward poor and uneducated farmers. "The root cause of the problem is the prevalent policy or attitude that a poor person—an uneducated inarticulate person of any race—is not worth the time and effort required to implement the federal farm programs as they were intended to be implemented." The loans, he concluded, went to the "affluent or middle class farmers," who repaid them and made the office look good. "They are not for the benefit of people who are still struggling to break the cycle of poverty."[41] Stevens's concise analysis went to the heart of FHA policy.

Numerous complaints flowed from the Mississippi Delta, including a petition in February 1971 signed by a dozen farmers claiming USDA discrimination that, added to a poor crop year, put them in jeopardy. Because USDA offices turned them away, these farmers resorted to local creditors, who became dissatisfied with "substantial payments on our debts" and threatened foreclosure. "This situation threatens total disaster," the complaint made clear, "total loss of land and capital goods." Johnnie Jenkins, for example, owned a 200-acre farm worth $500 an acre, but the Production Credit Association was pressuring him with foreclosure unless he paid off a $6,000 balance despite his pledge to apply a $3,500 subsidy check to the balance. To survive, Jenkins needed at least $15,000. Two farmers carried crop insurance but received no federal compensation for their crop failures. "In a 3-county area—Issaquena, Sharkey, and Washington—there are dozens of Black farm families in similar conditions," the complaint revealed. The complainants labeled USDA equal opportunity efforts "farcical and fraudulent" and offered that other federal programs "are little or no better." What was needed, they insisted, were more African American decision makers in local offices.[42]

Continuing crises in Greene County, Alabama, illustrated interlocking policies that worked against African Americans. Shirley D. Webb's experience in the county ASCS office typified the hateful treatment meted out to African American USDA workers. She began work in March 1966, a partition separating her from white workers, and several new white hires were paid more than Webb and took higher positions for which she was qualified. At Christmas, she was excluded from name drawing for presents. One of the new hires expressed relief that she would not be required to call her Mrs. Webb, and another, after a visit from one of Webb's friends, a home-demonstration agent, "sprayed lysol around Mrs. Webb's desk." Farmers and several ASCS committeemen loudly used the word "nigger" in Webb's hearing, and James Smith, county ASCS director, went along with such talk. Webb listed many instances of discrimination, but she endured insults because she wanted to keep her job. In October 1970, she sent a three-page letter to William Seabron, and in January 1971, she warned him that "unless someone from Washington or someone not involved here locally with this office handle this matter it will become a white-wash and nothing meaningful will get reported." Shortly afterward, she was fired, but festering nastiness continued, and Seabron's office ordered an audit that recommended she be rehired and promoted and that James Smith be transferred. Richard J. Peer, the chief of compliance and enforcement, reported in May 1971 that Shirley Webb's firing was still a source of concern and outrage in the black community. When the auditors met with federal employees and those who received services from them, Peer found that "this white group reflected negative thinking." Two Alabama members of his team representing the Extension Service and the FHA, he reported, "have indicated that their working relationship within the State inhibits their objective approach to the assignment."[43]

A month later, in June, Peer released the audit of Greene County USDA offices. "There are no blacks on the ASCS County Committee, the Soil Conservation District Board of Supervisors, or the Board of Directors of the REA Cooperative," the audit began, although two of the three FHA county committeemen were black. Ninety-five percent of black rental housing (1,586 homes) had no plumbing, while 33 percent of black-owned homes (423) had plumbing. Extension had finally pledged to provide information on its programs. "Black farmers, for the most part," the audit reported, "have simply been overlooked by SCS, ASCS and the Service Forester of the Alabama Forestry Commission in favor of the large white landowners." The county had been losing farm population, and between 1939 and 1964, the number of black and white farms had decreased by 59 percent. "It has been alleged that many black tenants were forced off the land when they began to exercise their right to vote," the audit revealed, and no black farmer served on USDA committees, shared equally in benefits, or had a voice in setting policy. The ASCS office was all white since Shirley Webb's termination, and the county Extension Service remained segregated. Housing continued to be one of the county's primary needs, and the audit criticized FHA loan policy. It recommended the election of qualified blacks to committees, better distribution of information to black farmers, collaboration with the local cooperative in promoting vegetable growing, and distribution of material on housing.[44]

Slowly Greene County agencies began hiring blacks, with the FHA hiring a black assistant county supervisor and the ASCS hiring a black program assistant. White extension workers remained stubbornly resistant to reform and refused to hire a white woman to work under a black supervisor. County extension chairman W. H. Jackson avoided a meeting called to discuss problems, claiming that such meetings would degenerate into "forums of dissension and abuse by certain minority leaders or would not be attended by the people intended to be reached." He said he did "not believe in such meetings." Black voting registration, however, was changing the political landscape. The election of an all-black county commission in 1969 opened the county to federal antipoverty funds. The county commission not only pushed for home loans, health clinics, and better food distribution but also lured industries to the county, leading to 800 new jobs that significantly raised the per capita income.[45] As black politicians won office in more counties, whites understood that both their public discrimination and their monopoly of federal funds were ending.

In a larger sense, Greene County epitomized how civil rights initiatives affected rural counties. Most county whites lived in and controlled Eutaw, the county seat, but the 1970 Greene County election resulted in African American control of county government. By then, however, the transformative combination of science and technology, biased USDA programs, flight from farms, and resegregation of education was reshaping the county. "Education level in the County is far below the national average," the audit reported, and "poverty and unemployment among the highest." Nearly all white children attended private schools, while black children went to public schools. USDA agencies lavished attention on larger farmers. Significantly, whites clung to economic power through control of banking, merchandising, and federal agricultural programs. As farmers abandoned cotton and turned to pasture, cattle, and forests, tenants, sharecroppers, and especially young people fled the county.[46]

When the FHA submitted its draft report on 1970 minority participation in loan programs, William Seabron found dubious statistics and confusing comparisons and sent it back severely marked up. Such statements as "compare favorably" and "participation . . . was very good," Seabron argued, were "too general and too vague." Comparisons were inexact, tables confusing, and statistics inaccurate, he complained. In October 1973, biweekly USDA staff meetings on civil rights were abandoned in favor of meeting only when something came up.[47] Discriminatory fingerprints covered many USDA memorandums and reports, but except for when it was detected by diligent eyes such as Seabron's, bias passed under the radar.

Among the interactions of many farmers with the FHA, Welchel Long's experience exposed the bigoted, petty, and unprofessional actions of the agency's staff. Long's father died when he was thirteen, leaving his mother, him, and his two younger brothers to tend the farm in Madison County, Georgia. After a year, the family moved to Athens, where his mother got public work and fourteen-year-old Welchel got a job paying $2.00 a week delivering beer and wine from a saloon to University of Georgia students. At seventeen, he tried farming and, for four years, rented land from an uncle before serving in the army during World War II. "We didn't have any chemistry in the public schools at that time, not much of anything, really," he recalled, and after his military service, he took supplementary high school courses and entered Tuskegee Institute on the GI Bill, graduating in 1950. Then he took a job teaching agriculture in high school,

*Welchel Long, 1987. Photograph by Lu Ann Jones. Courtesy of*
*National Museum of American History, LJ 87-17112-2.*

where he witnessed halting integration until he retired in 1978. His wife,
Emma D. Long, taught school for thirty-five years.[48]

When the historian Lu Ann Jones asked Welchel Long about his experiences with the FHA, his memory flooded with dozens of cases of
black farmers who experienced everything from rudeness to blatant
discrimination, and he blamed the FHA for eliminating all but 2 of the
county's 400 black farmers since he arrived in 1952. When Jones asked
him what happened to the black farmers, Long indicated that many left
the area and some stayed behind to work in the granite quarries or at the
firms that made tombstones. "We call this tombstone territory." Echoing
other observers, he said the FHA would lend a black farmer enough to
buy an $800 tractor, but "they wouldn't let 'em have a dime to buy fer-
tilizer, seeds or nothing else. So all they let him do was get in trouble,
ruining his record credit, and they would sell him out." One year, Long
got a $12,500 FHA loan and used $3,000 to plant 500 acres of soybeans.
When he asked for the remainder for herbicides and machinery repairs,
the FHA impounded the money. When the district FHA supervisor visited
him in October and asked about his crop, Long pointed to his field and
said, "Weeds eat it up." The supervisor declared that he should get the remainder of the funds and apply herbicides, but Long told him, "You can't

kill weeds when they get as high as your head." The FHA, he judged, had little confidence in African American farmers and was tightfisted with loans. "Somehow they figured a black man could farm without money." When an ambitious black farmer wanted to buy land, the FHA would declare the acreage too large or too small, any excuse to deny a loan.[49]

Long's affidavit filed with his complaint to the FHA in 1971 chronicled in detail his farming and teaching careers and the petty and base behavior of Elbert County FHA supervisor Thomas K. Wilson. Long began teaching agriculture at Bowman High School in 1952 at a time when most black farmers were sharecroppers. He encouraged farmers to save money and buy land and equipment, and he invited county supervisor Wilson to speak to farmers about loans for land and machinery. "We don't have any of that kind of money on hand now," Wilson would plead, "or we are out of that type of money." Farmers would find a tract of land for sale, Long reported, but it "would never pass F.H.A. rules and regulations." In 1954, Long bought a tractor and equipment and did custom work for black farmers, financing his farm work through local banks, gins, and hardware stores.[50]

In 1960, he expanded his operation, but the local bank cooled on farm loans and suggested that Long apply for an FHA loan. In 1961, Wilson told Long that he could not qualify for a loan because he made too much money at his teaching job. "At that time I was making less than $6,000 per yr.," he said of his teacher's salary. Long continued to farm and did well until 1965, when his crop failed due to insect infestation and drought and he went $11,000 in debt. In 1966, he made another appeal to Wilson for a loan but got the same answer. Meanwhile, the bank was threatening foreclosure unless he paid off his debt, so he again appealed to Wilson, who met with the bank vice president, a fertilizer dealer, and a farm-supply dealer. Wilson agreed to arrange a loan but, after an extended delay, again told Long he made too much money to qualify. Long was now on the debt treadmill and only made enough to pay off expenses for his current crop, not his long-term debts. All of Long's creditors suggested that he secure an FHA loan, but Wilson continually delayed giving him an application. Finally, he filed an application, and Wilson offered him a $13,000 operating loan, but Long, suspecting the smaller loan was bait to fail, argued that he still had machinery and real estate debts and needed more. Wilson promised he would send Long's appeal to the state office. After another extensive delay, Long learned he did not get the amount he requested. He then talked with Simon Hunter, the black member of the FHA committee, and learned that all members had supported his application.

Long appealed to Senator Herman Talmadge's office, which arranged an interview with the state FHA director. When the director abruptly shoved the file across the desk, Long observed only two signatures and asked why the decision had been made without the approval of Simon Hunter, the black member. At that point, Long remembered, the director's assistant "flew into a rage and shouted nigger nigger I'm so tired of pushing niggers." He did not apologize.[51]

Long then complained to the FHA civil rights office, although he had been cautioned that he might lose his teaching job as a result. Dean N. Moser investigated the case and found few records of black applications in Wilson's office. Moser dragged out the investigation, and when Long complained, a Washington FHA staffer suggested that Long do his own investigation. "It seems to me that there is a systematic unwritten policy or conspiracy on the part of the National Administration to let this trend continue," Long judged, "to drag out complaints and charges with the hope that they will just disappear." He never did get a copy of the investigation. "They finally found that there wasn't nothing wrong with what they did," he resignedly concluded.[52]

Simon Hunter, who served on the Elbert County FHA committee from 1968 to 1971, complained that often he was not advised of meetings. In one case, Thomas Wilson told him he had tried to contact him by phone, but Hunter and his wife had been at home all day and the phone did not ring. The meeting, Hunter later learned, concerned a black farmer's application that was denied. Hunter said that Wilson arrived at meetings with decisions already made. "The Supervisor would bring an application in from a Black applicant," Hunter recalled, "and without any discussion, deliberation or information, tell the committee that he won't qualify, especially if it involved any type loan except housing." In Hunter's three years on the committee, no black farmer received a farm ownership loan. "I am sure that discrimination existed and that there were differences in lending practices of white and black citizens."[53]

Welchel Long had strong support from community whites. In April 1971, Russell Daniel Jr., owner of the Allis Chalmers tractor and farm-implement dealership in Athens, wrote a strong letter of support, observing that Long had been his "valuable customer" for twenty years and had paid his bills until his crops failed several years earlier. He analyzed Long's indebtedness and strongly suggested that Wilson had given Long the "run around." Both Simon Hunter and Horace Seymour on the FHA county committee had supported the loan. In an October 1972 letter to President Richard Nixon, Daniel stressed Long's discriminatory treatment by the state FHA office, including the epithets, and concluded that it "pretty well sums up the attitude of the F.H.A. employees with whom Mr. Long has tried to deal."[54]

In August 1971, Kit S. Yon, a writer and business manager at the University of Georgia, summed up Long's treatment after he submitted his loan application. "Subsequently he was told, 'it has not been approved,' 'the committee hasn't yet looked at it,' 'you were turned down' and other things." He urged Berkeley Burrell, president of the National Business League, to take action, and he contacted a White House adviser with ties to the FHA. Yon declared that Long's case was "without a doubt, the clearest case of colossal inefficiency, bureaucratic foot dragging and total indifference to the plight of a human being that will ever be recorded in the ledger of time."[55]

The letters of support, including one from the White House, and Long's history of mistreatment were sent to George C. Knapp in the FHA Office of the Administrator. Someone in the office made a phone call to Long's residence on September 1, 1971, asked one of his children to confirm his address, and then talked with Kit Yon, who, according to Knapp, "was unable to provide additional supportive information." Given Yon's strong letter, there was not much more that he could have offered. Despite overwhelming evidence of discrimination, Knapp informed OEO director Jerome Shuman that he would take no further action unless prompted by Shuman's office. In August 1971, Secretary Clifford Hardin had appointed Shuman, a cum laude graduate of Howard University law school and a professor of law at Georgetown University, to head a new Office of Equal Opportunity. In the reorganization, William Seabron became deputy director and the new office absorbed his staff and duties. Shuman's investigative report evaded charges of discrimination and instead faulted Long. It reviewed his credit, recorded his crop failures and increasing debt, jumbled his requests for loans to demonstrate his ineligibility, incorrectly stated that the county FHA committee met to review his application, claimed that his teaching salary made him ineligible, and judged that "we cannot conclude that racial discrimination was a factor in the denial of Mr. Long's loan application."[56]

In 1978, after Thomas Wilson had left Elbert County, the new supervisor asked Welchel Long to serve on the FHA committee, and a surprised Long told him, "You don't want me," and mentioned his previous trouble. Finally, he consented to serve. He learned that his experience seeking a loan was not exceptional and that the supervisor had enormous power. No matter what the committee ruled, "if the supervisor or the state people decided that you weren't going to get the money, you didn't get it, I don't care how well qualified [you were]. They were doing things their way. They didn't go by the books." When Lu Ann Jones interviewed Long in April 1987, he had retired from his teaching position and no longer farmed. He was a repository of information on FHA loans, the decline of farms, and the halting integration of schools.[57]

Long's case vividly demonstrated the power of county FHA supervisors to grant or withhold loans. Wilson's pattern of discrimination seemed well known to many whites in the community, and they complained when Long, a hardworking man who had paid his bills until his crops failed, was denied a loan that he deserved. Wilson in effect ignored the county committee by inviting only selected members to meetings, arriving with the decisions already made, and offering the committee selective information to justify his decisions. The bias at the state and federal levels seemed even more dismissive and rude. In Shuman's office, only a few cursory phone calls served as an investigation, and despite evidence supporting Long, Shuman backed Wilson. At every level, Welchel Long had encountered rude, nonresponsive, and prejudiced FHA staff. Still, Wilson kept his position, again demonstrating how willing the FHA was to tolerate discrimination.

No matter how powerful Long's support was, appeals could move only through tiers of FHA county, state, and federal offices. There was no provision for review by an administrative-law judge at least slightly removed from the bureaucracy. Much as county ASCS committees that ruled on points of law that were formerly settled in courts put farmers at a disadvantage, FHA bureaucrats sitting in judgment of regulations chained a farmer to FHA personnel. Farmers faced vexingly cumbersome federal regulations that governed USDA programs, but the former avenue of redress, decisions based on precedent from courts and judges, was moved to county committees and state and federal USDA offices. Even starting an appeal was out of the question for most small farmers, and USDA officials invariably supported county staffers. This shift in law deserves exhaustive historical analysis.[58]

In November 1971, the U.S. Commission on Civil Rights upbraided the Nixon administration for failing to enforce civil rights laws and criticized the president for issuing vague and unclear statements. The commission pilloried USDA claims that it was improving, for, it stressed, citizens were entitled to equal rights immediately. "No one can get greatly excited about progress that is made after he is dead," commission chair Reverend Theodore M. Hesburgh wryly observed. Even what seemed to be well-intended concern over the loss of black farms only resulted in sterile studies. In the spring of 1972, Deputy Assistant Secretary Alfred L. Edwards called to Jerome Shuman's attention "the plight of Black small land owners who are losing their land through a variety of foreclosure devices." Since there was "not sufficient hard information" about such foreclosures, Edwards suggested "information gathering activity." It was a civil rights issue, Edwards insisted, and should be funded by Shuman's office.[59] The petty debate over funding yet another unnecessary study epitomized how USDA civil rights action had been reduced to paperwork.

Usually USDA rhetoric hid the agrigovernment agenda, which supported large farmers at the expense of small producers, but a 1972 incident pulled aside the curtain. Secretary Clifford Hardin appointed fifteen staffers to a young executives committee, whose May 1972 report, "New Directions for U.S. Agricultural Policy," suggested changing the definition of "farm" from an operation of ten acres selling $50 of products to an operation selling $5,000 or more. The committee not only defined small farmers out of agriculture but also denigrated farming as a way of life, seeing it instead as a business. In 1970, the committee discovered, there were 2.9 million farms, but only 1.4 million had sales of $5,000 or more. Since the larger operations accounted for 95 percent of cash receipts from farming, the committee recommended, they should receive USDA support. The others should wither away, the dispossessed farmers set adrift to find public work. Marginalizing small farmers was nothing new at the USDA, for during World War II, sentiments flowed through hallways about useless noncommercial farmers, and after the war, agricultural policy was tailored for large producers who could not only afford the science and technology but also profit from experiment-station research and land-grant school scholarship. The USDA's young executives championed agribusiness and ignored what could be the end of a vibrant tradition of independent families on the land. Included in the 780,000 small farms to be undefined were 46 percent of all remaining black farms.[60]

The young executives' simplistic reasoning ignored the role the FHA would play given its mandate to help the most needy farmers, or how the Extension Service would deploy its agents, or even the acreage allotments, already commodified, that would be seized from farmers with less than the magic $5,000 sales figure. AAA legislation created allotments in 1933, and almost mysteriously, over time they accrued financial value, increasing the price of farmland and ultimately being bought, sold, and rented. The young executives were blinded by the glare of heavy metal and swooning from chemical visions. Shocked at what could become a public relations disaster, Hardin distanced himself from their much-too-explicit new directions.[61]

In many ways, the young executives' scenario had been accomplished. From 1959 to 1966, for example, 75 percent of the farms that failed had gross sales per year of less than $5,000, and 75 percent were smaller than 180 acres. "In the vast majority of instances," USDA agricultural economist Walter W. Wilcox judged, "these farms were far too small for an efficient family farming operation." Indeed, public work contributed 50 percent of income on farms with sales less than $5,000. Stressing the role of mechanization in erasing jobs, Wilcox tabulated that from 1963 to 1968, "farm numbers dropped 560,000 while the number of farm workers decreased by over 1.7 million." Meanwhile, larger farms with sales of $20,000 or more increased 62 percent, from 325,000 to 527,000, and the number with $40,000 or more in sales nearly doubled. He predicted that the trend toward larger farms would continue. Wilcox was unsure what to make of these vast structural changes or of the fact that the USDA's farm programs had created "this rather dramatic change in farm structure in the past 9 years." In 1972, the USDA's young executives gloated over the agribusiness transformation and urged it along. Perhaps because farm failures had become USDA policy, Secretary Freeman and his staff had seemed remarkably unconcerned that in the 1950s 169,000 farms failed annually or that 124,000 failed annually between 1960 and 1965.[62] Of course, the direction had been set as far back as the New Deal and had been implemented with vigor, but culling farms never went fast enough to suit agri-government zealots.

*It has always been my dream to own and farm my own land.*
—*Timothy Pigford*

# 9

# THE END GAME

When the *Pigford* class-action suit was filed in 1997, some of the African American farmers involved traced discrimination back to the early 1980s, when, as Judge Paul L. Friedman observed, the USDA "disbanded its Office of Civil Rights and stopped responding to claims of discrimination." Since USDA discrimination went back beyond the two-year statute of limitations, in 1998, Congress extended the statute for class-action members to cover farmers who had filed complaints after January 1, 1981.[1] That year thus marked a dividing line between farmers who might gain compensation for discrimination and those who could not. Farmers entangled in the web of discrimination described in the first eight chapters of this book, then, were unable to pursue restitution.

Meanwhile, the history of the civil rights movement in popular memory increasingly took on a mythology of heroes, great achievements, and monuments to success. Like faked compliance reports, the civil rights success story edited out conflict, a construction that undermined the movement's legitimacy and scope. Erasing the important role of SNCC, CORE, COFO, the NAACP, the NSF, and the SCLC; ignoring U.S. Commission on Civil Rights reports; contemptuous of congressional oversight; destroying complaints; and compiling mountains of useless investigations and compliance reports, the USDA denied its unsavory history. As recent suits have revealed, discrimination against African American, women, Native American, and Hispanic farmers persisted even after the *Pigford* decision at the turn of the twenty-first century.

As Welchel Long observed in the 1980s, it made little difference whether or not an African American served on a county FmHA committee ("FmHA" replacing "FHA" in 1974 to distinguish it from such federal agencies as the Federal Housing Authority and the Federal Highway Administration) for county supervisors had the final say on loans and whites controlled county offices. As the term "civil rights" degenerated into a rhetorical tool under the Nixon administration, USDA bureaucrats spoke of equal opportunity even as they discriminated in hiring and continued policies that drove black farmers from the land. The cumbersome compliance machinery indicated fair treatment for minorities while camouflaging inequities in borrowing, acquiring information, and program participation. Highly critical reports from the U.S. Commission on Civil Rights, the NSF, the Southern Regional Council, and congressional hearings revealed the swamp of discrimination festering in the agency. In "The Decline of Black Farming in America," issued in 1982, the commission reported that FmHA officials in eastern North Carolina denied loan applications to blacks, loaned less money than requested, sometimes withheld even that, accelerated payment schedules, and made negative reports to businesses and thus cut off other sources of credit.[2] It was a familiar scenario, one that had become hardwired into FmHA loan programs.

The commission's 1982 report on FmHA discrimination very much resembled its 1965 exposé, but in the intervening seventeen years, over 150,000 black farmers had left the land. Only some 33,000 remained. The report included a complaint filed in U.S. District Court in 1979 arguing that the drastic decline in Mississippi's black farms resulted in part from "a racially discriminatory policy and practice in awarding, supervising and servicing farm loans." Between 1954 and 1974, the complaint revealed, the percentage of black farmers in Mississippi declined from 46.8 percent to 15.2 percent, black farmers lost 140,881 acres a year, and acreage in black farms fell from 18 percent to 5.8 percent. FmHA county committees in Marshall and Leflore Counties, the report continued, "are staffed with persons who are biased against Black and/or small farmers," and the Marshall County committee loaned to "large financially secure White farmers who do not qualify for such loans." A 1990 study reported that between 1910 and 1987, the number of black farms in Alabama declined from 110,443 to 1,828, and a similar trend occurred throughout the South. From 1979 to 1980, black members of FmHA county committees declined from 7.2 percent to 4.3 percent, endangering even tokenism.[3]

The commission's 1982 analysis found discrimination in all FmHA loan programs. Black farmers received only 1.9 percent of home ownership loans, and "Others," including Hispanics, Native Americans, and Asians, claimed 4.1 percent. In Alabama, loan amounts for black farmers dropped from an average of $27,811 in 1979 to $10,769 in 1981, while over the same years, white averages rose from $47,057 to $64,664. The number of operating loans granted to minorities dropped from 5,287 in 1971 to 3,024 in 1981. The same pattern held true for disaster, economic-emergency, soil-and-water, and limited-resource loans. USDA payments thus continued to favor farmers who tilled 500 or more acres. Ward Sinclair, who diligently covered agricultural issues for the *Washington Post*, reported in November 1981 that 10 percent of the largest farmers took in nearly half of the USDA's $2 billion in payments. Even though there was a $40,000 limit on what a farmer could receive in federal agricultural payments, Sinclair discovered 279 cases where farmers evaded the rule.[4] Imaginative farmers disbursed holdings among family members and easily evaded such restrictions.

Under the Reagan administration, not only did USDA civil rights enforcement drag to a halt but also agencies sharpened their punitive edge. Isidoro Rodriguez, who headed the Office of Minority Affairs, released a memo in February 1983 proposing to strip civil rights guidelines that were "contrary to the administration's direction." Ethnic and women's groups, he reasoned, had benefited financially and politically from civil rights policies but failed to support Republicans, while whites were impatient with what they saw as favoritism. Since Rodriguez took office in 1981, civil rights investigations of USDA programs fell from 90 a year to 0, and compliance field reviews fell from 92 to 1. He earned praise from USDA administrators for heroically returning $475,000 in unspent office funds and for cutting ten staff members. After Rodriguez was placed on a ten-day leave as his memo on eviscerating civil rights guidelines was being reviewed, his office staff charged him with bullying and with decimating civil rights work, and he was later fired. Reporter Ward Sinclair learned that Rodriguez was also under investigation for collecting unemployment benefits when he was employed.[5]

In April 1983, the commission confronted USDA secretary John R. Block with his dismal civil rights record. Commission chair Clarence M. Pendleton Jr. pointed out that the USDA had "dismantled" its already weak civil rights enforcement machinery. Over a year earlier, the commission had reprimanded Block for failing to extend loans to black farmers, and members of the U.S. House of Representatives Agriculture Committee had upbraided Block for not directing the FmHA to increase the number of such loans. Pendleton put much of the blame for lax enforcement on the fired Isidoro Rodriguez, who, it seemed, perfectly embodied the Reagan administration's attitude toward civil rights.[6]

A 1984 House Judiciary Subcommittee hearing sounded like a scratchy old record from two decades earlier except that one of the witnesses, Timothy Pigford, would over the next decades lead a class-action suit that, like the *Strain* case, would have enormous implications. Attorneys from eastern North Carolina who worked with rural issues and black farmers testified to chilling USDA abuses. John W. Garland, director of Legal Services of the Coastal Plains from 1979 through 1983, testified about North Carolina black farmers who sought legal assistance in dealing with the FmHA. A July 1980 USDA Civil Rights Office investigation had found substantial discrimination in Gates and Hertford County FmHA offices, where, among other failings, staffs made biased real estate appraisals, delayed approving loans for blacks, had no deferred-payment schedules for blacks, and required some black farmers to voluntarily liquidate as a loan condition. "They issued a finding of no discrimination, in face of their own compliance review that found eight or nine verifiable differences in treatment between white and black farmers," Garland testified. "There was no enforcement." Garland noted not only the FmHA's discriminatory treatment but also "the blatant attempts to silence critics and retaliation against farmers who try to seek redress." Indeed, he pointed out, the hearing room would have been filled with witnesses had they not been intimidated. "Farmers who file claims against USDA or FmHA officials do so at the peril of losing their land and a way of life that has been theirs for generations," and he listed several farmers who had lost their land after filing a complaint. Black farmers also revealed that FmHA officials "taunted them and denied them assistance because they filed a discrimination complaint."[7]

Garland mentioned political intrusion in a case that involved David H. Harris Jr., executive director of the Land Loss Prevention Project in Durham, North Carolina. The OEO, Harris learned, had interviewed six black farmers who owned 150 or more acres each and were served by the Hertford/Gates County FmHA office. None had received information on the emergency loan program, and when one farmer inquired about hardship loans, the FmHA office told him there was no such loan and advised him to get an off-farm job. The OEO completed its report in July 1980, but the black farmers did not learn of it until October 1981, and then only because their attorney filed a Freedom of Information request. In February 1981, Harris filed a supplemental complaint charging that the FmHA had retaliated against the farmers. Despite overwhelming evidence, the OEO found no racial discrimination. Harris began preparations for a federal case, but before he could file suit, North Carolina FmHA state director Larry Godwin enlisted Republican U.S. senator John East's office, which imaginatively claimed that Harris engaged "in a campaign aimed at political harassment against FmHA." Senator East's intervention had the desired effect, for the lawsuit was never filed. "Larry Godwin's letter to Senator East was a clear abuse of his authority and was an attempt to neutralize the farmers' legal representatives," Harris made clear. "Harassing the attorneys had the same chilling effect upon the farmers as if the farmers themselves had been harassed."[8]

Garland argued that local autonomy and lack of accountability led to FmHA offices "being run like plantations." Supplying a broader perspective on the eastern North Carolina context, he added that he had fought in Vietnam "ostensibly to insure that the Vietnamese farmer could freely own his land and cultivate the soil and not be displaced by the communists." In Nicaragua, he added, the United States supported the Myskito Indians against dispossession. What reaction, he mused, would the United States have if the USSR began displacing "a religious or ethnic minority group from their land." Suggesting a parallel, he stressed that no action had been taken to rectify discrimination against black farmers in the United States. "They were foreclosed. They were sold out. They were threatened, they were taunted. They were intimidated. They were harassed." In 1980, black farmers filed eighty-five OEO complaints about discrimination in FmHA loans, to no avail.[9]

The next witness, Timothy Pigford, grew up on an eastern North Carolina farm. His father died when he was four years old, and his uncle managed the farm until he lost it. After attending college, Pigford took a job in 1973 at the Hercules Chemical Plant, and three months later, he found a 100-acre farm on sale for $47,000. "It has always been my dream to own and farm my own land," he testified. Landownership, of course, helped stabilize African American communities, and during the 1960s, black landowners often provided invaluable aid to civil rights workers. Pigford's experience with the FmHA resonated with that of many other young black men who wanted a farm and thousands of older black farmers who were denied FmHA loans. Pigford applied for an FmHA loan, was approved, but then was informed that he should get a loan from the Federal Land Bank and the FmHA would make up the difference. He could not afford the land bank's collateral requirement, and the FmHA did not advise Pigford of other loan plans or his right to appeal. "I didn't know my rights and I am not proud of that," he admitted, "but I figured FmHA was the agency to help me—I trusted them."[10]

In 1976, Pigford began renting land and each year applied for an FmHA operating loan. He endured the office's demeaning advice to get out of farming and realized that in eight years he had spent enough on rent to pay for the farm he wished to buy in 1973. Meanwhile, he attended seminars at agricultural colleges, pestered extension agents in three counties for advice, and learned that without capital such advice was useless. Black farmers understood technology, he insisted, they simply lacked the capital to purchase it. The FmHA offered worthless advice, such as assuring him that his 50 horsepower tractor was adequate to till the 280 acres he rented. Extension Service and FmHA agents discouraged his interest in soil conservation and diversification. "I've wanted to expand into vegetable crops, hogs, and get into tobacco which I currently rent out," he insisted, but the FmHA would not listen to his ideas. To carry on, he rented from fourteen different landlords. "None of them will sign anything," meaning he had no contracts but operated on word of mouth. When he found a 170-acre farm for $110,000 in 1976, the county FmHA supervisor judged he did not have enough equipment for the farm. Eight years later, he complained, "They're telling me I have too much equipment to farm 280 acres."[11]

For eight years, the Bladen County FmHA supervisor refused to loan Pigford money to buy a farm. "I feel like if I was white I would not have had to struggle since 1973 to own a piece of land," he judged. He had joined an organization of black farmers in North and South Carolina whose operating loans came late, who were not told of useful loan programs, and who puzzled over complex loan forms and received no help from FmHA offices. He analyzed the granting of farm loans and learned that county committees "are often made up of big farmers and people with political ties or business interest in controlling who gets and keeps loans." Of North Carolina's 236 county committee members in 1983, only 10 were black. "Big farmers have expanded because the government has continually pushed production, production, production—FmHA dished the money out," he observed. "But Black farmers can't get even adequate equipment and now they can't find land to rent because it's so expensive."[12]

In Pigford's mind, it was "very hard for a young black farmer or any black farmer in Bladen County or in eastern North Carolina to obtain proper operating funds for ownership and to buy equipment." Even to get an operating loan, Pigford had to make as many as four trips to Elizabethtown, thirty miles from his home, and pretend to heed poor advice that could ruin him. The FmHA encouraged Pigford to rent a tractor, even though the rental payments were more than installments on a new tractor. They denied him a loan for a combine that he needed to harvest his corn and soybeans and advised him to use a custom operator. "They come out to the farm and ride out there in that air-conditioned car," he scoffed. "They get out, it's hot, and they say 'well, the crops look all right,' and they get back in their car and head back to Elizabethtown." At the same time, Pigford saw young white men entering farming and sucking up the loan money that he believed should go to minority farmers. He had attended high school and cropped tobacco with these young white men, but "here I am 11 years later still struggling." It had gotten to the point that when he went to the FmHA office, he testified, "I just carry me a box of Maloxes and set it down there on the desk and tell him let's go at it."[13]

John Garland rejoined the discussion and stressed that FmHA county administrators, who held an appointed position, were imperious and would break off a check-signing meeting with a black client and abruptly leave for a luncheon appointment. They were rude, arrogant, and irresponsible, Garland added. As their testimony came to an end, Garland and Pigford shared their evaluations of Democratic and Republican administrations. Garland saw no difference, but Pigford allowed that under the Reagan administration, "there is more stonewalling. There is more paper, more regulations, more delay. There seems to be the attitude, if you want something to fail, you just bog it up in paperwork."[14]

As Garland and Pigford agreed, both Democratic and Republican administrations tolerated discrimination against African Americans, but during the Reagan administration, the complaint process froze as the USDA OEO flagrantly ignored issues dealing with minority farmers. Alma R. Esparza, director of the office, testified at the 1984 hearing, mixing boilerplate platitudes with promises to reduce the growing pile of discrimination complaints. She could not recall how many black farmers remained, how fast they were declining, how many went out of business every year, or how many suffered bankruptcies. She also had no idea how many blacks were on FmHA county committees. Interestingly, her supplemental statement revealed 187 blacks, 77 Hispanics, 36 Native Americans, and 18 Asians out of 6,803 FmHA committee members. The figures for the ASCS indicated that out of 8,897 committee members, there were 28 blacks, 54 Hispanics, 33 Native Americans, and 23 Asians.[15]

Congressman Don Edwards scathingly portrayed FmHA county supervisors as "apparently an 'old boy' network, in the past chiefly white, who have about as dismal a reputation of any group of civil servants in America." Edwards also pointed to an article by the *Washington Post*'s Ward Sinclair that claimed that Esparza's office had investigated only 6 out of 400 complaints. She did not dispute the figure. Sinclair, an astute observer of agricultural policy whose penetrating series of articles had prompted the hearing, had interviewed Esparza in July 1984, and although she had promised diligent enforcement, her office had stopped all discrimination investigations and sent all complaints back to local offices. Nor had she issued reports on minority participation in USDA programs, conducted compliance reviews, or updated discrimination guidelines. Esparza was the fifth person to head the minority affairs office since 1981, and clearly her work on the Reagan transition team and two executive posts in the Defense Department had not prepared the San Antonio, Texas, resident for her for civil rights position. William C. Payne Jr., who had for years been on the U.S. Commission on Civil Rights staff before moving to the USDA's minority affairs office, made a formal complaint to the OIG charging that Esparza had halted civil rights enforcement. She then demoted Payne, who had outstanding performance ratings and awards. Inspector General John V. Graziano found that the office suffered from inept management, poor morale, questionable personnel practices, and failure to carry out its civil rights mission. Even Esparza's handpicked internal task force reported that the office was a shambles.[16]

Other witnesses at the September 1984 hearing reiterated the FmHA's sorry performance, cataloging discrimination and incompetence. At times, it seemed that bias was the only consistent FmHA program. Former FmHA southeast regional director Arthur Campbell testified for the Rural Study Group, an organization that worked to improve the lives of low-income rural families. Campbell astutely observed that the FmHA had not tracked minority loans, had kept secret a review on integration in the multifamily-housing program, and had stonewalled a Freedom of Information request, ultimately "denying the very existence of the report." In 1983, the number of ownership loans to black farmers "was the lowest ever recorded since such statistics based on race were collected," Campbell added. The OEO sent complaints back to the staff member who was the subject of the complaint, and as a result, Campbell charged, "the field officer, often the accused in such instances, now is the investigator, the judge and jury." Campbell leveled a damning judgment, charging the FmHA with "insensitivity, a level of ineptitude unequalled anywhere else, a calculated system of delay."[17] Despite startling testimony from Garland, Pigford, and Campbell, Congress seemed helpless to intervene, and the

USDA's civil rights office and the USDA at large celebrated the post–civil rights era.

By the 1980s, scholars and attorneys increasingly focused on land loss among African Americans, and they discovered that more than USDA chicanery entered the equation. Many black landowners died without drawing up a will, and the property went to all eligible heirs, usually immediate family. Any heir could ask for his or her share, forcing sale of the property, or they could sell their share to someone outside the family, for example, a prosperous white landowner, who could then demand a partition sale. Other families failed to pay taxes and lost their land. Whites watched closely for opportunities to secure African American farms and, as many FmHA cases showed, stooped to vicious methods to take them.[18] The fierce opposition to legal-assistance attorneys in part reflected white hostility not only to interference in their semi-legal and often illegal schemes to take black property but also to challenges to their control of county agriculture.

*Washington Post* reporter Ward Sinclair continued his pursuit of USDA discrimination. In May 1986, he reported that the FES fired Edith P. Thomas, its ranking African American employee, for complaining that nutrition funding went directly to white land-grant schools. Supervisors informed her that "they wanted the black schools to seek aid from the white schools rather than from the USDA headquarters," another example of requiring African Americans to be supplicants. Since starting at the USDA in 1980, Thomas had filed ten complaints about racial slurs and discrimination, and she pressed these cases along with challenging her punitive firing and in November agreed to a confidential settlement with the USDA. She was rehired and given a two-year assignment at George Mason University. Thomas, who held a Ph.D., had formerly taught at Indiana University and the University of North Carolina. Sinclair also discovered that the Animal and Plant Health Inspection Service attempted to fire a black employee for complaining about an equal opportunity violation, and then it delayed a scheduled compliance report. To prevent leaks, it enforced a rule channeling information releases through the division head. The Federal Grain Inspection Service was forced to abandon a rule that permitted supervisors to receive satisfactory performance ratings, in Sinclair's words, "if they had been found guilty of no more than three discrimination complaints."[19]

On November 5, 1987, Congressman Don Edwards again chaired hearings on equal opportunity in USDA programs, a predictable rerun of the earlier hearing. Given the migratory habits of the USDA's affirmative action office, lack of enforcement was unsurprising. In 1978, it had moved from the Office of Personnel to the Office of Equal Opportunity, and then it bounced back to the Office of Personnel in 1981. Investigations moved from the Office of the Inspector General in 1982 to contractors, only to end up back in the Office of Personnel in 1984. Peter C. Myers, deputy secretary of agriculture, stressed Secretary of Agriculture Richard Lyng's June 1986 directive pledging civil rights enforcement, an increasingly contemptuous secretarial rite. Myers's testimony and charts chronicled hundreds of idling complaints. The categories grew far beyond complaints of discrimination toward African Americans: reprisals, 492; race/black, 338; age, 280; sex/female, 213; sex/male, 138; handicapped, 101; Hispanics, 59; race/white, 32; sexual harassment, 14.[20]

June K. W. Kalijarvi, a Washington attorney who specialized in civil rights and constitutional litigation, had observed no improvement in USDA civil rights enforcement for over a decade. While Secretary Lyng and other top managers might have good intentions, the professional staff that perpetuated discrimination, she testified, "were there before this administration and they will be there after this administration, and, frankly, they don't care." To document the heavy weight of bureaucratic inertia, she added, "If you work your way through this entire system, it can take 11 years." Employees who filed EEO complaints, she revealed, "are going to be identified for the rest of their tenure in Agriculture as troublemakers."[21] Kalijarvi located the reproductive center of discrimination not in leadership but among the professional staff, which could silence or eliminate civil rights advocates.

In November 1987, a coalition of legal groups headed by the North Carolina Association of Black Lawyers filed a class-action complaint against the USDA. The number of FmHA loans to black farmers, it revealed, declined from 5.4 percent in 1980 to 1.5 percent six years later, and the number of loans to Native Americans went from 4.1 percent to 1 percent. The number of black and Native American farmers declined by 23.2 percent between 1978 and 1982, the number of women farmers fell by 15.7 percent, while the number of white farmers dropped by 9.8 percent. The complaint of this North Carolina coalition foreshadowed the class-action suits that would emerge in the 1990s.[22]

Attempting to improve the USDA's fifty-second out of fifty-eight Equal Employment Opportunity Commission ranking in minority employment, Secretary Lyng in April 1988 announced a five-year plan that called for more women in higher positions; an increase in black, Hispanic, Native American, and Pacific Islander hires; and training programs to acquaint managers with these goals. Surprisingly, the Office of Advocacy and Enterprise strongly reprimanded Mississippi's FmHA for violating equal rights laws by attempting to block a compliance review and for ignoring Lyng's five-year objectives. Not surprisingly, the review found that the FmHA had discriminated against black farmers, that it had failed to monitor local civil rights objectives, that it did not comply with equal opportunity and affirmative action standards for employees, and that blacks and women were underrepresented on ASCS staffs. Despite this positive move, the weight of discrimination still crushed most equal rights initiatives throughout the South.[23]

In July 1990, African American Mississippi congressman Mike Espy, who in three years would become secretary of agriculture in the Clinton administration, summed up before a congressional committee the enduring complaints that had followed the FmHA. He mentioned the same complaints that had plagued black farmers for generations and suggested an agreed-upon FmHA strategy to end minority farming. David Harris, who had weathered Senator East's intervention as executive director of the Land Loss Prevention Project, footnoted Espy's analysis. "Now we are not just talking about benign neglect, a failure to enforce equal opportunity guidelines," Harris clarified, "but we are also talking about blatant acts of discrimination on the part of local Farmers Home Administration officers against minority farmers." These acts were much the same as reported in the 1965 and 1982 U.S. Commission on Civil Rights reports, and "those same acts and practices are in existence to this very day." Harris revealed that between 1982 and 1987, the number of black-operated farms in North Carolina dropped from 4,413 to 2,640, more than 40 percent, and the number of Native American farms fell from 911 to 653, more than 28 percent.[24]

More and more, FmHA discrimination complaints from women, Native Americans, and Hispanics boiled from county offices. The FmHA denied a Native American farmer a loan to enter the poultry business, claiming he lacked experience; he had twelve years' experience in the poultry business. Female clients, Harris continued, reported that FmHA supervisors told them that women should not be in farming. The U.S. Commission on Civil Rights report, he said, prompted the USDA to set up a task force. The task force's report, he chided, admitted some problems but assured the commission that the USDA would address the "temporary difficulties."[25] By discriminating against minority farmers and women, the FmHA punished the very farmers it was created to aid, and no memo or directive proclaiming civil rights had any effect on its discriminatory policies.

Randi Ilyse Roth, a staff attorney for the Farmers' Legal Action Group, Inc., testified that in some FmHA offices, "the decision making personnel clearly discriminate against black farmers on the basis of race" and added that the FmHA "consciously and systematically chooses not to effectively enforce the Nation's civil rights laws." In Lee County, Arkansas, Roth discovered that the FmHA office computed crop yields one way for white farmers and another way for black farmers. The office denied bias and claimed that it was "only a communication gap."[26]

Roth appended several affidavits to her statement. Betty Puckett, a Louisiana farmer, had worked since 1986 for the Louisiana Interchurch Farm Crisis Coalition helping farmers with credit problems. "For the first few years of my work," she recalled, "I was not convinced that black farmers were being hurt by active race discrimination within FmHA." Puckett changed her mind: "I know now that there is race discrimination in FmHA's administration of the farm loan programs." The worst cases, she argued, were in Avoyelles Parish, where the supervisor turned down black farmers' loan requests for reasons that he would never have given to white clients. "I have not seen any evidence that the state or national offices of FmHA have taken any steps or have any programs to correct this discrimination within FmHA," she concluded. Olly Neal, an attorney in Lee County, Arkansas, represented small farmers, black and white. He gave examples of discrimination. "The only time I have ever succeeded in getting a loan approved for a black farmer," he recalled, "was in a case in which I had a personal relationship with the FmHA county supervisor." Payments were not timely, collateral was figured inaccurately, and appraisals were weighted to the disadvantage of black farmers. In Woodruff County, he had witnessed a county supervisor "blatantly lie in an appeal hearing regarding a black farmer's loan: he said he had never seen the farm, when in fact he had personally done an on-site appraisal of the farm."[27]

Ben Burkett, a fourth-generation farmer, represented the Federation of Southern Cooperatives/Land Assistance Fund, Interfaith Action for Economic Justice, and the National Family Farm Coalition. He farmed in Forrest County, Mississippi, near Hattiesburg. In his work with hundreds of black farmers, Burkett often witnessed FmHA discrimination. After chronicling several cases, he reasoned that the FmHA "tries to break the spirit of farmers and discourages them from asking for or applying for assistance to solve their problems." Thirty-nine-year-old Walter Carroll of Perry County, for example, worked for other farmers, so Burkett suggested he apply for a socially disadvantaged FmHA loan of $14,000 and purchase a twenty-acre farm and equipment. The FmHA office first denied that there was such a loan, but even after Burkett persisted, it turned down Carroll's loan application. Carroll appealed but missed a crop year because of "this time consuming, unnecessary and discriminatory burdensome procedure." If the FmHA was interested in working with young African American farmers, Burkett testified, it would have embraced a farmer like Walter Carroll. There were thousands of such cases, he asserted, and even when farmers won appeals, the FmHA often failed to implement the decision.[28]

Since the AAA's inception during the New Deal, the county committees and office staffs of the AAA, ASCS, and FSA have minded the interests of prosperous farmers, but they have also looked after their own. In 1995, the nonprofit Environmental Working Group analyzed federal funds that USDA farmer-elected committee members and office employees received and discovered that over the previous decade, the figure was a startling $2.3 billion. While the often-criticized county-committee system forbade members to vote on their own applications, the clubby committee atmosphere provided ample opportunities. The study revealed that committee members over the past decade had received on average $14,743 per year, employees had been given $8,212, while farmers outside the office took $7,334. Both Farm Bureau and USDA officials defended the committee system with the tired rhetoric that farmer-elected committees were ethically pure and best represented the interests of farmers. The Farm Bureau, always defending, actually embodying, agribusiness, lamely offered that committee members were often prominent farmers with large operations but, of course, did not explain how committee members and office staffs cornered such a large share of federal payments. Although the study raised serious questions about the ethical caliber of committee members and office staffs, the system endured.[29] Since county agricultural committees seldom preserved records, even their public business remained clouded, and whatever deals might have been struck have remained secret.

Thirty-one years after Orville Freeman's memorandum on civil rights, President Bill Clinton's Secretary of Agriculture Dan Glickman appointed a Civil Rights Action Team to investigate the department's civil rights efforts and make recommendations for change. In February 1997, the team issued a monumental report, "Civil Rights at the United States Department of Agriculture: A Report by the Civil Rights Action Team." A group of black farmers, including Timothy Pigford, demonstrated at the White House in December 1996, asking for fair treatment by the USDA, and then they filed suit against Secretary Glickman, "asking for an end to farm foreclosures and restitution for financial ruin they claimed was brought on by discrimination."[30]

The USDA action team, meanwhile, held twelve listening sessions in January 1997 and heard "often emotionally charged" testimony condemning USDA employees and county committeemen for "bias, hostility, greed, ruthlessness, rudeness, and indifference." The report stated that "minority, socially disadvantaged, and women farmers charged that USDA has participated in a conspiracy to acquire land belonging to them and transfer it to wealthy landowners." Farmers eagerly testified to USDA discrimination. One witness observed that county officials were "short on moral rectitude and long on arrogance and sense of immunity." County committees, a Mississippi farmer judged, could "send you up the road to fortune, or down the road to foreclosure." The Office of the General Counsel, the action team learned, denied compensation to farmers who experienced discrimination and was "even hostile" to civil rights. Tellingly, performance ratings for loan officers were based on acres served and low default rates, "a performance management system that rewards service to large, financially sound producers while working against small and minority farmers."[31]

The action team, citing a Government Accountability Office report, found that the promises of equal employment opportunity had been broken. A quarter of USDA offices in the 101 counties with the most minority farmers had no minority employees, and although another quarter had minority executive directors, most offices hired blacks for less-skilled jobs. Again and again, the report observed that women and minorities were nearly invisible in agricultural policy. "In 1994, 94 percent of all county committees had no female or minority representation." Witnesses berated the "complaints processing system which, if anything, often makes matters worse," and even when discrimination was proven, "USDA refuses to pay damages." There were 495 pending complaints, and half were two years old or older. "USDA does not place a priority on serving the needs of small and limited-resource farmers," the report stressed, "and has not supported any coordinated effort to address this problem."[32]

Meanwhile, the *Pigford* class-action case moved through the courts, and on April 14, 1999, Judge Paul L. Friedman handed down his opinion in *Timothy Pigford v. Dan Glickman*. As mentioned in chapter 1, Friedman's first words, "Forty acres and a mule," resonated with broken promises from the Reconstruction era, but he highlighted the amazing accomplishments of 900,000 black farmers and the black ownership of approximately 16 million acres of farmland by the 1920s. The judge reviewed the sad chronicle of discrimination that by the mid-1990s left only 18,000 black farms with 3 million acres. This drastic decline was not an accident. Friedman stressed that USDA officials had "denied, delayed or otherwise frustrated" black loan applications. He drew from the action team's listening sessions to mention Alvin E. Steppes, a Lee County, Arkansas, farmer who was denied an FmHA operating loan in 1986, which resulted in the loss of his farm. Calvin Brown from Brunswick County, Virginia, also applied for an operating loan, was strung along for a month, and then was told there was no record of his application. By the time he received funds, the planting season had passed. Black farmer George Hall of Greene County, Alabama, was the only farmer in his county not to receive disaster payments for crop losses in 1994. Such discrimination, Friedman argued, had decimated the black farm population. USDA staff in 1983, he continued, "simply threw discrimination complaints in the trash without ever responding to or investigating them." Even in the rare cases when discrimination was found, "the farmer never received any relief."[33]

Judge Friedman realized that many black farmers in the suit did not have documentation, in part because the USDA had not acted on complaints or even maintained them. The Consent Decree set up a two-tier mechanism. Farmers lacking complete documentation could get "a virtually automatic cash payment of $50,000, and forgiveness of debts owed to the USDA." Farmers with documentary evidence "have no cap on the amount they may recover," and they could opt out of the class and seek compensation in separate suits. Friedman went into substantial detail on documentary evidence of discrimination. Because only a complex mechanism could estimate the costs of discrimination, monitors would supervise the process. Friedman carefully spelled out guidelines, dealt with objections, and set the resolution in motion. Yet he acknowledged that African Americans had an abiding distrust of the USDA, and he quoted attorney J. L. Chestnut, who argued in a 1999 hearing that mistrust "reaches all the way back to slavery." Black farmers desired that the court maintain jurisdiction "in perpetuity," for otherwise "USDA will default, ignore the lawful mandates of this Court, and in time march home scot-free while blacks are left holding the empty bag again." Friedman could not guarantee that discrimination would end, but he observed that the billions of dollars due farmers for discrimination would be a grim reminder to USDA bureaucrats. Plus, he stressed, it would demonstrate that "the USDA is not above the law." In February 2010, the Obama administration announced a $1.25 billion settlement with African American farmers.[34]

A series of investigative reports by *Washington Post* correspondents Dan Morgan, Gilbert M. Gaul, and Sarah Cohen in 2006 and 2007 documented outrageous USDA payments and widespread inefficiency. A Texas asphalt contractor collected $1,300 in "direct payments" on the eighteen-acre lot where he built his dream home because the land once grew rice. Real estate agents used the payments on old rice land to lure customers to what they called "cowboy starter kits," lots large enough to pasture a horse. Having the land assessed as agricultural, in one case, changed the property tax from $3,000 to $55. Working farmers, even those who did well, still received government payments. In one case, an Eastern Shore, Maryland, farmer grossed $500,000 for his corn crop and received another $75,000 from the USDA, a "deficiency payment." Texas cattle farmers received up to $40,000 in drought payments whether or not there was a drought. When the space shuttle *Columbia* disintegrated over Texas in 2003, President George W. Bush declared a disaster in order to obtain all debris from the accident, and farmers successfully applied for relief. Henderson County, Texas, cattle farmers, for example, received $433,000. After an earthquake in 2001, Washington State farmers who had no damage on their farms collected disaster payments. In July 2007, Sarah Cohen reported, the USDA "distributed $1.1 billion over seven years to the estates or companies of deceased farmers." Responding to a Government Accountability Office inquiry, local USDA offices justified such payments as necessary, Cohen learned, because of "staff shortages and competing priorities."[35]

In the decade after the *Pigford* decision, the USDA slipped back into its old discriminatory and inefficient ways. In 2008, a Government Accountability Office official faulted USDA statistics on civil rights complaints and questioned the credibility of compliance reports. When Secretary of Agriculture Tom Vilsack took office in 2009, he inherited 11,000 unprocessed civil rights complaints, several class-action suits, and complaints of discrimination among the 113,000 USDA employees. He instituted civil rights training, beefed up the compliance-review process, and commissioned an independent study on civil rights problems and solutions. His goal, he said, was "changing the culture at USDA." When the $8 million study by the Jackson Lewis consulting firm arrived in May 2011, not surprisingly, it documented discrimination against minorities and women.[36]

As monitors sorted through thousands of claims, three other suits charged USDA discrimination. In November 1999, Native American plaintiffs filed suit against the USDA, and in 2001, the court certified the plaintiffs as a class and the *Keepseagle* case progressed to the discovery stage. The USDA contested every claim. As in the *Pigford* case, the court extended the time frame for discrimination back to 1981. After reviewing eighty depositions and tens of thousands of documents, the court found that non–Native American FmHA administrators and committees, often resorting to cultural stereotypes, used subjective criteria in deciding whether to grant loans that caused Native Americans losses of some $608 million from 1981 to 2007. The court relied on a formula to determine losses since the FmHA did not keep adequate records of these transactions. Loan guidelines were vague and contained expressions such as "reasonable," "some experience in making management decisions on the farm," "community loan rates," "hands on supervision," and even "family farm," none of which lent themselves to specificity. There was blatant discrimination that included denigrating Native American farmers and underrating their yields. One South Dakota Native American involved in the suit served on an FmHA county committee. A white member told him he had no business serving on the committee, and another member called him a "damn Indian." When Native American applications came up, a member would say, "not another Prairie Nigger." Employees in FmHA offices insulted Native Americans by calling them "injuns" and telling them to go back to the reservation, and one person blatantly said he had been sent to an office "to sell you people out."[37] The settlement, negotiated over ten months, established a $760 million fund to pay damages to class members, extinguished Native Americans' outstanding loans, and initiated reforms in the FmHA loan program.[38]

Hispanics also sued the USDA. *Garcia v. Vilsack*, filed in 2000, is still pending. The *Garcia* suit was denied class status and has bounced around the courts for a decade without settlement. Women also sued, and *Love v. Vilsack* did not achieve class status either and was consolidated with the *Garcia* case. The 2007 Census of Agriculture showed 306,209 farms operated by women. The U.S. Supreme Court declined to hear the appeal of the combined cases. On February 25, 2011, the Obama administration offered to settle the Hispanics' and women's complaints for $1.3 billion, but both groups claimed they deserved more.[39]

Just as the civil rights movement of the 1960s created in its wake awareness of discrimination against women, Native Americans, Hispanics, Asians, gays, and other minorities, the *Strain* and *Pigford* cases opened avenues to address discrimination aimed at the broader minority farm population. No financial settlement could redress the humiliation, discrimination, and losses, but at last the USDA paid, not in firings and a purge of racists but at least in dollars, a currency long denied to African American farmers. Compensation arrived too late to save hundreds of thousands of farms lost by minority and women farmers, including, of course, those outside the 1981 window. The 3.5 million farmers who left the land after World War II did not fit the USDA modernist vision that materialized in the New Deal years.

At last, the USDA did fire someone for a purportedly racist statement. On July 19, 2010, when Secretary of Agriculture Tom Vilsack unceremoniously dismissed Shirley Sherrod, Georgia state director of rural development for the USDA, alarmed civil rights leaders and USDA executives, not to mention the White House and the press, excoriated her for what turned out to be a severely edited video constructed by conservative commentator Andrew Breitbart to appear as if Sherrod was making a racist remark. Instead of displaying racism, as Breitbart claimed, Sherrod's full remarks concerned overcoming potential prejudice in helping a white couple save their farm. The USDA, the White House, and even civil rights leaders spoke and acted before reviewing either the complete speech or Shirley Sherrod's biography. Her life resonates with civil rights activists and African American farmers. She was born Shirley Miller in Baker County (Bad Baker), Georgia, in 1948 and was seventeen years old when her father was murdered. Even before she attended Fort Valley State University, she was active in the civil rights movement and met her future husband, SNCC leader and pastor Charles Sherrod, while working on the Albany campaign. Later she earned a master's degree in community development at Antioch University. In 1969, the couple were among the founders of the 5,700-acre New Communities collective farm, which endured opposition from white neighbors (who claimed they were communists), challenges from Georgia governor Lester Maddox, and three years of drought. A promised OEO grant did not come through, and when they requested an emergency loan during the drought, the FHA supervisor told them they "would get one over his dead body." The land was sold in 1985. New Communities veterans joined plaintiffs in the *Pigford* suit and received $8.2 million for lost farmland and $4.2 million for loss of income, and the Sherrods were awarded an additional sum for anguish. The rush to dismiss Shirley Sherrod based on the edited video contrasts remarkably with the government's reaction to the thousands of USDA bureaucrats who denied African American farmers loans, jobs, acreage, information, and courtesy, none of whom were dismissed and few reprimanded. Her firing also demonstrates remarkable historical amnesia. Apparently no one involved in the decision to fire Shirley Sherrod understood the historical significance of USDA discrimination, and no tombstone marks its final resting place.[40]

# NOTES

*Abbreviations*

ACES Papers   Alabama Cooperative Extension Service Papers, Record Group 71, Archives and Manuscripts Department, Auburn University, Auburn, Ala.

Aderhold Papers   O. C. Aderhold Papers, Record Group 1, Office of the President, 1960–67, Hargrett Rare Book and Manuscript Library, University of Georgia, Athens

Colvard Papers   Dean W. Colvard Presidential Papers, Mitchell Memorial Library, Mississippi State University, Mississippi State

Delta Ministry Papers   Delta Ministry Papers, Special Collections, Mitchell Memorial Library, Mississippi State University, Mississippi State

Giles Papers   William Lincoln Giles Presidential Papers, Mitchell Memorial Library, Mississippi State University, Mississippi State

HRAHC   Homie Regulus Archives and Heritage Collection, H. A. Hunt Memorial Library, Fort Valley State University, Fort Valley, Ga.

LBJL   Lyndon Baines Johnson Library and Museum, Austin, Tex.

MML, MSU   Mitchell Memorial Library, Mississippi State University, Mississippi State

NAACP Records   Records of the National Association for the Advancement of Colored People, Manuscript Division, Library of Congress, Washington, D.C.

NAL   National Agricultural Library, Beltsville, Md.

NARA   National Archives and Records Administration, Archives II, College Park, Md.

CFASCS, RG 145   Central Files, Records of the Agricultural Stabilization and Conservation Service, Record Group 145

CFFPD, USCCR, RG 453   Chronological Files, February 1964–December 1965, Federal Programs Division, U.S. Commission on Civil Rights, Record Group 453

CFLID, USCCR, RG 453   Chronological Files, February 1964–December 1965, Liaison and Information Division, U.S. Commission on Civil Rights, Record Group 453

CFOSDCC, USCCR, RG 453   Chronological Files, Office of the Staff Director, Congressional Correspondence, 1968–76, U.S. Commission on Civil Rights, Record Group 453

CFOSDPRS, USCCR, RG 453   Chronological Files, Office of the Staff Director,

Publications, Reports, and Studies, 1973–76, U.S. Commission on Civil Rights, Record Group 453

CFOSDSP, USCCR, RG 453    Chronological Files, Office of the Staff Director, Records Relating to Special Projects, 1960–70, U.S. Commission on Civil Rights, Record Group 453

CFSS, USCCR, RG 453    Chronological Files, Records Relating to Surveys and Studies, 1958–62, U.S. Commission on Civil Rights, Record Group 453

GC 1906–76, SOA, RG 16    General Correspondence, 1906–76, Records of the Secretary of Agriculture, Record Group 16

NSF Papers    National Sharecroppers Fund Papers, Walter Reuther Library, Wayne State University, Detroit, Mich.

OHPJFKL    Oral History Program, John F. Kennedy Presidential Library and Museum, Boston, Mass.

OHSA    Oral History of Southern Agriculture, Archives Center, National Museum of American History, Washington, D.C.

Philpott Papers    Harry M. Philpott Papers, Archives and Manuscripts Department, Auburn University, Auburn, Ala.

Russell Papers    Richard Russell Papers, Richard Russell Library, University of Georgia, Athens

SCCHR Papers    South Carolina Commission on Human Relations Papers, South Caroliniana Library, University of South Carolina, Columbia

Seabron Papers    William M. Seabron Papers, Walter Reuther Library, Wayne State University, Detroit, Mich.

Smith Papers    Lillian Smith Papers, Hargrett Rare Book and Manuscript Library, University of Georgia, Athens

SNCC Papers    *Student Non-Violent Coordinating Committee Papers*, microfilm, 73 reels (New York, 1982), Martin Luther King Jr. Center, Atlanta, Ga.

SOHC    Southern Oral History Collection, University of North Carolina, Chapel Hill

Talmadge Papers    Herman Talmadge Papers, Richard Russell Library, University of Georgia, Athens

Taylor Papers    William L. Taylor Papers, in author's possession

## *Chapter 1*

1. Orville L. Freeman, Secretary's Memorandum No. 1572; "Racial Bias in Workings of U.S. Farm Aid Criticized by Federal Civil Rights Unit," *Wall Street Journal*, March 1, 1965, clipping, box 4255, GC 1906–76, SOA, RG 16, NARA; U.S. Commission on Civil Rights, *Equal Opportunity in Farm Programs: An Appraisal of Services Rendered by Agencies of the United States Department of Agriculture* (Washington, 1965).

2. Rod Leonard to secretary, January 3, 1964, box 4120, Farm Program, GC 1906–76, SOA, RG 16, NARA; transcript, Orville Freeman Oral History Interview IV, November 17, 1988, by Michael L. Gillette, internet copy, LBJL.

3. "Agriculture: A Hard Row to Hoe," *Time*, April 6, 1963, http://www.time.com /time/magazine/article/0,9171,830035-1,00.html; transcript, Orville Freeman Oral History Interview I, February 14, 1969, by T. H. Baker, internet copy, LBJL; Krissah Thompson, "USDA Chief Details Agency Efforts to Improve Record on Civil Rights," *Washington Post*, February 16, 2010, A11.

4. *Pigford v. Glickman*, 185 F.R.D. 82 (D.D.C. 1999), 85-86; U.S. Bureau of the Census, *1969 Census of Agriculture*, section 1, Summary Data, table 3, "Farm Operators— Tenure and Characteristics, 1969, 1964, and 1959," in volumes for Louisiana, South Carolina, Tennessee, Alabama, North Carolina, Georgia, Arkansas, Virginia, and Mississippi; William C. Payne Jr. to author, February 12, 1992, in author's possession.

5. Ray Marshall and Allen Thompson, "Status and Prospects of Small Farmers in the South," Southern Regional Council, Atlanta, Ga. (1976), 13-14, 23, 26, 61-64. On the relations between African Americans and whites in rural Hancock County, Georgia, and white patrons, see Mark Schultz, *The Rural Face of White Supremacy: Beyond Jim Crow* (Urbana, 2005), 97-130.

6. Walter W. Wilcox to William C. Black, May 31, 1968, box 4815, Farming, GC 1906-76, SOA, RG 16, NARA; Bruce J. Reynolds, "Black Farmers in America, 1865-2000: The Pursuit of Independent Farming and the Role of Cooperatives," USDA Rural Business Cooperative Service, Research Report 194 (October 2002), table 3, "Farm Operators in the U.S. by Race, 1900 to 1997," 24. On property owning, see Loren Schweninger, *Black Property Owners in the South, 1790-1915* (Urbana, 1990), 143-237; Schweninger, "A Vanishing Breed: Black Farm Owners in the South, 1651-1982," *Agricultural History* 63 (Summer 1989): 41-57; and Peggy G. Hargis, "Beyond the Marginality Thesis: The Acquisition and Loss of Land by African Americans in Georgia, 1880-1930," *Agricultural History* 72 (Spring 1998): 241-62.

7. Deborah Fitzgerald, *Every Farm a Factory: The Industrial Ideal in American Agriculture* (New Haven, 2003), 157-83, quote on 6; James C. Scott, *Seeing Like a State: How Certain Schemes to Improve the Human Condition Have Failed* (New Haven, 1998), 193-222, 286-87, quote on 196. "Modernism" and "modernity," of course, have expansive meanings. See "Historians and the Question of 'Modernity,'" *American Historical Review* 116 (June 2011): 577-751.

8. For an overview of the USDA, see Wayne D. Rasmussen and Gladys L. Baker, *The Department of Agriculture* (New York, 1972). For a concise discussion of the policies that shaped the New Deal and its implementation, see Paul K. Conkin, *A Revolution Down on the Farm: The Transformation of American Agriculture since 1929* (Lexington, 2008), 51-76.

9. Wendell Berry, *The Unsettling of America: Culture and Agriculture* (New York, 1977), 31. On Berry's ideas, see Jeffrey Filipiak, "The Work of Local Culture: Wendell Berry and Communities as the Source of Farming Knowledge," *Agricultural History* 85 (Spring 2011): 174-94.

10. Pete Daniel, *Breaking the Land: The Transformation of Cotton, Tobacco, and Rice Cultures since 1880* (Urbana, 1985), 65; Rasmussen and Baker, *Department of Agriculture*, 226. For an overview of USDA programs, see Ruth R. Harkin and

Thomas R. Harkin, "'Roosevelt to Reagan': Commodity Programs and the Agriculture and Food Act of 1981," *Drake Law Review* 31 (1981–82): 499–517. On how government policies bypassed African Americans, see Ira Katznelson, *When Affirmative Action Was White: An Untold History of Racial Inequality in Twentieth-Century America* (New York, 2005).

11. See Pete Daniel, "Going among Strangers: Southern Reactions to World War II," *Journal of American History* 77 (December 1990): 886–911.

12. Herschel C. Ligon to Orville Freeman, January 5, 1965, box 4295, Farm Programs, folder 6; Joseph W. Barr to Freeman, December 18, 1968; Freeman to Barr, December 24, 1968 (quote), box 4815, Farming, GC 1906–76, SOA, RG 16, NARA. On tax-loss farming, see Charles Davenport, "Farm Losses under the Tax Reform Act of 1969: Keepin' 'Em Happy Down on the Farm," *Boston College Law Review* 12 (1971): 319–50, http://lawdigitalcommons.bc.edu/bclr/vol12/iss3/2; Hoy F. Carman, "Tax Loss Farming: A Perennial Problem," *California Agriculture* (December 1978): 12–14; and Sally Hanlon, "Joint Economic Committee Investigates Tax-loss Farming," *Tax Notes*, May 5, 1986, 443. On farmer complaints about nonfarmers, see Pete Daniel, *Lost Revolutions: The South in the 1950s* (Chapel Hill, 2000), 54–55.

13. Alan E. Fusonie, "John H. Davis: Architect of the Agribusiness Concept Revisited," *Agricultural History* 69 (Spring 1995): 343–44.

14. Conkin, *Revolution Down on the Farm*, 97–99.

15. David Westfall, "Agricultural Allotments as Property," *Harvard Law Review* 79 (April 1966): 1188–89; Neil D. Hamilton, "Legal Issues Arising in Federal Court Appeals of ASCS Decisions Administering Federal Farm Programs," *Hamline Law Review* 12 (1989): 633–48; Brainerd S. Parrish, "Cotton Allotments: Another 'New Property,'" *Texas Law Review* 45 (March 1967): 734–53.

16. Lu Ann Jones, interview with Tom B. Cunningham, Darlington, S.C., January 19, 1987; interview with Leslie Ardoin, Mamou, La., May 19, 1988, OHSA; Berry, *Unsettling of America*, 92–93.

17. Lu Ann Jones, interview with Florentine Daniel, Franklin, La., May 27, 1988, OHSA.

18. James R. Wimberly to Herman Talmadge, ca. June 12, 1969, box 90, folder 6; Donald Wane to Talmadge, August 4, 1970, box 135, folder 7, Senatorial V, Administration A, category files, Talmadge Papers; C. E. Sheffield to Richard Russell, April 23, 1970, box 4, folder 8, Senatorial Papers, XV, General, Russell Papers.

19. On white southern opposition to civil rights, see Sarah H. Brown, "The Role of Elite Leadership in the Southern Defense of Segregation, 1954–1964," *Journal of Southern History* 77 (November 2011): 827–64.

20. Mildred Bond and Medgar Evers, confidential report on Presley Flakes, January 20, 1956; Bond and Evers, confidential report on Jake Tanner, January 20, 1956, Mississippi Pressures, series IIA, box 422, NAACP Records (emphasis in original); Lu Ann Jones, interview with Henry Woodard, Tunica, Miss., October 5, 1987, OHSA.

21. "Population of Dawson Is 55 Per Cent Negro," *Washington Post*, June 8, 1958,

A12; Robert E. Lee Baker, "Death and Violence Terrorize Negroes of Georgia Town," *Washington Post*, June 8, 1958, A1 (quote).

22. "Investigation of Terrell County, Ga.," confidential (1958), Reprisals, Georgia General, 1956–65, series III, box 277, NAACP Records. See Danielle L. McGuire, *At the Dark End of the Street: Black Women, Rape, and Resistance—A New History of the Civil Rights Movement from Rosa Parks to the Rise of Black Power* (New York, 2010).

23. Testimony of Francis Joseph Atlas, September 27, 1960, in *Hearings before the U.S. Commission on Civil Rights Held in New Orleans, Louisiana, September 27, 1960* (Washington, 1961), 24–27.

24. Clarence A. Laws to Roy Wilkins, March 30, 1961 (first four quotes); F. J. Atlas to Wilkins, April 30, 1961 (fifth quote); Wilkins to Atlas, May 8, 1961, Reprisals, Louisiana General, 1960–65, series III, box 277, NAACP Records. See also "Elections, Registration—Louisiana, *U.S. v. Baxter Deal et al.*," *Race Relations Law Reporter* 6 (1961–62): 474–76.

25. Lela Turner to Dear Sir, November 20, 1962; Turner to National Sharecroppers Fund, January 25, 1963, box 41, Appeals for Help, folder 25, NSF Papers.

26. Winson Hudson and Constance Curry, *Mississippi Harmony: Memoirs of a Freedom Fighter* (New York, 2002), 53–64, quote on 82; Winson Hudson to Robert C. Weaver, July 31, 1961, Mississippi Pressures A–L, 1956–64, series III, box 230, NAACP Records.

27. "Surplus Food Programs," n.d., reel 20, frame 144; Charles Cobb and Charles McLaurin, memo, November 19, 1962, reel 38, frames 135–36 (first quotation); James Forman to Orville Freeman, December 6, 1962, reel 10, frame 651, *SNCC Papers* (second quotation); Greta de Jong, "Staying in Place: Black Migration, the Civil Rights Movement, and the War on Poverty in the Rural South," *Journal of African American History* 90 (Fall 2005): 390–91.

28. Elmer Neufeld to Leo Driedger, January 11, 1961; John A. Morsell to Wesley T. Frost, February 21, 1961; Gloster B. Current to Roy Wilkins and Randolph White, June 15, 1961, Reprisals, Fayette County, Tenn., 1961–63, series III, box 280, NAACP Records; Anne Braden to Herman H. Long, March 28, 1963, box 111; Fay Bennett to Lewis W. Jones, November 5, 1962, box 28; Claudius A. Turner, "Political Difficulties of Negroes in Haywood County, Tennessee," June 14, 1952, box 111, Race Relations Department, United Church Board for Homeland Ministries Archives, Amistad Collection, Tulane University, New Orleans, La.

29. Minutes of the Delta Council Labor Committee, Indianola, March 27, 1961; Meeting of the Delta Council Board of Directors, Cleveland, November 7, 1962, box 32, folder 10, Delta and Pine Land Company Records, MML, MSU; Hodding Carter III, "The Negro Exodus from the Delta Continues," *New York Times Magazine*, March 10, 1968, 26.

30. De Jong, "Staying in Place," 392–95; Daniel, *Lost Revolutions*, 217.

31. Josephine Ripley, "Title VI: Door to New South," *Christian Science Monitor*, November 12, 1964, 1.

32. National Sharecroppers Fund to Orville Freeman, August 29, 1963, copy in USDA, series III, box 144, NAACP Records.

33. Ira Kaye to Alice Spearman, November 9 (first quote), 20 (second quote), 1961, box 25, SCCHR Papers.

34. John Salter, "Exploratory Report re. Economic Destitution of Rural and Urban Negro Families in the Delta Region of the State of Mississippi, 1963," Mississippi Pressures, General, 1963–65, series III, box 231, NAACP Records.

## Chapter 2

1. Minutes of meeting, February 13, 1964, box 1, CFFPD, USCCR, RG 453, NARA. See also Donald S. Safford to F. Peter Libassi, February 13, 1964, ibid.

2. Dean W. Determan to Marian P. Yankauer, March 6, 1964, ibid.; Larry J. Hackman, recorded interview of Harris Wofford, February 3, 1969, 146–51, OHPJFKL.

3. Transcript, Aaron Henry Oral History Interview, September 12, 1970, by T. H. Baker, internet copy, LBJL.

4. Richard M. Shapiro and Donald S. Safford, interview with Rev. W. H. Hall and Joseph Wiley Register, Valdosta, Ga., March 2, 1964, box 1, CFLID, USCCR, RG 453, NARA.

5. Richard M. Shapiro and Donald S. Safford, interview with James Register, Valdosta, Ga., March 2, 1964, ibid.

6. Richard M. Shapiro and Donald S. Safford, interview with Andrew Blakeney, Valdosta, Ga., March 2, 1964, ibid.

7. Richard M. Shapiro and Donald S. Safford, interview with George Miller, Valdosta, Ga., March 3, 1964, ibid.

8. Richard M. Shapiro and Donald S. Safford, interview with Rev. W. H. Hall, Valdosta, Ga., March 2, 1964, ibid.

9. Richard M. Shapiro and Donald S. Safford, interview with Freeling Scarborough, Valdosta, Ga., March 4, 1964, ibid. Earl Anderson, the office manager in Brooks County, gave much the same account of long-serving committee members and a black janitor. Richard M. Shapiro and Donald S. Safford, interview with Earl Anderson, Quitman, Ga., March 10, 1964, ibid.

10. Mary B. Gibbs to Orville Freeman, September 15, 1964; John A. Thomas to Dear Sir, October 20, 1964, box 288, Tobacco, CFASCS, RG 145, NARA.

11. E. A. Weinheimer to Marvin L. McLain, April 22, 1960; True D. Morse to Weinheimer, May 12, 1960, box 3432, Cotton, Acreage Allotments and Marketing Quotas, GC 1906–76, SOA, RG 16, NARA. See also Stanton Brown Jr. to Olin Teague, August 12, 1960; Marvin L. McLain to Teague, September 1, 1960, ibid. On discriminatory ASCS handling of acreage, see Elmo Holder, "Discrimination in A.S.C.S.," n.d. [1965], reel 20, frame 144, SNCC Papers.

12. Richard M. Shapiro and Donald S. Safford, "Meeting of the Lowndes County Committee (All White) of the Agriculture Stabilization and Conservation Service," Valdosta, Ga., March 5, 1964, box 1, CFLID, USCCR, RG 453, NARA.

13. Marian P. Yankauer, interview with James Hunt, May 12, 1964, and notes of telephone conversations with Hunt, May 13, 1964, ibid.; Sue Cronk, "She's Traveled Far in Civil Rights: From the Picket Line to a Government Office," *Washington Post*, March 19, 1964, C5.

14. Richard M. Shapiro and Donald S. Safford, interview with Cornelius Wallace McIver and William L. Whittle, Thomasville, Ga., March 6, 1964, box 1, CFLID, USCCR, RG 453, NARA.

15. Richard M. Shapiro and Donald S. Safford, interview with L. Monroe Jackson and Henry L. Carr, Quitman, Ga., March 6, 1964; interview with Sarah Martin Clark, Thomasville, Ga., March 6, 1964, ibid.

16. Richard M. Shapiro and Donald S. Safford, interview with J. B. Stevens, Quitman, Ga., March 3, 1964, ibid.

17. Richard M. Shapiro and Donald S. Safford, "Overnight Accommodations at Camp John Hope, March 7, 1964," March 23, 1964, ibid.

18. Richard M. Shapiro and Donald S. Safford, report on Sharecroppers Fund Conference, Fort Valley State College, Fort Valley, Ga., March 7, 8, 1964, ibid.

19. Ibid.

20. Richard M. Shapiro and Donald S. Safford, interview with D. D. Slappey, March 7, 1964, ibid.

21. Richard M. Shapiro and Donald S. Safford, interview with Dr. C. L. Ellison, Fort Valley, Ga., March 7, 8, 1964, ibid.

22. Richard M. Shapiro and Donald S. Safford, interview with James Mays, Fort Valley, Ga., March 7, 1964, ibid.; Penny Patch, "Sweet Tea at Shoney's," in Constance Curry et al., *Deep in Our Hearts: Nine White Women in the Freedom Movement* (Athens, 2000), 145-46.

23. Richard M. Shapiro and Donald S. Safford, interview with Daniel W. Young, Monroe, Ga., March 7, 1964, box 1, CFLID, USCCR, RG 453, NARA.

24. Richard M. Shapiro and Donald S. Safford, interview with Thomas L. Delton, Fort Valley, Ga., March 9, 1964, ibid. See also Marian P. Yankauer, conversation with William Seabron, May 27, 1964, ibid.

25. F. Peter Libassi to Leslie W. Dunbar, March 22, 1965, box 2, CFFPD, USCCR, RG 453, NARA; Marian P. Yankauer to Liaison and Information Division, May 28, 1964, box 1, CFLID, USCCR, RG 453, NARA.

26. Richard M. Shapiro and Donald S. Safford, interview with John H. Rollins, Fort Valley, Ga., March 7, 1964, box 1, CFLID, USCCR, RG 453, NARA.

27. Richard M. Shapiro and Donald S. Safford, interview with Earl Anderson, Quitman, Ga., March 10, 1964, ibid.

28. T. T. Williams, interview with Bernard L. Collins, Montgomery, Ala., May 11, 1964, ibid.; "Keys to Rural Community Development: The 1890 Land Grant Universities Approach," in *Proceedings of the 45th Annual Professional Agricultural Workers Conference* (Tuskegee, 1987), iii-iv.

29. T. T. Williams, interview with Bernard L. Collins, Montgomery, Ala., May 11, 1964, box 1, CFLID, USCCR, RG 453, NARA.

30. Richard M. Shapiro and Donald S. Safford, "Report of Field Trip to Alabama and Mississippi from May 10, 1964, through May 16, 1964, Agricultural Stabilization and Conservation Service," June 3, 1964, ibid.

31. T. T. Williams, interview with George Parris, Montgomery, Ala., May 11, 1964, ibid.

32. Ibid.

33. T. T. Williams, interview with Frank Jackson, Eutaw, Ala., May 12, 1964, ibid.

34. "Alabama Extension Service," box 8, CFSS, USCCR, RG 453, NARA.

35. T. T. Williams, interview with Charles Scott, Selma, Ala., May 11, 1964, box 1, CFLID, USCCR, RG 453, NARA.

36. T. T. Williams, interview with William Ammons, Jackson, Miss., May 13, 1964, box 2, ibid.; affidavit of William E. Ammons (confidential), May 24, 1966, box 8, Extension Service, Director and Associate Director, A79-39, Colvard Papers.

37. T. T. Williams, interview with Jasper Davis, Prentiss, Miss., May 14, 1964, box 1, CFLID, USCCR, RG 453, NARA.

38. T. T. Williams, interview with Al Johnson, Prentiss, Miss., May 14, 1964, ibid.

39. Marian P. Yankauer, conference with Howard Bertsch, May 15, 1964, ibid.; Layne R. Beaty, recorded interview with Howard Bertsch, July 9, 1964, OHPJFKL. According to NSF staffer Mike Kenney, Bertsch was still showing "hostility and defensiveness" in December 1966. See Mike Kenney to Jac Wasserman, January 20, 1967, box 30, Mike Kenney Reports and Correspondence, folder 5, NSF Papers.

40. Marian P. Yankauer, conference with Donald Oberle and Lawrence Washington, May 22, 1964, box 1, CFLID, USCCR, RG 453, NARA. See also Valerie Grim, "Black Participation in the Farmers Home Administration and Agricultural Stabilization and Conservation Service, 1964–1990," *Agricultural History* 70 (Spring 1996): 321–36; and "Howard Bertsch, 59, Farmers Home Aide," *New York Times*, November 6, 1969.

41. Marian P. Yankauer, conference with Donald Oberle and Lawrence Washington, May 22, 1964, box 1, CFLID, USCCR, RG 453, NARA; Ocie Smith to Jac Wasserman, June 16, 1964, box 40, Ocie Smith, folder 41; L. S. James, report to Fay Bennett, June 16–July 16, 1964, box 40, South Carolina, folder 40, NSF Papers.

42. Marian P. Yankauer, interview with Joseph L. Matthews, Albert Bacon, and J. Neil Raudabaugh, May 12, 1964, box 1, CFLID, USCCR, RG 453, NARA. On Bacon, see "Albert S. Bacon Heads up Georgia Soil Conservation," *Atlanta Daily World*, December 13, 1945, 6; and Cornelius V. Troup, *Distinguished Negro Georgians* (Dallas, 1962), 22–23.

43. Marian P. Yankauer, interview with John W. Banning, May 12, 1964, box 1, CFFPD, USCCR, RG 453, NARA.

44. Marian P. Yankauer, "The Negro Farmer and the Agriculture Department," attached to F. Peter Libassi note, May 1, 1964, box 20, CFOSDSP, USCCR, RG 453, NARA.

45. Marian P. Yankauer to Liaison and Information Division, June 3, 1964, box 1, CFLID, USCCR, RG 453, NARA; press release, Lloyd Davis, series IX, subseries 3, box 9.3/1, Special Collections, NAL.

46. Jac Wasserman to Fay Bennett, July 25, 1964, box 40, Jac Wasserman, folder 7,

NSF Papers; Joseph M. Robertson obituary, *Washington Post*, June 11, 1996. On Bennett, see obituary, *New York Times*, December 29, 2002, 34; and interview with Fay Bennett Watts by Amelia Fry and Ingrid Scobie, Helen Gahagan Douglas Project, vol. 3, University of California, Berkeley, 1981.

47. Marian P. Yankauer, telephone conversation with Albert Bacon, May 18, 1964, box 1, CFLID, USCCR, RG 453, NARA.

48. Richard M. Shapiro to F. Peter Libassi, August 25, 1964, box 1 (quotes); William C. Payne to Marian P. Yankauer, October 29, 1965; Walter B. Lewis to William Seabron, November 5, 1965, box 3, CFFPD, USCCR, RG 453, NARA.

49. Richard M. Shapiro and Donald S. Safford, interview with Arthur Britton Sr., June 11, 1964, box 1, CFLID, USCCR, RG 453, NARA; Ocie Smith to Jac Wasserman, June 16, 1964, box 40, Ocie Smith, folder 41, NSF Papers.

50. Richard Shapiro and Ronald M. McCaskill, interview with W. M. Bost, August 4, 1964, box 1, CFLID, USCCR, RG 453, NARA.

51. James Robinson to Marian P. Yankauer, "Detailed Study of Extension Personnel to Serve Negro and White Rural Population in All Mississippi Counties," August 14, 1964, box 1, CFFPD, USCCR, RG 453, NARA.

52. James Robinson to Marian P. Yankauer, August 11, 1964, ibid.

*Chapter 3*

1. The Greensboro sit-ins are documented in numerous civil rights works. For the larger context, see William H. Chafe, *Civilities and Civil Rights: Greensboro, North Carolina, and the Black Struggle for Freedom* (New York, 1980); and Jeffrey A. Turner, *Sitting In and Speaking Out: Student Movements in the American South, 1960–1970* (Chapel Hill, 2010).

2. Clayborne Carson, *In Struggle: SNCC and the Black Awakening of the 1960s* (Cambridge, 1981), 20–29; Barbara Ransby, *Ella Baker and the Black Freedom Movement: A Radical Democratic Vision* (Chapel Hill, 2003), 241–47.

3. Pete Daniel, *Lost Revolutions: The South in the 1950s* (Chapel Hill, 2000), 91–175; Daniel, "Rhythm of the Land," *Agricultural History* 68 (Fall 1994): 1–22.

4. Carson, *In Struggle*, 79.

5. "Special Report—Federal Programs Project," n.d. [Fall 1965], reel 38, frame 129; Gren Whitman, "Report on the ASCS Election for 1965," n.d. [Fall 1965], reel 37, frames 1086–89, *SNCC Papers*.

6. Greta de Jong, "Staying in Place: Black Migration, the Civil Rights Movement, and the War on Poverty in the Rural South," *Journal of African American History* 90 (Fall 2005): 387–409. See also James N. Gregory, *The Southern Diaspora: How the Great Migration of Black and White Southerners Transformed America* (Chapel Hill, 2005).

7. Robert Earl Martin, "Negro-White Participation in the A.A.A. Cotton and Tobacco Referenda in North and South Carolina: A Study in Differential Voting and Attitudes in Selected Areas," Ph.D. diss., University of Chicago, 1947, 258. On the FES,

see Wayne D. Rasmussen, *Taking the University to the People: Seventy-five Years of Cooperative Extension* (Ames, 1989).

8. F. Wainwright Blease to Victor B. Phillips, July 6, 1965, box 321, CFASCS, RG 145, NARA.

9. For an overview of the ASCS committee system, see Neil D. Hamilton, "Farmers' Rights to Appeal ASCS Decisions Denying Farm Program Benefits," *South Dakota Law Review* 29 (1983–84): 282–302.

10. Martin, "Negro-White Participation in the A.A.A. Cotton and Tobacco Referenda," 259–63, 266–67, 270, quotes on 266. In Wilson County, all of the committeemen and most owners were members of the Farm Bureau.

11. E. A. Jaenke to William L. Taylor, June 20, 1967, box 20; Joseph Hajda, "Historical Background of the Farmer Committee System"; "Separate Statement by Morton Grodzins," box 2, CFOSDSP, USCCR, RG 453, NARA; *Fulford v. Forman*, 245 F. 2d 145 (1957), 151, 152 (n. 20).

12. On Fitzgerald, see *Wall Street Journal*, July 12, 1977, 7; *Hartford Courant*, August 16, 1961; and *New York Times*, July 12, 1977. On Godfrey, see transcript, Horace Godfrey Oral History Interview, October 31, 1968, by Thomas H. Baker, internet copy, LBJL.

13. Acting deputy, DASCO, to Horace Godfrey, January 27, 1965, box 320, CFASCS, RG 145, NARA. See also DASCO to chairman, Arkansas state committee, December 16, 1964; Godfrey to Orville Freeman, January 5, 1965; and Ray Fitzgerald to chairman, Arkansas state committee, January 23, 1964, ibid.

14. Affidavit of Bill D. Fowler, January 17, 1964; affidavit of Oris Nix, January 9, 1964, box 293, Administration, ibid.

15. State ASC committee to Ray Fitzgerald, January 9, 1964; state executive director to DASCO, January 17, 1964; DASCO to chairman, Arkansas ASC state committee, January 22, 1964; Mark S. Glover to DASCO, March 30, 1964; Fitzgerald to chairman, ASCS, Arkansas, April 2, 1964; Clifford Brummit to Orville Freeman, October 1, 1965, box 321, ibid. (quote).

16. Separate statement by Morton Grodzins, copy in box 2, CFOSDSP, USCCR, RG 453, NARA (emphasis in original). See also Pete Daniel, "The Legal Basis of Agrarian Capitalism: The South since 1933," in Melvyn Stokes and Rick Halpern, eds., *Race and Class in the American South since 1890* (Providence, 1994), 79–110.

17. Karen Sorlie Russo, "Farm Client Beware of ASCS Mysteries," *Compleat Lawyer* (Winter 1991): 57–60; "ASCS Operations in Twenty-six Alabama Counties," May 1967, box 39, CFOSDSP, USCCR, RG 453, NARA.

18. "The Federal Agricultural Stabilization Program and the Negro," *Columbia Law Review* 67 (1967): 1121–36, first quote on 1126, second quote on 1128, third quote on 1135. On the complexity of appeal, see C. Michael Malski, "Agricultural Law: Appealing Agricultural Allotments," *Mississippi Law Journal* 41 (Summer 1970): 422–43; and David Ray James, "The Transformation of Local, State, and Class Structures and Resistance to the Civil Rights Movement in the South," Ph.D. diss., University of Wisconsin, 1981.

19. Bob Moses to Fay Bennett, November 2, 1962, box 41, Appeals for Help, folder 25, NSF Papers. On Moses, see Carson, *In Struggle*, 46; and John Dittmer, *Local People: The Struggle for Civil Rights in Mississippi* (Urbana, 1994), 102.

20. Charles Cobb and Charles McLaurin, "Preliminary Survey on the Condition of the Negro Farmers in Ruleville, Mississippi, at the Close of the Cotton Season," November 19, 1962, reel 38, frames 135–36, *SNCC Papers*.

21. Carson, *In Struggle*, 96–108.

22. Howard Zinn, *SNCC: The New Abolitionists* (Boston, 1964), 10; Wesley C. Hogan, *Many Minds, One Heart: SNCC's Dream for a New America* (Chapel Hill, 2007), quote on 149, 170–76.

23. John Lewis to Editor, *New York Times*, May 7, 1964, copy in reel 42, frame 609, *SNCC Papers*; Hogan, *Many Minds*, 77–78.

24. Muriel Tillinghast, "Depending on Ourselves," in Faith S. Holsaert, Martha Prescod Norman Noonan, Judy Richardson, Betty Garman Robinson, Jean Smith Young, and Dorothy M. Zellner, *Hands on the Freedom Plow: Personal Accounts by Women in SNCC* (Urbana, 2010), 250–57, first quote on 251, second quote on 252, third and fourth quotes on 253, fifth quote on 255.

25. Mary E. King, "Getting Out the News," 342, ibid.

26. Jane Stembridge to Lillian Smith, May 9, 1964, 1283A, box 4, Smith Papers.

27. Jane Stembridge to Lillian Smith, May 23, June 23, 24, 1964, ibid.

28. Chana Kai Lee, *For Freedom's Sake: The Life of Fannie Lou Hamer* (Urbana, 1999), 85–102; Carson, *In Struggle*, 126; Hogan, *Many Minds*, 190–96; Chris Myers Asch, *The Senator and the Sharecropper: The Freedom Struggles of James O. Eastland and Fannie Lou Hamer* (Chapel Hill, 2008), 210–13, quote on 212.

29. Dittmer, *Local People*, 315–18.

30. Ibid., 134–37. On Gulfside, see Andrew W. Kahrl, *The Land Was Ours: African American Beaches from Jim Crow to the Sunbelt South* (Cambridge, 2012), 52–85; and Elaine DeLott Baker to author (email), November 7, 2011, in author's possession.

31. See Elaine DeLott Baker, "The 'Freedom High' and 'Hardliner' Factions of the Student Nonviolent Coordinating Committee (SNCC): A Reexamination" (1994), in author's possession; Baker and Francesca Polletta, "The 1964 Waveland Memo and the Rise of Second-Wave Feminism," paper given at 2009 OAH convention, in author's possession; Penny Patch, "Sweet Tea at Shoney's," in Constance Curry et al., *Deep in Our Hearts: Nine White Women in the Freedom Movement* (Athens, 2000), 154–55.

32. Carson, *In Struggle*, 133–52; Hogan, *Many Minds*, 197–218; Dittmer, *Local People*, 331–32; Jennifer Jensen Wallach, "Replicating History in a Bad Way? White Activists and Black Power in SNCC's Arkansas Project," in Jennifer Jensen Wallach and John A. Kirk, eds., *Arsnick: The Student Nonviolent Coordinating Committee in Arkansas* (Fayetteville, 2011), 69–84; R. Hunter Morey, "Cross Roads in COFO," Clay County Civil Rights Movement Material, acc. no. 175, MML, MSU.

33. Charles Cobb to Fay Bennett, November 6, 1963, box 30, Mississippi, folder 19, NSF Papers; Special Report: Federal Programs Project, n.d., reel 38, frame 129, *SNCC Papers*.

34. J. B., "Some Thoughts on the ASCS Elections: COFO Worker after 1964 Elections," reel 20, frame 449; Mike Kenney to Hi Everyone, July 15, 1964, reel 71, frames 66–67, *SNCC Papers*.

35. Dittmer, *Local People*, 190–93. For an account of the Mileston co-op in 1967, see Myron Cook, "Mileston Cooperative, Holmes County, Mississippi," November 21, 1967, box 58, Myron Cook folder, NSF Papers.

36. Lu Ann Jones, interview with Howard Taft Bailey, Lexington, Miss., October 16, 1987, OHSA. On self-defense, see Emilye Crosby, "'It Wasn't the Wild West': Keeping Local Studies in Self-Defense Historiography," in Emilye Crosby, ed., *Civil Rights from the Ground Up: Local Struggles, A National Movement* (Athens, 2011), 194–255.

37. Mike Kenney, "Report of the ASC Nominating Meeting in Durant," November 12, 1964, reel 71, frame 1, *SNCC Papers*.

38. Anonymous, "My Trip to Washington," n.d., box 30, Mississippi, folder 19, NSF Papers.

39. C. M. Gordy to DASCO, December 1, 1964, box 320, CFASCS, RG 145, NARA; Mike Kenney, "Preliminary ASC Election Report," December 4, 1964, reel 91, frame 3, *SNCC Papers*.

40. Mike Kenney, Christmas 1964, reel 71, frame 35, *SNCC Papers*.

41. Kenneth Birkhead to Thomas Hughes, April 30, 1965, box 4255, Civil Rights, GC 1906–76, SOA, RG 16, NARA; Marian P. Yankauer to F. Peter Libassi, March 9, 1965, box 2, CFFPD, USCCR, RG 453, NARA.

42. Benton County *Freedom Train* 2, no. 1 (October 6, 1964), reel 40, frames 38–39, *SNCC Papers*; Aviva Futorian to C. W. Sullivan, n.d. (ca. November 13, 1964), box 320, CFASCS, RG 145, NARA; Futorian, telephone conversation with Marian P. Yankauer, November 12, 1964, box 2, CFFPD, USCCR, RG 453, NARA.

43. C. W. Sullivan to Aviva Futorian, n.d., box 320, CFASCS, RG 145, NARA.

44. Patch, "Sweet Tea at Shoney's," 133–70, quote on 139; Penny Patch, "The Mississippi Cotton Vote," in Holsaert et al., *Hands on the Freedom Plow*, 403–6.

45. *COFO News*, December 11, 1964, loaned by Elaine DeLott Baker; Marian P. Yankauer, telephone conversation with Benjamin Graham of COFO, Batesville, Panola County, Miss., November 16, 1964, box 2, CFFPD, USCCR, RG 453, NARA.

46. Patch, "Mississippi Cotton Vote," 406–7; *COFO News*, December 11, 1964.

47. Patch, "Mississippi Cotton Vote," 408–9.

48. Ray Fitzgerald to Joseph M. Robertson, November 20, 1964; Aviva Futorian to C. W. Sullivan, n.d.; Sullivan to Futorian, n.d.; Fitzgerald to James May, November 20, 1964, box 320, CFASCS, RG 145, NARA.

49. F. Wainwright Blease to DASCO, April 2, 1965, ibid.; Jenny Irons, *Reconstituting Whiteness: The Mississippi State Sovereignty Commission* (Nashville, 2010), 162.

50. Elaine DeLott Baker to mother and dad, n.d. [1964] (quote); affidavit of Baker, December 15, 1964, loaned by Elaine DeLott Baker.

51. F. Wainwright Blease to DASCO, December 16, 1964; statement of George Raymond before F. Phillip Snowden and John F. Eakins, January 7, 1965, Canton, Miss.,

box 320, CFASCS, RG 145, NARA; "U.S. to Probe Bias in Miss. Farm Unit Poll," *New York Amsterdam News*, December 12, 1964, 7; Dittmer, *Local People*, 187–88.

52. George Raymond to Orville L. Freeman, January 9, 1965; Marvin Rich to Freeman, February 17, 1965; Walter L. Bieberly to Rich, March 1, 1965, box 320, CFASCS, RG 145, NARA.

53. Horace Godfrey to Philip A. Hart, January 16, 1965, ibid.

54. C. W. Sullivan to director, south-central area, February 25, 1965, ibid.

55. Walter L. Bieberly to DASCO, March 30, 1965, box 320, CFASCS, RG 145, NARA; Marian P. Yankauer, telephone conversation with Richard Fitch, February 16, 1965; Yankauer to Samuel J. Simmons, April 9, 1965, box 2, CFFPD, USCCR, RG 453, NARA.

56. F. Wainwright Blease to DASCO, April 2, 1965; USDA press release, April 2, 1965, box 320, CFASCS, RG 145, NARA.

57. W. H. Forsyth Jr. to C. W. Sullivan, May 4, 1965; Forsyth to Orville Freeman, May 5, 1965, ibid.

58. DASCO to William Seabron, May 25, 1965, ibid.

59. Ray Fitzgerald to W. H. Forsyth Jr. and attached notes, May 27, 1965, box 320, ibid.

60. "Agriculture Acts on Race Inequity," *Washington Post*, June 1, 1965, A4; "William M. Seabron Dies, Ex-Agriculture Official," *Washington Post*, December 31, 1980, B8; "3 Negroes Appointed to State Farm Units," *Washington Post*, April 3, 1965, A4.

61. Department of Agriculture Civil Rights Assignments, n.d., box 4454, Civil Rights, GC 1906–76, SOA, RG 16, NARA; Pete Daniel, interview with William L. Taylor, Washington, D.C., February 21, 2006, SOHC; Richard M. Shapiro to F. Peter Libassi, April 27, 1965; William Payne to Libassi, April 26, 1965; Libassi to Taylor, May 11, 1965, box 2, CFFPD, USCCR, RG 453, NARA.

62. "Official Bias in Workings of U.S. Farm Aid Is Criticized by Federal Civil Rights Unit," *Wall Street Journal*, March 1, 1965, copy in box 4255, Civil Rights, GC 1906–76, SOA, RG 16, NARA; "Calls U.S. Unfair to Negroes," *Chicago Tribune*, March 1, 1965, 5; *Equal Opportunity in Farm Programs: An Appraisal of Services Rendered by Agencies of the United States Department of Agriculture* (Washington, 1965).

63. Transcript, Horace Godfrey Oral History Interview, October 31, 1968, by Thomas H. Baker, internet copy, LBJL.

64. Marcus B. Braswell to Eugene W. Bayol, April 7, 1965; Ray Fitzgerald to Braswell, May 14, 1965, box 321, CFASCS, RG 145, NARA.

65. Marcus B. Braswell to Eugene W. Bayol, April 7, 1965, box 321, CFASCS, RG 145, NARA.

66. H. C. McShan to Horace Godfrey, March 31, 1965; Ray Fitzgerald to McShan, April 13, 1965, box 321, Compliance; John B. Vance to DASCO, May 2, 9, 1966, box 418, Employment, CFASCS, RG 145, NARA.

67. "ASCS Operations in Twenty-six Alabama Counties," May 1967, box 39, CFOSDSP, USCCR, RG 453, NARA. See also "The Agricultural Stabilization and Conservation Service in the Alabama Black Belt," exhibit no. 2, in *Hearings before the*

*United States Commission on Civil Rights, Montgomery, Ala., April 27–May 2, 1968* (Washington, 1969). These reports provide excellent summaries of ASCS programs and the discrimination that county committees allowed.

68. J. H. Spence [?], "Shaw, Mississippi: New Sounds in the Delta," n.d., box 5, CFOSDSP, USCCR, RG 453, NARA; J. Todd Moye, *Let the People Decide: Black Freedom and White Resistance Movements in Sunflower County, Mississippi, 1945–1986* (Chapel Hill, 2004), 151–54.

69. "Shaw, Mississippi," box 5, CFOSDSP, USCCR, RG 453, NARA (quotes); Moye, *Let the People Decide*, 154; Dittmer, *Local People*, 364.

70. Lawrence Guyot to Orville Freeman, November 22, 1965; William Seabron to Kenneth M. Birkhead, December 15, 1965, box 4254, Civil Rights, GC 1906–76, SOA, RG 16, NARA.

71. Bessie Mims to James O. Eastland, April 22, 1965, box 321, CFASCS, RG 145, NARA. See also Kenneth M. Birkhead to Eastland, May 25, 1965, ibid.

72. Mary Anne McGee to Phil M. Landrum, April 23, 1965, ibid.

73. Sandra S. Greene to Orville Freeman, July 13, 1965; Donald L. Gillis to Greene, July 29, 1965, ibid.

74. John L. McClellan to Orville L. Freeman, May 27, 1965; Edith Dickey Moses to Horace Godfrey, June 9, 1965, ibid.

75. Horace Godfrey to Edith Dickey Moses, June 17, 1965, ibid.; "Agriculture Acts on Race Inequity," *Washington Post*, June 1, 1965, A4; William Payne to Richard Shapiro, May 26, 1965, box 2, CFFPD, USCCR, RG 453, NARA.

76. F. Wainwright Blease to DASCO, June 14, 1965, box 321, CFASCS, RG 145, NARA.

77. Horace Godfrey to director of personnel, USDA, July 29, 1965, ibid.; transcript, Horace Godfrey Oral History Interview, October 31, 1968, by Thomas H. Baker, internet copy, LBJL.

78. C. W. Sullivan to director, south-central area, June 13, 1966; F. Wainwright Blease to DASCO, June 16, 1966; John B. Vance to DASCO, June 23, 1967, box 418, Employment; DASCO to Horace Godfrey, November 29, 1967, box 476, Reports, CFASCS, RG 145, NARA. For job issues in 1965, see Thomas Hughes to Orville Freeman, December 23, 1965, box 4254, Civil Rights, GC 1906–76, SOA, RG 16, NARA.

79. See M. John Bundy and Allen D. Evans, "Racial Discrimination in USDA Programs in the South: A Problem in Assuring the Integrity of the Welfare State," *Washington Law Review* 45 (1970): 738–46.

## Chapter 4

1. Thomas R. Hughes to Orville Freeman, May 26, 1965, box 4255, Civil Rights, GC 1906–76, SOA, RG 16, NARA; John Dittmer, *Local People: The Struggle for Civil Rights in Mississippi* (Urbana, 1994), 341–45.

2. Fay Bennett to Jac Wasserman, August 5, 1964, box 40, Jac Wasserman, folder 7, NSF Papers.

3. Walter B. Lewis to F. Peter Libassi, July 27, 1965; Lewis to William L. Taylor,

September 15, 1965, box 3, CFFPD, USCCR, RG 453, NARA; U.S. Commission on Civil Rights, *Equal Opportunity in Farm Programs: An Appraisal of Services Rendered by Agencies of the United States Department of Agriculture* (Washington, 1965).

4. William L. Taylor to Orville Freeman, August 20, 1965; Thomas R. Hughes to Horace Godfrey, September 14, 1965; Godfrey to Hughes, September 22, 1965, box 4454, Civil Rights, GC 1906–76, SOA, RG 16, NARA; Marian P. Yankauer to Walter Lewis, June 9, 1965; William C. Payne to Yankauer, July 21, 1965, box 2, CFFPD, USCCR, RG 453, NARA; DASCO to Merwin W. Kaye, September 29, 1966, box 418, County Committee, CFASCS, RG 145, NARA; Pete Daniel, interview with William L. Taylor, Washington, D.C., February 21, 2006, SOHC.

5. William Seabron to Merwin W. Kaye, September 28, 1966, box 4454, Civil Rights, GC 1906–76, SOA, RG 16, NARA; Ray Fitzgerald to Kaye, September 29, 1966, box 418, County Committee, CFASCS, RG 145, NARA.

6. C. W. Sullivan to DASCO, August 11, 1965, and attached questions for August 27, 1965, meeting; William H. Booth to Ray Fitzgerald, August 13, 1965; Louisiana state ASCS director to F. Wainwright Blease, August 10, 1965, box 321, CFASCS, RG 145, NARA.

7. Karel M. Weissberg, "A Study of the Freedom Movement in Cotton County, Mississippi," honors thesis, Radcliffe College, January 3, 1966, loaned by Elaine DeLott Baker.

8. "ASCS Elections," September 13, 1965, reel 60, frames 657–58; "SNCC Program: ASCS Elections, 1965," reel 60, frames 659–61; memo to Friends of SNCC, November 5, 1965, reel 20, frames 451–52, *SNCC Papers*. See also "SNCC Program: ASCS Elections, 1965," reel 20, frames 471–73; "The Agricultural Stabilization and Conservation Service: What the ASCS Is," reel 20, frames 424–31, ibid.

9. Barbara Brandt, "We Weren't the Bad Guys," in Faith S. Holsaert, Martha Prescod Norman Noonan, Judy Richardson, Betty Garman Robinson, Jean Smith Young, and Dorothy M. Zellner, *Hands on the Freedom Plow: Personal Accounts by Women in SNCC* (Urbana, 2010), 434–35; "Acreage Allotment Program," reel 20, frame 423; "The Agricultural Stabilization and Conservation Service," reel 20, frame 424; "ASCS Background Information," B1, no. 1, February 1965, reel 20, frames 425–31; "ASCS Background Information," B1, no. 5, February 1965, reel 20, frames 431–38; "Do You Plant Cotton?," reel 20, frame 440; "A.S.C.S. Organizers Handbook," reel 20, frames 404–16, *SNCC Papers*.

10. F. Wainwright Blease to Victor B. Phillips, July 6, 1965, box 321, CFASCS, RG 145, NARA; Joel Horowitz to Orville Freeman, September 9, 1965, box 12, CFOSDSP, RG 453, NARA. See also "Proposal by Rural Advancement Fund of the National Sharecroppers Fund, Inc. for Massive ASCS Election Campaign in Alabama and Other Southern States," box 32, Rural Advancement Fund, folder 26, NSF Papers.

11. Harold Ickes to USDA, July 15, 1965; William Seabron, memo, August 2, 1965; Seabron to Ickes, August 9, 1965; F. Wainwright Blease to DASCO, August 10, 1965, box 320, CFASCS, RG 145, NARA.

12. Acting director, south-central area, to DASCO, August 13, 1965; Ray Fitzgerald

to Horace Godfrey, August 17, 1965; F. Wainwright Blease to DASCO, August 18, 1965, ibid. See also Louisiana state executive director to F. Wainwright Blease, October 29, 1965, box 321, ibid.

13. "Report on ASCS Election in St. Landry Parish, Summer 1965," box 53, Reports, folder 2, NSF Papers.

14. Ocie Lee Smith to Jac Wasserman, October 22, 1965, box 60, Ocie Smith, folder 10, ibid.; William Seabron to J. William Howell, December 13, 1965, box 4254, Civil Rights, GC 1906–76, SOA, RG 16, NARA.

15. Ralthus Hayes, Susan Lorenzi, and Henry Lorenzi, "Report on the 1965 ASCS Community Committee Election in Holmes County, Mississippi," n.d., box 28, ASCS, folder 42, NSF Papers; Dittmer, *Local People*, 191.

16. Jack Anderson, "Trip to Mississippi," attached to Lester P. Condon to John A. Schnittker, November 24, 1965, box 4254, Civil Rights, GC 1906–76, SOA, RG 16, NARA; Jenny Irons, *Reconstituting Whiteness: The Mississippi State Sovereignty Commission* (Nashville, 2010), 166–69.

17. Unita Blackwell and Annie Devine, "Congressman Resnick's Visit to Mississippi," n.d., reel 20, frames 420–21, *SNCC Papers*; Joseph Y. Resnick to Orville Freeman, December 4, 1965, and attached report and affidavit of Willie Holmes, November 28, 1965, box 4254, Civil Rights, GC 1906–76, SOA, RG 16, NARA.

18. Unita Blackwell and Annie Devine, "Congressman Resnick's Visit to Mississippi," n.d., reel 20, frames 420–21, *SNCC Papers*.

19. Horace Godfrey to Kenneth M. Birkhead and attachments, January 7, 1966, box 4490, Farm Program; Ray Fitzgerald to William Seabron, December 6, 1965; F. Wainwright Blease to Fitzgerald, December 6, 1965, box 4254, Civil Rights, GC 1906–76, SOA, RG 16, NARA.

20. Ralthus Hayes, Susan Lorenzi, and Henry Lorenzi, "Report on the 1965 ASCS Community Committee Election in Holmes County, Mississippi," n.d., box 28, ASCS, folder 42, NSF Papers. Ray Fitzgerald relied on a report by regional ASCS director F. Wainwright Blease to sanitize McWilliams's discrimination. See Fitzgerald to William Seabron, December 6, 1965; Blease to Fitzgerald, December 6, 1965, box 4254, Civil Rights, GC 1906–76, SOA, RG 16, NARA.

21. Ralthus Hayes to Charles M. Cox, August 31, 1966 (emphasis in original); William Seabron to Hayes, September 9 (draft), 13, 1966; Joseph Searles to Victor Phillips, September 14, 1966; Seabron to Hayes, September 30, 1966, box 4454, Civil Rights, GC 1906–76, SOA, RG 16, NARA.

22. Kenneth Birkhead to undersecretary, October 19, 1966, ibid.

23. "Friends of SNCC Newsletter Number One," n.d. [November 1965], reel 20, frames 464–65, *SNCC Papers*.

24. Untitled report, August 20, 1965, reel 65, frames 332–33, ibid.; Howard Zinn, *SNCC: The New Abolitionists* (Boston, 1964), 12.

25. Ada Holliday to Department of Agriculture, November 26, 1965, reel 65, frame 323, *SNCC Papers*.

26. Ibid.

27. F. Wainwright Blease to DASCO, December 10, 1965; Ray Fitzgerald to Kenneth M. Birkhead, December 16, 1965, box 320, CFASCS, RG 145, NARA; Lester P. Condon to Ada Mae Holliday, July 25, 1966, box 4455, Civil Rights, GC 1906–76, SOA, RG 16, NARA.

28. William Seabron to Ada Mae Holliday, July 25, 1966, box 4455, Civil Rights, GC 1906–76, SOA, RG 16, NARA. See also Victor B. Phillips to Orzell Billingsley Jr., April 21, 1966, box 418, County Committee, CFASCS, RG 145, NARA.

29. Barbara Brandt, "Notes on a Meeting of the ASCS State Board, the State ASCS Civil Rights Advisory Committee, plus Invited Guests," November 9, 1965, reel 20, frame 456; Brandt to Judy Richardson, Elizabeth Sutherland, and Ben Smith, October 29, 1965, reel 20, frame 385, *SNCC Papers*; press release, Victor B. Phillips, September 27, 1965, series IX, subseries 3, box 9.3/1, Special Collections, NAL. See also Barbara Brandt, "A Report on a Meeting at the ASCS State Office," box 5, Delta Ministry Papers.

30. Jan Maedke to ASCS, December 12, 1966; affidavit of Walter McGloithan, December 12, 1966, box 4629, Civil Rights, GC 1906–76, SOA, RG 16, NARA.

31. Elmo Holder to National Sharecroppers Fund, May 28, 1965, box 31, SNCC folder, NSF Papers; "SNCC Program: ASCS Elections, 1965," reel 60, frames 659–61, *SNCC Papers*; John Liutkus to Ray Fitzgerald, November 7, 1965, box 321, Program, CFASCS, RG 145, NARA.

32. Jac Wasserman to James Mays, June 7, 1965, box 31, SNCC folder, NSF Papers; Elmo Holder to B. L. Collins, July 19, 1965 (quotes), box 321, CFASCS, RG 145, NARA. On Alabama ASCS elections, see Susan Youngblood Ashmore, *Carry It On: The War on Poverty and the Civil Rights Movement in Alabama, 1964–1972* (Athens, 2008), 140–46, 205–8; and Hasan Kwame Jeffries, *Bloody Lowndes: Civil Rights and Black Power in Alabama's Black Belt* (New York, 2009), 126–31.

33. B. L. Collins to E. W. Bayol, July 21, 1965; Bayol to Collins, August 9, 1965, box 321, CFASCS, RG 145, NARA; Collins to Bayol, August 24, 1965, reel 20, frame 384, *SNCC Papers*.

34. Affidavit of Stokely Carmichael before Edward E. Reed Jr., OIG, USDA Atlanta office, Selma, Ala., March 11, 1966, reel 20, frames 460–61, *SNCC Papers*.

35. "SNCC Program: ASCS Elections, 1965," reel 20, frames 471–73; "First Wins in '64," *The Voice*, December 20, 1965, reel 20, frame 457, *SNCC Papers*; John Lewis to Horace Godfrey, November 30, 1965, box 321, Program; William Seabron to Stokely Carmichael, June 21, 1966, box 418, Program, CFASCS, RG 145, NARA.

36. "SNCC Program: ASCS Elections, 1965," reel 20, frames 471–73; Gren Whitman, "Report on the ASCS Elections for 1965," n.d. [Fall 1965], reel 37, frames 1086–89, *SNCC Papers*; John Lewis to Horace Godfrey, November 30, 1965, box 321, Program, CFASCS, RG 145, NARA. On preparations for the Hale County election, see the undated report of Cleophus Hobbs and Terry Shaw, reel 18, frames 51–52, *SNCC Papers*.

37. "'Bad Baker' Workers Cite ASCS Steal," *The Voice*, December 20, 1965, reel 20, frame 459, *SNCC Papers*; Fred Anderson to William H. Booth, October 18, 1965; ex-

ecutive director, Georgia state ASCS office, to director, southeast area, October 19, 1965, box 4254, Civil Rights, GC 1906–76, SOA, RG 16, NARA. See also the discrimination complaint from Isaac Simkins taken by William Seabron by telephone on October 20, 1965; Seabron to Frank J. Brechenser, October 28, 1965, ibid. See also Ray Fitzgerald to Isaac Simkins, February 10, 1966, box 418, Program, CFASCS, RG 145, NARA.

38. William Seabron to Rodney Leonard, December 3, 1965, box 4259, Committees, Civil Rights Advisory, GC 1906–76, SOA, RG 16, NARA.

39. "Equal Opportunity in Farm Programs," Georgia State Advisory Committee, box 40, CFOSDSP, USCCR, RG 453, NARA.

40. Horace Godfrey to chairmen, ASC state committees, et al., January 27, 1966; director, southeast area ASCS, to chairmen, ASC state committees, February 28, 1966, box 418, County Committee, CFASCS, RG 145, NARA.

41. Horace Godfrey to Kenneth Birkhead, January 7, 1966, box 4490, Farm Program; Birkhead to undersecretary, December 7, 1966, box 4458, Civil Rights Advisory, GC 1906–76, SOA, RG 16, NARA.

42. Aaron Henry to Lyndon Johnson, February 12, 1966, box 4456, Civil Rights, ibid.; Dittmer, *Local People*, 120–21. See also Julian J. Keiser to Orville Freeman, June 7, 1966; William Seabron to Keiser, June 21, 1966, box 4455, Civil Rights, GC 1906–76, SOA, RG 16, NARA.

43. William Seabron to Orville Freeman, February 15, 1966, box 4456, Civil Rights, GC 1906–76, SOA, RG 16, NARA.

44. Jac Wasserman to Ray Fitzgerald, May 5, 1966; acting DASCO to Wasserman, May 11, 1966, box 418, County Committee, CFASCS, RG 145, NARA.

45. John Zippert to DASCO, June 17, 1966; Ray Fitzgerald to Zippert, June 29, 1966, box 418, Program, CFASCS, RG 145, NARA.

46. William Seabron to Charles M. Cox, May 25, 1966, box 4456; Seabron[?] (fragment, no second page and signature) to Stokely Carmichael, June 10, 1966; Carmichael to Seabron, July 25, 1966; Seabron to Carmichael, August 15, 1966, box 4455, Civil Rights, GC 1906–76, SOA, RG 16, NARA; Eric Wentworth, "U.S. Refuses to Delay Ala. Farm Unit Voting," *Washington Post*, August 5, 1966, A2; Paul Valentine, "Federal Judge Voids Negro Farmers' Suit," *Washington Post*, October 29, 1966, 2; "Negro Candidates Lose in Alabama," *New York Times*, November 9, 1966, 25.

47. "Lawsuit Claims Discrimination in Federal Agricultural Programs in Alabama," press release, July 27, 1966; William L. Taylor memorandum for the commission, August 11, 1966, box 20, CFOSDSP, USCCR, RG 453, NARA.

48. Miriam Wasserman, "Farm Elections: White Power in the Black Belt," typescript, box 28, ASCS, folder 38, NSF Papers; Wasserman, "White Power in the Black Belt," *New South* (Winter 1967): 27–36; affidavit of Peter Agee, August 1966, box 2, CFOSDSP, USCCR, RG 453, NARA. On the preparations for the 1966 elections, evictions, and rural conditions in Alabama, see Ashmore, *Carry It On*, 201–8.

49. Zinn, *SNCC*, 40; Clayborne Carson, *In Struggle: SNCC and the Black Awaken-*

*ing of the 1960s* (Cambridge, 1981), 206–43; Stembridge interview, December 5, 1996, SOHC; Dorothy M. Zellner, "My Real Vocation," in Holsaert et al., *Hands on the Freedom Plow*, 325.

50. Virgil Dimery to Orville Freeman, July 27, 1966; DASCO to William Seabron, August 13, 1966, box 418, Program, CFASCS, RG 145, NARA.

51. Ray Fitzgerald to Fay Bennett, June 30, 1966, ibid.

52. Ray Fitzgerald to Horace Godfrey, June 28, 1966, County Committee, ibid.

53. Jac Wasserman to Fay Bennett, July 25, 1966, box 30, Wasserman, folder 2; Mike Kenney to Bennett and Wasserman, July 27, 1966, box 30, Mike Kenney, folder 6, NSF Papers.

54. "Proposal by Rural Advancement Fund of the National Sharecroppers Fund, Inc. for a Massive ASCS Election Campaign in Alabama and Other Southern States," n.d., box 30, folder 26; Mike Kenney to Gary Piper, August 22, 1966, box 30, Mike Kenney, folder 6, NSF Papers (emphasis in original); Ashmore, *Carry It On*, 201, 206,

55. Ralph S. Tyler III to Orville Freeman, August 4, 1966, box 418, Program, CFASCS, RG 145, NARA.

56. Ashmore, *Carry It On*, 208–52, 278; Greta de Jong, "Staying in Place: Black Migration, the Civil Rights Movement, and the War on Poverty in the Rural South," *Journal of African American History* 90 (Fall 2005): 399–400; "Aid Group Decries Plight of Farmer," *New York Times*, July 8, 1969, 26. See also Will D. Campbell, "Staying the Course," *World: The Journal of the Unitarian Universalist Association* 8 (September/October 1994): 30–35, loaned by Elaine DeLott Baker.

57. Mike Kenney to Ray Fitzgerald, August 1, 1966; Fitzgerald to Kenney, August 5, 1966; Kenney to Fitzgerald, August 12, 1966; Fitzgerald to Kenney, August 18, 1966, box 418, County Committee, CFASCS, RG 145, NARA.

58. Mike Kenney to Fay Bennett, September 9, 1966, box 30, Mike Kenney, folder 6; Kenney to Bennett, September 23, 1968; Kenney memo to file, October 19, 1966, box 30, Mike Kenney, folder 5, NSF Papers.

59. Mike Kenney to Jac Wasserman, November 12, 1966, box 30, Mike Kenney, folder 6, NSF Papers; John B. Vance to Ray Fitzgerald, September 1, 1966, Program; F. Wainwright Blease to DASCO, October 18, 1966, box 418, County Committee, CFASCS, RG 145, NARA.

60. William Seabron to Aviva Futorian, September 7, 1966; Futorian to Seabron, October 24, 1966; Seabron to Futorian, November 14, 1966, box 418, Compliance, CFASCS, RG 145, NARA.

61. Vernon White et al. to William Seabron, November 8, 1966; Seabron to Victor Phillips, November 18, 1966, box 4454, Civil Rights, GC 1906–76, SOA, RG 16, NARA.

62. "General Outline of the Need to Organize in the Poor White Communities in the South," n.d. [1966], box 30, Mike Kenney, folder 5, NSF Papers; affidavit of Shirley Mesher, August 1966, box 2, CFOSDSP, USCCR, RG 453, NARA.

63. Shirley Mesher, "Agricultural 'Progress' Report," n.d., box 2, CFOSDSP, USCCR, RG 453, NARA.

64. Vincent O'Connor to Ray Fitzgerald, October 21, 1966; Jac Wasserman to Fitzgerald, October 27, 1960; O'Connor to Fitzgerald, November 2, 1966; Fitzgerald to O'Connor, November 2, 1966, box 418, CFASCS, RG 145, NARA.

65. Vincent O'Connor to Donald Jelinek, November 2, 1966, box 47, Arkansas, folder 13, NSF Papers. See also Jac Wasserman to Ray Fitzgerald, October 7, 1966; F. Wainwright Blease to DASCO, November 28, 1966, box 418, County Committee, CFASCS, RG 145, NARA. On O'Connor, see Jennifer Jensen Wallach, "Replicating History in a Bad Way? White Activists and Black Power in SNCC's Arkansas Project," 78, and on SNCC's work in urban Arkansas, see Randy Finley, "Crossing the White Line: SNCC in Three Delta Towns, 1963–1967," 54–68, both in Jennifer Jensen Wallach and John A. Kirk, eds., *Arsnick: The Student Nonviolent Coordinating Committee in Arkansas* (Fayetteville, 2011).

66. F. Wainwright Blease to DASCO, November 28, 1966, box 418, County Committee, CFASCS, RG 145, NARA.

67. John B. Vance to William Seabron, November 2, 1966, box 418, Program, ibid.

68. Victor B. Phillips to L. C. Holm, December 1, 1966, ibid.

## Chapter 5

1. Geer Morton to Mike Kenney, n.d. [December 1966?], box 53, Mike Kenney, folder 32; Kenney to Fay Bennett, November 18, 1966, box 30, Mike Kenney, folder 5; Kenney to Bennett, May 12, 1967, box 30, Mike Kenney, folder 4, NSF Papers.

2. Geer Morton and Ted Seaver to William Seabron, February 9, 1967, box 476, Program, CFASCS, RG 145, NARA.

3. Geer Morton, "How to Lose an ASCS Election Right and Left," February 27, 1967, box 53, ASCS, folder 40, NSF Papers (emphasis in original).

4. Orville Freeman to Charles C. Diggs, September 12, 1966, box 4454, Civil Rights; Kenneth Birkhead to undersecretary, December 7, 1966, box 4458, National Citizens Committee on Civil Rights, GC 1906–76, SOA, RG 16, NARA; Gilbert Ware to Walter B. Lewis, January 13, 1967, box 36, CFOSDSP, USCCR, RG 453, NARA.

5. DASCO to Horace Godfrey, February 24, 1967, box 4628, Civil Rights; William Seabron to Orville Freeman, March 6, 1967, box 4629, Civil Rights, GC 1906–76, SOA, RG 16, NARA.

6. Gilbert Ware to Walter B. Lewis, January 13, 1967, box 36, CFOSDSP, USCCR, RG 453, NARA.

7. Horace Godfrey to Orville Freeman, March 29, 1967, and attached "Negro Owners and Operators," box 476, Reports, County Committee; John B. Vance to Ray Fitzgerald, April 24, 1967, box 476, Compliance, CFASCS, RG 145, NARA. See also Victor B. Phillips to William Seabron, May 5, 1967, box 476, Directives, ibid.; and Karel M. Weissberg, "A Study of the Freedom Movement in Cotton County, Mississippi," honors thesis, Radcliffe College, January 3, 1966, loaned by Elaine DeLott Baker.

8. Horace Godfrey to William Seabron, March 27, 1967; Ray Fitzgerald to Godfrey,

April 12, 1967; John B. Vance to Fitzgerald, April 24, 1967, box 476, Compliance; J. Wainwright Blease to DASCO, April 27, 1967, box 476, Employment, CFASCS, RG 145, NARA.

9. Ray Fitzgerald to Horace Godfrey, April 12, 1967 (first and third quotes); Godfrey to William Seabron, March 27, 1967 (second quote), box 476, Compliance, ibid.

10. J. Wainwright Blease to chairmen, state ASCS committees, May 9, 1967, box 476, Civil Rights, Equal Opportunity; director, south-central area, to DASCO, March 22, 1967; Ray Fitzgerald to Horace Godfrey, March 22, 1967, box 476, Employment, ibid.

11. John B. Vance to DASCO, May 29, 1967, and attached "ASCS Attendance at Selma, Alabama, Civil Rights Meeting, May 26–27, 1967," box 476, Reports, ibid.

12. Beth Wilcox to Horace Godfrey, May 31, 1967; Godfrey to Wilcox, June 27, 1967, box 476, Program, ibid.

13. Victor B. Phillips to Don Looper, June 8, 1967; Phillips to Jac Wasserman, June 14, 1967, box 476, Civil Rights, Equal Opportunity, ibid.

14. James Mays, "Trip to Batesville, Mississippi," June 26–28, 1967, box 49, Various Field Reports, folder 34, NSF Papers.

15. Ibid.; Jennifer Gong and David Mallie, memo to NSF, n.d. [July 1967], box 30, David Mallie, folder 15, ibid.; William Seabron to Horace Godfrey, July 17, 1967, box 476, Civil Rights, CFASCS, RG 145, NARA. On Robert Miles, see Elaine DeLott Baker, "They Sent Us This White Girl," in Constance Curry et al., *Deep in Our Hearts: Nine White Women in the Freedom Movement* (Athens, 2000), 274–78.

16. C. M. Gordy to DASCO, August 8, 1967; John B. Vance to Alan Durick, August 15, 1967, box 476, County Committee, CFASCS, RG 145, NARA.

17. Ray Fitzgerald to Kenneth M. Birkhead, September 28, 1967; William Seabron to Fitzgerald, October 23, 1967, ibid.

18. Philip C. Beach to director, southeast area, October 31, 1966, box 418, County Committee, ibid.

19. Philip C. Beach to DASCO, September 29, 1967, box 527, Macon County Appeal, ibid.

20. Mary Ellen Gale, "No Way to Win It," *Southern Courier*, September 30–October 1, 1967, copy in ibid.

21. Ray Fitzgerald to Jack Puterbaugh, September 29, 1967; R. E. Bond to Fred M. Acuff, September 29, 1967, ibid.

22. Donald A. Jelinek to Jack Bridges, November 17, 1967, box 4793, Committees, GC 1906–76, SOA, RG 16, NARA.

23. Notes of phone conversation with Leary Whatley, Macon County, Ala., December 7, 1967, box 527, Macon County Appeal, CFASCS, RG 145, NARA.

24. J. H. M. Henderson to Jack Bridges, October 5, 1967; Alan Durick to Ray Fitzgerald, October 26, 1967; Mary Ellen Gale, "Farmers Challenge Vote," *Southern Courier*, November 18–19, 1967, copy; Henderson to Leary Whatley, December 14, 1967; Bridges to Henderson, December 12, 1967; Bridges to Ellis Hall et al., December 14, 1967; Ellis Hall et al. to Fitzgerald, December 27, 1967; Mary Ellen Gale,

"Macon Farmers Lose Vote Case," *Southern Courier*, December 23–24, 1967 (Jelinek quote), copy in ibid.

25. Ray Fitzgerald to chairman, Alabama state ASCS committee, December 27, 1967, ibid.

26. Mary Ellen Gale, "'You Got to Try' Says Farmer," *Southern Courier*, January 27–28, 1968, copy in ibid.

27. Mary Ellen Gale, "Farmers Visited, Lose Case," *Southern Courier*, February 24–25, 1968, copy; statistics on county ASCS committee landownership, attached to John B. Vance to Ray Fitzgerald, June 27, 1968, box 527, Reports, ibid.

28. Mary Ellen Gale, "Farmers Visited, Lose Case," *Southern Courier*, February 24–25, 1968, copy; Ray Fitzgerald to J. H. M. Henderson et al., August 30, 1968, ibid.

29. Alan Durick to Mrs. Mead, October 8, 1968, box 527, Committees, ibid.

30. Beth Wilcox, "Not Even an Alternate," *Southern Courier*, September 30–October 1, 1967, copy in box 527, Macon County Appeal; Macon County ASCS committee and community committee (1968), box 527, Reports, ibid.

31. Robert B. Fitzpatrick to William Seabron, January 24, 1968; Seabron to Ray Fitzgerald, February 13, 1968; Estelle Fine, "Campaigns Fail in Miss.," *Southern Courier*, September 30–October 1, 1967, copy in box 527, Macon County Appeal, ibid.

32. William Seabron to Clifton Haralson, January 26, 1968; Estelle Fine, "Campaigns Fail in Miss.," *Southern Courier*, September 30–October 1, 1967 (quotes), copy in ibid.

33. William Seabron to Jan Maedke, December 19, 1967, box 476, Program, CFASCS, RG 145, NARA.

34. William Seabron to Ray Fitzgerald, February 13, 1968, box 527, Program, ibid.

35. C. M. Gordy to DASCO, August 8, 1967, box 476, County Committee; "Number of Negroes Elected as ASCS County and Community Committeemen in 1966 and 1967"; "ASCS County Full-time Employment of Negroes by State for Periods Ending December 31, 1966 and December 1967," box 527, Reports, ibid.

36. Acting director, south-central area, through Ray Fitzgerald to Horace Godfrey, January 23, 1968; C. W. Sullivan to director, south-central area, January 26, 1968, box 527, Civil Rights, Equal Opportunity, ibid.

37. Nick Kotz, "Negroes' Plight on Eastland's Plantation," *Des Moines Register*, February 25, 1968; "A Poor, Hard Life on Eastland Plantation," *Des Moines Register*, February 26, 1968; "Machine Shift Adds to Plight of Negroes," *Des Moines Register*, n.d., copies in box 5; Kotz, "Notes on Interview with Senator Eastland," April/May 1967, box 5, Delta Ministry Papers; William Seabron to Orville Freeman, and attachment, March 13, 1967, box 4628, Civil Rights, GC 1906–76, SOA, RG 16, NARA. See also Chris Myers Asch, *The Senator and the Sharecropper: The Freedom Struggles of James O. Eastland and Fannie Lou Hamer* (Chapel Hill, 2008).

38. Citizens Advisory Committee on Civil Rights, recommendations to secretary; minutes of March 14–15 meeting in Tampa, Florida (quotes), box 527, Committees; Alabama ASC State Advisory Committee (1968), Reports, CFASCS, RG 145, NARA.

39. Horace Godfrey to Orville Freeman, May 15, 1968, box 527, Program, ibid.

40. Victor B. Phillips to Horace Godfrey et al., June 11, 1968, box 527, Civil Rights, Equal Opportunity, ibid.

41. John B. Vance to Ray Fitzgerald, June 27, 1968, Reports; Horace Godfrey to Tom Hughes, June 24, 1968, Committees, ibid.

42. Horace Godfrey to William Seabron, July 31, 1968, Committees, ibid.

43. Horace Godfrey to William Seabron, July 17, 1968, Reports, ibid.; CEC to Richard Russell, January 11, 1968, box 73, folder 2, Senatorial Papers, XCV, General, Russell Papers.

44. Horace Godfrey to William Seabron, July 31, 1968; Godfrey to John A. Schnittker, August 16, 1968, box 527, Committees, CFASCS, RG 145, NARA.

45. Statement of C. H. Erskine Smith (exhibit no. 2), April 27, 1968, in *Hearings before the United States Commission on Civil Rights, Montgomery, Ala., April 27–May 2, 1968* (Washington, 1969).

46. William Seabron to Carol B. Kummerfeld, September 6, 1968, box 527, Reports; Victor B. Phillips to Seabron, November 5, 1968, box 527, Program; Charles Cox to William Galbraith, March 26, 1969, box 578, County Committee, CFASCS, RG 145, NARA.

47. Orville Freeman to William Seabron, January 12, 1969; Seabron to Freeman, January 15, 1969; Horace Godfrey to Seabron, January 17, 1969, box 2, folder 11, Personal Correspondence, January–March 1969, Seabron Papers; Bruce J. Reynolds, "Black Farmers in America, 1865–2000: The Pursuit of Independent Farming and the Role of Cooperatives," USDA Rural Business Cooperative Service, Research Report 194 (October 2002), 24.

48. Maxine W. Lacy to secretary of agriculture, February 20, 1969; William E. Galbraith to Lacy, March 12, 1969, box 578, County Committee, CFASCS, RG 145, NARA.

49. Ed Edwin, interview with Earl L. Butz, January 15, 1968, Dwight D. Eisenhower Papers, Eisenhower Presidential Library, Abilene, Kans.

50. Percy R. Luney to Frank B. Elliott, June 3, 1971; William Seabron to Kenneth E. Frick, July 12, 1971, box 5363, Committees, GC 1906–76, SOA, RG 16, NARA. See also Jim Ward, "The Questions and Problems of Image," February 4, 1970, box 6, Office of the Administrator, Annual Reports, ASCS Historical Records, 1933–69, CFASCS, RG 145, NARA.

51. "Federal Agency Failure to Implement and Enforce Federal Civil Rights Law," March 1973, box 1, CFOSDCC, USCCR, RG 453, NARA.

52. Ward Sinclair, "Old-Boy Network Still Haunts Agriculture's Problem Child," *Washington Post*, September 21, 1987, A13.

## Chapter 6

1. Louis R. Harlan, *Booker T. Washington: The Wizard of Tuskegee, 1901–1915* (New York, 1983), 206–12; Lu Ann Jones, *Mama Learned Us to Work: Farm Women in the New South* (Chapel Hill, 2002), 139–69; Dwayne Cox, "Alabama Farm Agents, 1914–1922," *Alabama Review* (October 1994): 285–304.

2. Pete Daniel, *Breaking the Land: The Transformation of Cotton, Tobacco, and Rice Cultures since 1880* (Urbana, 1985), 12–18.

3. Doxey A. Wilkerson, *Agricultural Extension Services among Negroes in the South* (Conference of Presidents of Negro Land Grant Colleges, 1942), 20–23, 32.

4. Juanita Y. DeVaughn to W. B. Hill, June 22, 1962, box 361, ACES Papers. See also Jones, *Mama Learned Us to Work*, 139–69; and Carmen V. Harris, "'Well I Just Generally Bes the President of Everything': Rural Black Women's Empowerment through South Carolina Home Demonstration Activities," *Black Women, Gender and Families* 3 (Spring 2009): 91–112.

5. Michael Piore to W. M. Bost, July 19, 1964; Bost to Piore, July 23, 1964, box 8, Colvard Papers; Bost to Piore, July 28, 1964; Piore to Lloyd H. Davis, July 28, 1964, box 34, Mississippi, folder 15, NSF Papers (emphasis in original). See also Craig Piper, interview with William M. Bost, July 10, 2002, MML, MSU.

6. Pete Daniel, interview with Robert Church, Fort Valley, Ga., February 20, 2007, SOHC.

7. Ibid.

8. Ibid.

9. Ibid.

10. William Seabron to Robert J. Pitchell, November 15, 29, 1965, box 4254, Civil Rights, GC 1906–76, SOA, RG 16, NARA.

11. W. M. Bost to Lloyd Davis, June 24, October 20, 1969; Bost memo to all extension workers, January 13, 1970, box 21, Giles Papers. See also Charles E. Bell Jr. to Lester P. Condon (Audit of Civil Rights Implementation in the Mississippi Cooperative Extension Service), August 29, 1968, ibid.

12. Pete Daniel, interview with Robert Church, Fort Valley, Ga., February 20, 2007, SOHC. See also James Robinson's report to Marian P. Yankauer on the Extension Service in twenty-five Georgia counties, August 11, 1964, box 1, CFLID, USCCR, RG 453, NARA.

13. Richard M. Shapiro and Donald S. Safford, "Report of Field Trip to South Carolina from April 12, 1964, through April 16, 1964"; Shapiro and Ronald M. McCaskill, interviews with Harold McNeill and Minnie Brown, Greensboro, N.C., August 12, 1964, box 1, CFLID, USCCR, RG 453, NARA.

14. W. Ralph Eubanks, *Ever Is a Long Time: A Journey into Mississippi's Dark Past* (New York, 2003), xii, 35–36.

15. William Seabron to Lloyd Davis, October 18, 1965, box 4254, Civil Rights, GC 1906–76, SOA, RG 16, NARA.

16. "Equal Opportunity in Farm Programs," Georgia State Advisory Committee, box 40, CFOSDSP, USCCR, RG 453, NARA. See also William Seabron to Robert J. Pitchell, November 15, 1965, box 4254, Civil Rights, GC 1906–76, SOA, RG 16, NARA; Richard M. Shapiro and Ronald M. McCaskill, interview with Pauline Moore and William C. Cooper, Greensboro, N.C., August 12, 1964, box 1, CFLID, USCCR, RG 453, NARA.

17. On Eberhardt, see E. T. York to John P. Duncan Jr., May 8, 1963; O. C. Aderhold

to York, April 23, 1963, and attachment, box 1, Aderhold Papers; "Equal Opportunity in Farm Programs," Georgia State Advisory Committee, box 40, CFOSDSP, USCCR, RG 453, NARA. See also "Agricultural Extension Service Integration to Downgrade Negroes," November 5, 1965, reel 20, frame 474, *SNCC Papers.*

18. Richard M. Shapiro and Donald S. Safford, report on Sharecroppers Fund Conference, Fort Valley State College, Fort Valley, Ga., March 7, 8, 1964, box 1, CFLID, USCCR, RG 453, NARA; William Seabron to Nyle C. Brady, July 20, 1965, box 4255, Civil Rights, GC 1906–76, SOA, RG 16, NARA.

19. William Seabron to Thomas Hughes, August 3, 1965, Civil Rights, GC 1906–76, SOA, RG 16, NARA.

20. C. O. Pearson to Orville Freeman, October 13, 1965, box 4541, Personnel 9-2, ibid.

21. Thomas Hughes to William Seabron, October 18, 1965, box 4254, Civil Rights, ibid.

22. W. W. Law to Orville Freeman, November 10, 1965; William Seabron to Law, November 16, 1965, ibid.

23. Annye H. Braxton to William Seabron, December 6, 1965, ibid.; affidavit of Fred R. Robertson, January 7, 1966, box 17, RG 533, Philpott Papers; William Taylor to Nicholas Katzenbach, March 8, 1966; Taylor to John Doar, December 19, 1967, CFOSDSP, USCCR, RG 453, NARA.

24. Pauline F. Smith to Orville Freeman, April 2, 1966, box 4456, Civil Rights, GC 1906–76, SOA, RG 16, NARA.

25. Lloyd Davis to W. M. Bost, July 19, 1966; Bost to Davis, July 29, 1966 (Smith quote); Davis to Bost, July 20, 1966; Bost to Davis, July 29, 1966 (Reed quote), box 21, Giles Papers. See also affidavit of William M. Bost, May 16, 1966; affidavit of William E. Ammons, May 24, 1966; affidavit of Bost, June 7, 1966, box 8, Colvard Papers.

26. W. M. Bost to Lloyd Davis, July 29, 1966, box 21, Giles Papers; Cooperative Extension Service, Minutes of the Executive Board, January 13, 1965, vertical file, University Archives, MML, MSU; Craig Piper, interview with William M. Bost, July 10, 2002, MML, MSU.

27. Kitty Galbraith, interview with Sadye Wier, November 12, 1980, Wier Papers, Special Collections, MML, MSU.

28. John Marszalek and George Lewis, interview with Sadye Wier, August 13, 1992, ibid.

29. Richard M. Shapiro to Walter B. Lewis, September 10, 1965, box 3, CFFPD, USCCR, RG 453, NARA.

30. Meeting of Citizens Advisory Committee on Civil Rights, December 2–3, 1965, box 4458, Citizens Committee on Civil Rights, GC 1906–76, SOA, RG 16, NARA.

31. "Board to Meet June 15–16 in Miami, Regional Action," June 1966, copy in box 47, RG 653, Philpott Papers.

32. Meeting of Citizens Advisory Committee on Civil Rights, February 23–24, 1966, box 4458, Committees, Civil Rights Citizens Advisory Committee, GC 1906–76,

SOA, RG 16, NARA; press release, July 19, 1962, series IX, subseries 3, box 9.3/1, Special Collections, NAL.

33. William Seabron to Robert J. Pitchell, March 5, 1966; Delores G. Morse to Seabron, July 11, 1966; Seabron to Morse, July 14, 1966; W. E. Skelton to Morse, August 3, 1966; Morse to Seabron, September 15, 1966; Seabron to Morse, October 6, 1966, box 4541, Personnel 9-2, Discrimination—Preferences, GC 1906–76, SOA, RG 16, NARA. On continuing discrimination, see Seabron to Lloyd Davis, May 11, 1966, box 4456, Civil Rights, ibid.

34. John W. Slusser to William Seabron, "Report on Trip to North Carolina, June 14–17, 1966," box 4455, Civil Rights; Orville Freeman to W. Marvin Watson, September 8, 1966, box 4483, Extension Work, ibid.

35. Fletcher F. Lassiter to William Payne, July 8, 1966; Walter B. Lewis to William Seabron, July 18, 1966, box 4455; Lewis to Seabron, August 24, 1966; Lassiter to Seabron, September 20, 1966; Seabron to Lassiter, November 3, 1966, box 4454, Civil Rights, ibid. See also Marian P. Yankauer, telephone conversation with Albert Bacon, May 18, 1964; Richard M. Shapiro to F. Peter Libassi, August 25, 1964, box 1; Yankauer to Lassiter, October 16, 1964; Yankauer, telephone conversations with Lassiter, February 1, 1965, box 2; Payne to Yankauer, October 29, 1965; Lewis to Seabron, November 5, 1965, box 3; Lewis to Lloyd H. Davis, June 17, 1965, box 2, CFFPD, USCCR, RG 453, NARA.

36. "Alabama Extension Service," box 8, Records Relating to Surveys and Studies, 1958–62, box 1, CFFPD, USCCR, RG 453, NARA; T. T. Williams, interview with Frank Jackson, Eutaw, Ala., May 12, 1964, box 1, CFLID, USCCR, RG 453, NARA. See also A. S. Bacon to C. D. Scott, August 7, 1964, box 361, ACES Papers. Vocational-education high school students could belong to either the white Future Farmers of America or the African American New Farmers of America, and in the mid-1960s, these organizations were negotiating a merger. See Dean W. Determan to F. Peter Libassi, June 9, 1964; Determan, interview with Mildred Reel, May 20, 1964, box 1, CFLID, USCCR, RG 453, NARA.

37. William Seabron to George L. Mehren, January 20, 1966, box 4456, Civil Rights, GC 1906–76, SOA, RG 16, NARA. For a statement of USDA compliance rules, see F. H. Hankes to Roland F. Ballou, May 19, 1966, box 418, Compliance, CFASCS, RG 145, NARA. For examples of title VI discussions, see William C. Payne to F. Peter Libassi, April 26, 1965; Payne to Richard M. Shapiro, May 13, 1965; Shapiro to Walter B. Lewis, May 26, 1965, box 2; Payne to the file, July 21, 1965, box 3, CFFPD, USCCR, RG 453, NARA.

38. William Seabron to George L. Mehren, January 20, 1966, box 4456; Seabron to John A. Schnittker et al., October 3, 1966; Elmer Mostow to Seabron, March 1, 1966, box 4454, Civil Rights, GC 1906–76, SOA, RG 16, NARA.

39. W. M. Bost to William L. Giles, August 26, 1966, box 21, Giles Papers.

40. Lloyd Davis to W. M. Bost, December 13, 1967, box 51, Giles Papers.

41. Secretary's Citizens Advisory Committee on Civil Rights to secretary, attached to Kenneth M. Birkhead to committee, December 28, 1966, box 4458, Committees,

Secretary's Citizens Advisory Committee on Civil Rights, GC 1906–76, SOA, RG 16, NARA. The committee members were Robert C. Edwards, Emmett Peter Jr., Elmer Ellis, John B. Evans, Charles G. Gomillion, Cecil Newman, Robert H. Montgomery, Lionel S. Steinberg, Vivian W. Henderson, Louis J. Twomey, Ralph E. McGill, Clay L. Cochran, Aaron Henry, Capus Waynick, J. J. Simmons, and Oscar Cavazos. See also Robert A. Carlson, "Cooperative Extension: A Historical Assessment," *Journal of Extension* 3 (Fall 1970): 10–15.

42. Daniel J. Zeigler to William Seabron, July 16, 1966; Seabron to Zeigler, July 22, 1966, box 4541, Personnel 9-2; Zeigler to Seabron, January 28, 1967; Seabron to Thomas Hughes, February 6, 1967, box 4711, Personnel 9-2, GC 1906–76, SOA, RG 16, NARA.

43. William Seabron to Jim Thornton, November 30, 1966, box 4537, Organization FES, ibid.

44. William Seabron to Lloyd Davis, February 9, October 31, 1967, box 4654, Extension Work 1, ibid.

45. William Seabron to Lloyd Davis, March 20, 1967, box 4654, Extension Work 1; Seabron to Orville Freeman, April 21, 1967, box 4628, Civil Rights, ibid.

46. William Seabron to Luke M. Schruben, July 24, September 28, 1967, Civil Rights, ibid.

47. Tom L. Lambert to J. William Howell, November 9, 1967, box 4653, Extension Work; William Seabron to Lester P. Condon, December 4, 1967, box 4627, Civil Rights, ibid.

48. Calvin L. Beale, "Notes of a Visit to the Cooperative Extension Office in Fayette County, Texas," box 4629, Civil Rights, ibid. (emphasis in original). See also William Seabron to Lloyd Davis, January 25, 1967, ibid.

49. Ruth W. Harvey to Orville Freeman, July 15, 1967, box 4627; C. L. Tapley to J. E. Wilmore, September 19, 1967, and enclosures; William Seabron to Frank J. Brechenser, October 16, 1967, box 4653, Extension Work, ibid.

50. E. V. Smith to Harry M. Philpott, October 13, 1965; Smith to Ralph Draughon, January 7, 1965; Philpott to L. H. Foster, February 25, 1966, box 52, RG 533, Philpott Papers; L. A. Potts to Thomas R. Hughes, September 14, 1965, box 4455, Civil Rights, GC 1906–76, SOA, RG 16, NARA.

51. O. C. Aderhold to Harmon W. Caldwell, May 1, 1963, box 1, Aderhold Papers. See also Cozy L. Ellison to USDA, November 8, 1963; Luke M. Schruben to Ellison, November 15, 1963, ibid.

52. C. C. Murray to O. C. Aderhold, November 27, 1963 (Eberhardt quote); Aderhold to John I. Spooner, April 3, 1964; "Statement of President O. C. Aderhold regarding Location of Negro State Staff Members of the Extension Service," n.d., attached to Aderhold to L. R. Siebert, May 18, 1964, ibid.; Cozy L. Ellison to C. V. Troup, June 23, 1964, box 16, Troup folder, HRAHC.

53. Joe Western, "Rural Civil Rights," *Wall Street Journal*, September 14, 1964, 1, 12.

54. L. W. Eberhardt Jr. to C. C. Murray, March 8, 1965; S. Walter Martin to O. C. Aderhold, April 14, 1965; Eberhardt to Aderhold, May 12, 1965, box 1, Aderhold

Papers; Cozy L. Ellison to W. W. E. Blanchet, August 15, 1967, box 16, HRAHC. See also Stephen J. Karina, *The University of Georgia College of Agriculture: An Administrative History, 1785-1985* (n.p., n.d.), in Hargrett Rare Book and Manuscript Library, University of Georgia, Athens.

55. Dorothy B. Webster, "Life's Story of William B. Hill," box 360, RG 71, ACES Papers.

56. W. B. Hill, "Extension Work with Limited Resource Families in Alabama," box 358, RG 71, ibid.

57. John W. Slusser to William Seabron, February 20, 1968; Orville Freeman to L. W. Eberhardt Jr., February 26, 1968, box 4783, Civil Rights; Eileen Hemphill to the files, August 8, 1968; Seabron to Norman Smith, August 15, 1968, box 4803, Extension Work, GC 1906-76, SOA, RG 16, NARA.

58. Ernest A. Turner to Orville Freeman, August 15, 1968, box 4782, Civil Rights; William Seabron to Freeman, December 19, 1967, box 4653, Extension Work, ibid.

59. William L. Giles to George L. Mehren, July 14, 1967; Russell I. Thackrey to Giles, July 18, 1967; Giles to Thackrey, July 25, 1967, box 31, Giles Papers. See also Jim Hightower, *Hard Tomatoes, Hard Times: A Report of the Agribusiness Accountability Project on the Failure of America's Land Grant College Complex* (Cambridge, 1973), 75-77.

60. William Seabron to George L. Mehren, October 20, 1967; Seabron to secretary et al., December 1, 1967, box 4627, Civil Rights; Seabron to secretary, March 18, 1968, box 4783, Civil Rights, GC 1906-76, SOA, RG 16, NARA.

61. U.S. Commission on Civil Rights, "The Mechanism for Implementing and Enforcing Title VI of the Civil Rights Act of 1964," July 1968, box 4947, Civil Rights, ibid.

62. William Seabron to Orville Freeman, March 18, 1968, box 4783, ibid.

63. Orville Freeman to William L. Giles, May 16, 1968; Giles to Freeman, June 3, 1968; Jamie L. Whitten to Giles, June 19, 1968; Giles to H. G. Carpenter, July 3, 1968; Clifford M. Hardin to Giles, July 23, 1970; Giles to Hardin, July 27, 1970, box 51, Giles Papers.

64. Ned D. Bayley to Orville Freeman, July 26, 1968, box 4782, Civil Rights, GC 1906-76, SOA, RG 16, NARA.

65. John C. Lynn to Herman Talmadge, May 22, 1968, Senatorial V, Administration A, category files, box 58, Talmadge Papers; Ned D. Bayley to Orville Freeman, August 22, 1968; Edward M. Shulman to Freeman, August 22, 1968, box 4782, Civil Rights, GC 1906-76, SOA, RG 16, NARA. See Emmett Peter Jr., "Keep 'Em Down on the Farm," *New Republic*, October 1968, copy in Lyle Schertz to Joe Robertson, October 28, 1968, box 4803, Extension Work, ibid. See also Jamie Whitten to William L. Giles, July 29, 1968; Giles to Whitten, August 1, 1968, box 51, Giles Papers. For a summary of USDA approaches to FES discrimination, see Bayley to Joseph Robertson, September 12, 1968, box 4782, Civil Rights, GC 1906-76, SOA, RG 16, NARA.

66. Mrs. A. L. Carson to Herman Talmadge, February 5, 1968, box 58, Senatorial V, Administration A, category files, Talmadge Papers.

67. William Seabron to Lloyd H. Davis, September 17, 1968, box 4803, Extension Work, GC 1906–76, SOA, RG 16, NARA.

68. U.S. Commission on Civil Rights, "The Mechanism for Implementing and Enforcing Title VI of the Civil Rights Act of 1964," July 1968, box 4947, Civil Rights, ibid.

69. Joseph M. Robertson and Ned D. Bayley to Clifford Hardin, February 5, 1969; William Seabron to secretary, February 6, 1969; Robertson to Hardin, March 12, 1969, and enclosure, "Responsibility Delegated to HEW for Enforcement of Title VI of the Civil Rights Act," box 4950, Civil Rights, ibid. (emphasis in original).

70. Lloyd H. Davis to secretary, February 18, 1969, ibid.

71. Mississippi State Advisory Committee of the U.S. Commission on Civil Rights, "Equal Opportunity in the Mississippi Cooperative Extension Service," June 1969; W. M. Bost to A. B. Britton Jr., August 18, 1969, box 21, Extension Service Director, Special Collections, MML, MSU.

72. W. M. Bost to Lloyd Davis, August 11, 1969; William L. Giles to Bost, August 26, 1969, box 21, Giles Papers; *Charles F. Wade v. Mississippi Cooperative Extension Service*, 372 F. Supp. 126 (1974).

73. Joseph M. Robertson to Clifford Hardin, March 12, 1969, box 4950, Civil Rights, GC 1906–76, SOA, RG 16, NARA. See also Robertson to Hardin, March 12, 1969, and enclosure, "Responsibility Delegated to HEW for Enforcement of Title VI of the Civil Rights Act," ibid. Greg Moses pointed out that the U.S. Commission on Civil Rights considered that "the office of the USDA's Assistant Secretary for Administration was a bureaucratic tool for blocking civil rights." See Greg Moses, "Apartheid in Texas Agriculture: A Biography of 'Affirmative Action,'" presented at the National Association for African American Studies, Houston, February 16, 1996, http://members .tripod.com/~gmoses/tcrr/apart3.htm. This indictment of USDA foot-dragging also furnishes insight into the Texas Extension Service's discrimination.

74. "Report of the Citizens Advisory Committee on Civil Rights," attached to William Seabron to Secretary Clifford Hardin, March 14, 1969, box 4956, Civil Rights, GC 1906–76, SOA, RG 16, NARA; Elmer Ellis to Hardin, December 16, 1968, box 2, folder 10, Seabron Papers.

75. J. Samuel Walker, *ACC Basketball: The Story of the Rivalries, Traditions, and Scandals of the First Two Decades of the Atlantic Coast Conference* (Chapel Hill, 2011), 184–85.

76. Joseph M. Robertson to Clifford Hardin, September 17, 1970, box 5166, Committees; William A. Carlson to Warren A. Blight, September 28, 1972; Frank B. Elliott to the files, October 3, 1972, box 5532, Committees; R. B. Wilson to Jose M. Villarreal, June 10, 1974, box 5818, Committees, GC 1906–76, SOA, RG 16, NARA; William Seabron to Ramon Montalvo, May 8, 1969, box 2, folder 12, Personal Correspondence, April–June 1969, Seabron Papers.

77. Richard J. Peer to William Seabron, November 3, 1969, box 4946, Civil Rights, GC 1906–76, SOA, RG 16, NARA; T. K. Martin to administrative council, academic department heads, and nonacademic supervisors, July 28, 1970, box 53, Giles Papers.

See also Joseph M. Robertson to Robert J. Brown, October 2, 1969, and enclosure, "USDA Civil Rights Activities," box 4947, Civil Rights, GC 1906–76, SOA, RG 16, NARA.

78. "Federal Agency Is Called Biased," *New York Times*, November 30, 1969, 39 (quote); Richard J. Peer to William Seabron, November 3, 1969, box 4946, Civil Rights, GC 1906–76, SOA, RG 16, NARA. See also "Agriculture Agency Said to Be Nurturing Racial Bias in South," *Wall Street Journal*, December 1, 1969, 5; and Don Kendall, "Hardin Acts on Rights Complaints," *Washington Post*, October 20, 1969, A3.

79. Clifford M. Hardin to Theodore H. Hesburgh, April 14, 1971, box 5363, Committees, GC 1906–76, SOA, RG 16, NARA; "Federal Agency Failure to Implement and Enforce Federal Civil Rights Law," March 1973, box 1, CFOSDCC, USCCR, RG 453, NARA.

80. Theodore M. Hesburgh to Earl L. Butz, January 4, 1972, box 5529, Civil Rights, GC 1906–76, SOA, RG 16, NARA.

81. "Minutes, Annual Meeting, Council on Higher Education in the Agricultural Sciences," Atlanta, Ga., January 7–8, 1971, box 48, RG 533, Philpott Papers. See also "Guidelines for Administration of Appropriations to 1890 Land-Grant Colleges and Tuskegee Institute," August 1971, box 90, RG 71, ACES Papers.

82. Hightower, *Hard Tomatoes, Hard Times*, 33, 226 (Hightower and Morrison quotes); "A Summary of *Hard Tomatoes, Hard Times: The Failure of the Land Grant College Complex*," box 38, RG 533, Philpott Papers; B. F. Smith to William L. Giles, June 5, 1972, box 17, Giles Papers; "Suit Seeks Ban on Aid to Land-Grant System," *Washington Post*, October 12, 1972, A2; Robert S. Catz, "Land Grant Colleges and Mechanization: A Need for Environmental Assessment," *George Washington Law Review* 47 (May 1979): 740–60; Howard S. Scher, Robert S. Catz, and Gregory H. Mathews, "USDA: Agriculture at the Expense of Small Farmers and Farmworkers," *Toledo Law Review* 7 (1975–76): 837–62; U.S. Congress, Office of Technology Assessment, *Technology, Public Policy, and the Changing Structure of American Agriculture* (Washington, 1986).

83. "The Butz Episode," *Christian Science Monitor*, October 4, 1976, 28; Grayson Mitchell, "Butz Helped Circumvent Rights Laws, Critics Say," *Los Angeles Times*, October 5, 1976, B1.

84. "Butz Denies Report of Bias in Job Hiring," *Chicago Tribune*, October 6, 1976, B8 (first quote); Jeff Prugh, "U.S. Farm Agency Accused of Racial Bias," *Los Angeles Times*, October 15, 1976, B2 (second and third quotes). See also "After Butz Exit," *Christian Science Monitor*, October 6, 1976, 28.

85. "Agriculture Department Forms New Rights Unit," *New York Times*, August 7, 1971, 33; Bert Keys Jr. to J. Phil Campbell, July 4, 1973, box 5672, Civil Rights; Distressed Employees of the Compliance and Enforcement Division to Earl Butz, March 29, 1974, box 5816, Civil Rights, GC 1906–76, SOA, RG 16, NARA. See also *Strain v. Philpott*, 331 F. Supp. 836 (1971); *Bazemore v. Friday*, 478 U.S. 385 (1986); *Wade v. Mississippi Cooperative Extension Service*, 372 F. Supp. 126 (N.D. Mississippi 1974); Joseph Brooks, "The Emergency Land Fund: A Rural Land Retention and Development Model," in Leo McGee and Robert Boone, eds., *The Black Rural Land-*

*owner—Endangered Species: Social, Political, and Economic Implications* (Westport, 1979), 120–22; Hezekiah S. Jones, "Federal Agricultural Policies: Do Black Farm Operators Benefit?" *Review of Black Political Economy* 22, no. 4 (1994): 25–50; and Patricia E. McLean-Meyinsse and Adell Brown Jr., "Survival Strategies of Successful Black Farmers," *Review of Black Political Economy* 22, no. 4 (1994): 73–83.

### Chapter 7

1. Reverend K. L. Buford to Fred Robertson, March 22, 1968, box 4849, Personnel 9-2, GC 1906–76, SOA, RG 16, NARA.

2. Pete Daniel, interview with Bertha Jones, Tuskegee Institute, Tuskegee, Ala., February 21, 2007, SOHC.

3. Ibid.

4. Ibid.

5. Ibid.

6. Pete Daniel, interview with Willie L. Strain, Tuskegee Institute, Tuskegee, Ala., February 21, 2007, SOHC.

7. G. W. Taylor to Willie L. Strain, June 22, 1959, box 360; A. P. Torrence to Strain, January 26, 1962, box 361, ACES Papers.

8. Pete Daniel, interview with Willie L. Strain, Tuskegee Institute, Tuskegee, Ala., February 21, 2007, SOHC.

9. Willie L. Strain to Robert R. Chesnutt, July 2, 1962, box 71, ACES Papers.

10. See *The Negro Farmer*, 1963–65 issues, in NAL.

11. *The Negro Farmer*, June 1965, 1, ibid.

12. Pete Daniel, interview with Willie L. Strain, Tuskegee Institute, Tuskegee, Ala., February 21, 2007, SOHC.

13. William Payne to Walter Lewis, October 25, 1965, box 3, CFFPD, USCCR, RG 453, NARA.

14. William Payne to Richard Shapiro, October 1, 1965, ibid.

15. David S. Cecelski, *Along Freedom Road: Hyde County, North Carolina, and the Fate of Black Schools in the South* (Chapel Hill, 1994), 8–9, 32, first quote on 9, second quote on 32. See also Joseph Crespino, *In Search of Another Country: Mississippi and the Conservative Counterrevolution* (Princeton, 2007), 173–204.

16. On Auburn University's actions after the Civil Rights Act of 1964, see Hoyt M. Warren, "Actions Associated with Implementation of the Civil Rights Act of 1964 in the Cooperative Extension Service, Auburn University," December 1972 (administratively confidential), copy in Archives and Manuscripts Department, Auburn University, Auburn, Ala.

17. Ibid., appendix A, "Selected Policy Statements and Announcements," 5.

18. Pete Daniel, interview with Willie L. Strain, Tuskegee Institute, Tuskegee, Ala., February 21, 2007; interview with Bertha Jones, Tuskegee Institute, Tuskegee, Ala., February 21, 2007, SOHC.

19. Pete Daniel, interview with Willie L. Strain, Tuskegee Institute, Tuskegee, Ala.,

February 21, 2007, ibid.; Martin E. Sloane to William Seabron, July 11, 1969, box 1, CFOSDSP, USCCR, RG 453, NARA.

20. Warren, "Actions Associated with Implementation," part II, "Individual Complaint Transformed into Class Action (Civil Action 840-E) Based upon Fifth and Fourteenth Amendments of the United States Constitution and the Civil Rights Act of 1964, Complaint Procedures," 1–8.

21. Ibid., "*Strain v. Hardin*," 11–19.

22. Ibid., "Answers to Complaint," 21–31.

23. Ibid., "Complaint Changed to *Strain, United States vs Philpott*," 33–43.

24. Pete Daniel, interview with Willie L. Strain, Tuskegee Institute, Tuskegee, Ala., February 21, 2007, SOHC. On Judge Frank M. Johnson's ideology, see Tony A. Freyer, ed., *Defending Constitutional Rights: Frank M. Johnson* (Athens, 2001), especially "The Sibley Lecture," 105–18. Susan Youngblood Ashmore provided this source. See also Jack Bass, *Taming the Storm: The Life and Times of Frank M. Johnson and the South's Fight on Civil Rights* (New York, 1993).

25. *Willie L. Strain v. Harry M. Philpott*, 331 F. Supp. 836 (1971), quotation on 839.

26. Ibid., 836–44.

27. Warren, "Actions Associated with Implementation," "Court Opinion and Decree," 90–103.

28. On Alabama's Farm Bureau Federation, see Wayne Flynt, *Alabama in the Twentieth Century* (Tuscaloosa, 2004), 56, 61, 72, 118–19; Warren, "Actions Associated with Implementation," "Promotion of Former Negro County Agents and Former Negro Home Demonstration Agents," 106–16, quote on 114; and David L. Norman to Frank B. Elliott, December 2, 1971, box 5352; Caroline F. Davis to Ned D. Bayley, January 21, 1972, box 5557, Extension Work; K. L. Buford to Parren J. Mitchell, January 24, 1973, box 5671, Civil Rights, GC 1906–76, SOA, RG 16, NARA. See also Richard Lyng to Frank C. Carlucci, April 12, 1972, box 5529, ibid.

29. Warren, "Actions Associated with Implementation," "Implementation of Decree in Regard to Specifically Named Individuals and Conditions," 138–46; ibid., "Strain Considers Radio-TV Position," 159–61.

30. Pete Daniel, interview with Willie L. Strain, Tuskegee Institute, Tuskegee, Ala., February 21, 2007, SOHC.

31. Ibid.

32. Carolyn R. Newton to Earl Butz, May 9, 1973; Edwin L. Kirby to Charles P. Ellington, May 31, 1973; Kirby to Newton, May 31, 1973, box 5671, Extension Service, GC 1906–76, SOA, RG 16, NARA.

33. T. Marshall Hahn Jr. to Earl L. Butz, December 27, 1972; Ned D. Bayley to Butz, January 15, 1973; Paul A. Vander Myde to Hahn, August 27, 1973, box 5671, Extension Service, GC 1906–76, SOA, RG 16, NARA.

34. *Bazemore v. Friday*, 751 F. 2d 662 (CA4 1984); *Bazemore v. Friday*, 478 U.S. 385 (1986), quotes on 395, 400 (emphasis in original). For discussion of the lower court decisions, see "*Bazemore v. Friday*: Salary Discrimination under Title VII," *Harvard*

*Law Review* 99 (January 1986): 655–67. For background on the case, see C. O. Pearson to Orville Freeman, October 13, 1965; William Seabron to Pearson, June 14, 1966, box 4541, Personnel 9-2, GC 1906–76, SOA, RG 16, NARA.

35. *Charles F. Wade v. Mississippi Cooperative Extension Service*, 372 F. Supp. 126 (1974), 138.

36. Ibid., 139, 146, 147. See "Proposed Action Plan of United States to Achieve Full Compliance in Employment Practices and Conduct of Programs of the MCES," December 12, 1972; "MCES Response to Proposed Action Plan of the United States to Achieve Full Compliance in Employment Practices and Conduct of Programs in MCES," December 12, 1972, box 20; Ned D. Bayley to William L. Giles, March 20, 1972, and attached, "Cooperative State Extension Services—Civil Rights—Plans for Compliance," February 25, 1972, box 21, Giles Papers. For more detail on the case, see Fred B. Smith to Jamie L. Whitten, July 21, 1972; Edward M. Shulman to Frank B. Elliott, August 11, 1972, box 5587, Civil Rights, GC 1906–76, SOA, RG 16, NARA. On implementing equal opportunity guidelines, see Edwin L. Kirby to Jerome Miles, November 8, 1973, box 5672, Extension Service, ibid.

37. Joseph R. Wright Jr. to undersecretary, January 11, 1974, box 5816, Extension Service, GC 1906–76, SOA, RG 16, NARA.

38. Ronald Gluck to Robert Dempsey, July 17, 1974; Joseph R. Wright Jr. to Robert W. Long, July 31, 1974, ibid.

39. "The Federal Civil Rights Enforcement Effort—1974, vol. VI, to Extend Federal Assistance, USDA Extension Service" (typescript), U.S. Commission on Civil Rights report, box 4, CFOSDPRS, USCCR, RG 453, NARA.

40. William Peterson to Earl Butz, October 9, 1974, box 5816, Extension Service, GC 1906–76, SOA, RG 16, NARA.

41. William Peterson to Earl Butz, October 9, 1974, box 5816; Miles S. Washington Jr. to Joseph R. Wright, April 10, 1975, box 5945, Extension Service, ibid. See also Roy D. Cassell to T. K. Cowden, September 25, 1975, ibid., for a synopsis of pending cases and the status of noncompliance.

42. See *Knight v. Alabama*, 787 F. Supp. 1030 (N.D. Ala. 1991); *Knight v. Alabama*, 14 F. 3d 1534 (11th Cir. 1994); *Knight v. Alabama*, 900 F. Supp. 272 (N.D. Ala. 1995); *Knight v. Alabama, aff'd*, 476 F. 3d 1219 (11th Cir. 2007), *cert. denied*, 127 S. Ct. 3014 (2007); and Pete Daniel, interview with Willie L. Strain, Tuskegee Institute, Tuskegee, Ala., February 21, 2007, SOHC.

## Chapter 8

1. Wayne D. Rasmussen and Gladys L. Baker, *The Department of Agriculture* (New York, 1972), 123–28; Howard A. Glickstein to William L. Taylor, September 3, 1968, Taylor Papers.

2. Karen Kubovec McIlvain, "Agricultural Law: FmHA Farm Foreclosures, An Analysis of Deferral Relief and the Appeals System," *Washburn Law Journal* 23

(1983–84): 287–91; Patrick C. Murphy, "Representing Farmers in Administrative and Court Proceedings Involving the United States Department of Agriculture," *South Dakota Law Review* 29 (1983–84): 275–81.

3. Joe Henry Thomas to Fay Bennett, January 25, 1963, box 41, Alabama Appeals for Help, folder 25, NSF Papers.

4. Richard M. Shapiro and Donald S. Safford, interview with Fred Amica, Marshville, Ga., March 9, 1964, box 1, CFLID, USCCR, RG 453, NARA; Georgia State Advisory Committee, "Equal Opportunity in Farm Programs," 45–52, Georgia State Advisory Committee, box 40, CFOSDSP, USCCR, RG 453, NARA.

5. Ira Kaye to Fay Bennett, February 9, December 18, 1962, February 19, 1963, box 46, Ira Kaye, folder 44, NSF Papers.

6. Ira Kaye to Fay Bennett, January 7, March 4, 1963; Bennett to Kaye, January 6, 1964, ibid.

7. Handwritten note to Fay Bennett, February 9, 1963; case history of Muldrew Burgess, March 16, 1963, box 41, South Carolina, folder 9; L. S. James to Edwin P. Rogers, December 31, 1963, box 48, FHA, folder 31, ibid.

8. Case history of Willis P. Canty, March 2, 1963, box 41, South Carolina, folder 9; L. S. James to Edwin P. Rogers, December 31, 1963, box 48, FHA, folder 31, ibid.

9. Case history of Mamie Deschamps, March 16, 1963, box 41, South Carolina, folder 9; L. S. James to Edwin P. Rogers, December 31, 1963, box 48, FHA, folder 31, ibid.

10. Case history of John H. Wheeler, March 16, 1963, box 41, South Carolina, folder 9; L. S. James to Edwin P. Rogers, December 31, 1963, box 48, FHA, folder 31, ibid.

11. L. S. James to Edwin P. Rogers, January 17, 1964, box 48, FHA, folder 31, ibid.

12. James Robinson and Donald S. Safford, interview with E. T. Fatheree, Washington, D.C., August 4, 1964, box 1, CFLID, USCCR, RG 453, NARA.

13. T. T. Williams, interview with John S. Currie, Jackson, Miss., May 13, 1964, ibid.

14. Richard M. Shapiro and Donald S. Safford, interview with W. C. Thigpen, Quitman, Ga., March 10, 1964, ibid. On Virginia's FHA, see Marian P. Yankauer, interview with James E. Walters, Petersburg, Va., July 22, 1964; interview with Leonard V. Shelton and Milton K. Brown, Richmond, Va., July 23, 1964, ibid.

15. Marian P. Yankauer to F. Peter Libassi, October 9, 1964, box 2; Yankauer, memo to the file, July 22, 1965, box 3, CFFPD, USCCR, RG 453, NARA.

16. William Seabron to J. William Howell, October 29, 1965, box 4254, Civil Rights, GC 1906–76, SOA, RG 16, NARA.

17. Affidavit of Willie Joe White, n.d. [July 1965], reel 18, frame 710, *SNCC Papers*.

18. Statement of Cato Lee, Lowndesboro, Ala., n.d.; statement of Threddie Stewart, Haynesville, Ala., n.d.; statement of Eugene People, Tylar, Ala., n.d., box 57, Alabama, folder 35, NSF Papers.

19. Robert A. Cook to Marian P. Yankauer, November 22, 1965, box 4254, Civil Rights, GC 1906–76, SOA, RG 16, NARA.

20. Orzell Billingsley Jr. and Harvey Burg to Robert C. Bamberg, July 20, 1965; "The Substance of Reverend McShan's Testimony Offered to Alabama State Advi-

sory Committee to U.S. Commission on Civil Rights," July 10, 1965, box 4255, ibid.; State Advisory Commission open meeting in Demopolis, Ala., July 10, 1965, box 39; "Evaluation of the Effect of FHA Policy in Greene County on the Negro Farm Families Residing There," box 2, CFOSDSP, USCCR, RG 453, NARA.

21. T. T. Williams, interview with Harry Means, Eutaw, Ala., May 12, 1964; Richard M. Shapiro and Donald S. Safford, "Report on Field Trip to Alabama and Mississippi from May 10, 1964 through May 16, 1964," box 1, CFLID, USCCR, RG 453, NARA.

22. T. T. Williams, interview with Adam White, Prentiss, Miss., May 15, 1964, ibid.

23. William Seabron to J. William Howell, March 21, 1967, box 4711, Personnel 9-2; Seabron to Mildred E. Meadows, January 2, 1968, box 4844, Personnel 9-2, GC 1906–76, SOA, RG 16, NARA.

24. Richard M. Shapiro and Donald S. Safford, "Report of Field Trip to Louisiana," June 8-12, 1964, box 1, CFLID, USCCR, RG 453, NARA.

25. William A. Tippins to John W. Slusser, September 1967(?), box 4627, Civil Rights, GC 1906–76, SOA, RG 16, NARA; Lu Ann Jones, interview with Henry Woodard, Tunica, Miss., October 5, 1987, OHSA; affidavit of Jerome Anthony, July 8, 1965, reel 18, frame 500, *SNCC Papers.*

26. William A. Tippins to John W. Slusser, September 1967(?), box 4627, Civil Rights, GC 1906–76, SOA, RG 16, NARA.

27. "Equal Opportunity in Farm Programs," 34–52, Georgia State Advisory Committee, box 40, CFOSDSP, USCCR, RG 453, NARA; Will Bacon to FHA, May 1970, box 5161, Civil Rights, GC 1906–76, SOA, RG 16, NARA.

28. William P. Mitchell to USDA, July 8, 1966, box 4455; Richard J. Peer to Joseph Robertson, November 20, 1970, box 5159, GC 1906–76, SOA, RG 16, NARA.

29. William Seabron to Howard Bertsch, December 5, 1966, box 4454, ibid.

30. Testimony of Robert C. Bamberg, April 29, 1968, 230–32, 235, 242, in *Hearing before the United States Commission on Civil Rights, Montgomery, Ala., April 29, 1968* (Washington, 1969). See also Susan Youngblood Ashmore, *Carry It On: The War on Poverty and the Civil Rights Movement in Alabama, 1964-1972* (Athens, 2008), 255–56.

31. William L. Taylor to Orville Freeman, May 20, 1968, copy in Taylor Papers; *Hearings before the U.S. Commission on Civil Rights, April 27–May 2, 1968* (Washington, 1969), exhibit no. 5, James T. Bonnen, "Progress and Poverty: The People Left Behind," table 1, "Gross Program Outlays for Federal Agricultural Programs in Alabama, by Agency, Fiscal Year 1967," 742.

32. Bob Mants to FHA, April 4, 1966, box 4455; William Seabron to Mants, November 22, 1966; Seabron to Howard Bertsch, November 17, 1966, box 4454, Civil Rights, GC 1906–76, SOA, RG 16, NARA.

33. William Seabron to Howard Bertsch, March 21, 1967, and attached, Kenneth L. Dean to Seabron, February 13, 1967, box 4628, Civil Rights, ibid.; Mrs. Edd W. Johnson to Herman Talmadge, n.d., stamped received December 23, 1969, box 135, folder 9, Talmadge Papers.

34. William Greider, "Whites Only Clubs Got U.S. Loans," *Washington Post*, August 14, 1970, A7; Laura Kolar, "'Selling' the Farm: New Frontier Conservation and the USDA Farm Recreation Policies of the 1960s," *Agricultural History* 86 (Winter 2012): 55–77.

35. Looby & Williams to Orville Freeman, February 2, 1967, box 4629, Civil Rights, GC 1906–76, SOA, RG 16, NARA. On the Tennessee FHA, see Ronald M. McCaskill and Donald S. Safford, interview with T. Cavendar, August 6, 1964, box 1, CFLID, USCCR, RG 453, NARA.

36. William Seabron to Floyd Higbee, February 9, 1967, box 4711, Personnel 9-2, GC 1906–76, SOA, RG 16, NARA.

37. William Seabron to Howard Bertsch, September 18, 1967, box 4627, Civil Rights, ibid.

38. Fay Bennett to William McElhannon, August 1, 1968; McElhannon to Bennett, August 24, 1968; Bennett to Howard Bertsch, September 13, 1968, box 48, FHA, folder 31, NSF Papers; Adrienne Petty, "I'll Take My Farm: The GI Bill, Agriculture, and Veterans in North Carolina," *Journal of Peasant Studies* 35 (October 2008): 760–63, quote on 757.

39. Ferguise Mayronne to William Seabron, June 24, 1968, box 4782, Civil Rights, GC 1906–76, SOA, RG 16, NARA.

40. M. John Bundy and Allen D. Evans, "Racial Discrimination in USDA Programs in the South: A Problem in Assuring the Integrity of the Welfare State," *Washington Law Review* 45 (1970): 746–84, first quote on 783, second quote on 746.

41. Francis B. Stevens to Clifford M. Hardin, September 1, 1970, box 5160, Civil Rights, GC 1906–76, SOA, RG 16, NARA.

42. Johnnie Jenkins Jr. et al. to Richard M. Nixon, February 22, 1971, box 5350, ibid.

43. Shirley D. Webb, statement, October 9, 1970; William Seabron to Webb, November 23, 1970, box 5163; Webb to Seabron, January 4, 1971, box 5350; Richard J. Peer, "Shirley D. Webb, Summary," box 5351; Peer to Seabron, May 17, 1971, box 5350, Civil Rights, ibid.

44. "Civil Rights Compliance Review of Greene County, Alabama," June 21, 1971, box 5351, Civil Rights, ibid.

45. Richard J. Peer to William Seabron, July 30, 1971, ibid.; Greta de Jong, "Staying in Place: Black Migration, the Civil Rights Movement, and the War on Poverty in the Rural South," *Journal of African American History* 90 (Fall 2005): 403.

46. "Civil Rights Compliance Review of Greene County, Alabama," June 21, 1971, box 5351, Civil Rights, GC 1906–76, SOA, RG 16, NARA.

47. William Seabron to James V. Smith, February 17, 1971, box 5350; Joseph R. Wright Jr. to Jerome Shuman, Leonard Greess, John A. Knebel, and S. B. Pranger, October 15, 1973, box 5671, Civil Rights, ibid. See Percy R. Luney to Joseph M. Robertson, February 22, 1971, box 5350, ibid., for a report on the ASCS.

48. Lu Ann Jones, interview with Welchel Long, Elbert County, Ga., April 16, 1987, OHSA.

49. Ibid.

50. Affidavit of Welchel Long, n.d., box 5559, Farm Credit 1–2, GC 1906–76, SOA, RG 16, NARA.

51. Ibid.

52. Lu Ann Jones, interview with Welchel Long, Elbert County, Ga., April 16, 1987, OHSA.

53. Simon Hunter to whom it may concern, July 31, 1972, box 5559, Farm Credit 1–2, GC 1906–76, SOA, RG 16, NARA.

54. Russell Daniel Jr., "Information concerning Welchel Long, Elberton, Georgia," April 19, 1971; Daniel to Richard M. Nixon, October 14, 1972, ibid.

55. K. S. Yon to Berkeley Burrell, August 12, 1971; Burrell to George Bell, August 25, 1971; Yon to Burrell, December 10, 1971, ibid.

56. George C. Knapp to Jerome Shuman, September 10, 1971, box 5325, Farm Credit 1; Shuman to George T. Bell, January 5, 1972, box 5559, Farm Credit 1–2, ibid.; "Agriculture Department Forms New Rights Unit," *New York Times*, August 7, 1971, 33.

57. Lu Ann Jones, interview with Welchel Long, Elbert County, Ga., April 16, 1987, OHSA.

58. On the appeal process, see McIlvain, "Agricultural Law: FmHA Farm Foreclosures," 287–308; and Patrick C. Murphy, "Representing Farmers in Administrative and Court Proceedings Involving the United States Department of Agriculture," *South Dakota Law Review* 29 (Spring 1984): 275–81.

59. Paul Delaney, "Rights Panel Again Assails Efforts by U.S. Agencies," *New York Times*, November 17, 1971, 1; Alfred L. Edwards to Jerome Shuman, April 11, 1972, box 5529, Civil Rights, GC 1906–76, SOA, RG 16, NARA.

60. Howard S. Scher, Robert S. Catz, and Gregory H. Mathews, "USDA: Agriculture at the Expense of Small Farmers and Farmworkers," *Toledo Law Review* 7 (Spring 1976): 843–51.

61. David Westfall, "Agricultural Allotments as Property," *Harvard Law Review* 79 (April 1966): 1180–1202; Brainerd S. Parrish, "Cotton Allotments: Another 'New Property,'" *Texas Law Review* 45 (March 1967): 734–53; Keith G. Meyer, "Potential Problems Connected with the Use of 'Crops' as Collateral for an Article 9 Security Interest," *Agricultural Law Journal* (1981–82): 115–68.

62. Walter W. Wilcox to Orville Freeman, February 19, 1968 (quotes); Wilcox to William C. Black, May 31, 1968, box 4815, Farming, GC 1906–76, SOA, RG 16, NARA.

## Chapter 9

1. *Pigford v. Glickman*, 185 F.R.D. 82 (D.D.C. 1999), 85–95, quote on 85.

2. U.S. Commission on Civil Rights, "The Decline of Black Farming in America" (February 1982), 85.

3. U.S. Commission on Civil Rights, "Decline of Black Farming in America," 88–91; Bruce J. Reynolds, "Black Farmers in America, 1865–2000: The Pursuit of Independent Farming and the Role of Cooperatives," USDA Rural Business Cooperative Ser-

vice, Research Report 194 (October 2002), 24; Robert Zabawa, Arthur Siaway, and Ntam Baharanyi, "The Decline of Black Farmers and Strategies for Survival," *Southern Rural Sociology* 7 (1990): 107.

4. U.S. Commission on Civil Rights, "Decline of Black Farming in America," 98–126; Ward Sinclair, "Fat-Cat Farmers Get Bulk of U.S. Aid, Study Finds," *Washington Post*, November 10, 1981, A2. See also Herbert H. Denton, "Study Finds Black Family Farmers Vanishing Group," *Washington Post*, February 10, 1982, A2.

5. Juan Williams and Ward Sinclair, "Agriculture's Minority Affairs Chief Would Purge Rights Rules," *Washington Post*, February 17, 1983, A22; Sinclair, "Laxity, Confusion Mar USDA Rights Program," *Washington Post*, July 10, 1984, A11; Sinclair, "Agriculture Aide Placed on Leave during Review of Rights Memo," *Washington Post*, February 18, 1983, A5.

6. Ward Sinclair, "USDA, Block Scored for Stance on Rights," *Washington Post*, April 7, 1983, A8.

7. U.S. Congress, House, Committee on the Judiciary, *Hearings before the Subcommittee on Civil and Constitutional Rights on Civil Rights Enforcement Record of the Department of Agriculture*, 98th Cong., 2nd sess. (Washington, 1984), September 26, 1984, testimony of John W. Garland, 83 (quotes); U.S. Commission on Civil Rights, "Decline of Black Farming in America," 71–120.

8. U.S. Congress, House, Committee on the Judiciary, *Hearings before the Subcommittee on Civil and Constitutional Rights on Civil Rights Enforcement Record of the Department of Agriculture*, 98th Cong., 2nd sess. (Washington, 1984), September 26, 1984, testimony of John W. Garland, 59; House Subcommittee on Government Information, Justice, and Agriculture, *Hearings on Civil Rights Enforcement Record of the Department of Agriculture*, 101st Cong., 2nd sess. (Washington, 1990), July 25, 1990, testimony of David H. Harris Jr., 36, quote on 38. See also U.S. Commission on Civil Rights, "Decline of Black Farming in America," 84–86.

9. House Subcommittee on Civil and Constitutional Rights, *Hearings on Civil Rights Enforcement Record of the Department of Agriculture*, statement of John W. Garland, 56–61, 84; U.S. Commission on Civil Rights, "Decline of Black Farming in America, 84–85.

10. House Subcommittee on Civil and Constitutional Rights, *Hearings on Civil Rights Enforcement Record of the Department of Agriculture*, statement of Timothy Pigford, 68.

11. Ibid., 70–72.

12. Ibid.

13. Ibid., 63–75, first quote on 63, second quote on 65, third quote on 75.

14. Ibid., testimony of John W. Garland and Timothy Pigford, 77–81, quote on 81.

15. Ibid., testimony of Alma R. Esparza, 95–123.

16. Ibid., 125, Edwards quote on 129; Ward Sinclair, "Laxity, Confusion Mar USDA Rights Program," *Washington Post*, July 10, 1984, A11; Sinclair, "Civil Rights Dispute Erupts in Agriculture," *Washington Post*, July 12, 1984, A19; Sinclair, "USDA Rights Programs Hit in Internal Report," *Washington Post*, August 25, 1984, A2; Sinclair, "In-

side: The USDA," *Washington Post*, September 21, 1984, A19. See also Sinclair, "Agriculture's Equal-Rights Office Hit on Stalled Investigations," *Washington Post*, July 28, 1984, A6.

17. House Subcommittee on Civil and Constitutional Rights, *Hearings on Civil Rights Enforcement Record of the Department of Agriculture*, testimony of Arthur Campbell, 159–62. For a discussion of legislation relevant to the FmHA, see Mark Keenan, "Food Security Act of 1985: FmHA Farm Program Reforms," *South Dakota Law Review* 31 (1986): 478; and Martha A. Miller, "The Role of the Farmers Home Administration in the Present Agricultural Crisis," *Alabama Law Review* 38 (Spring 1987): 587–623.

18. For a discussion of heir and tax sales and a review of the substantial literature on the subject, see Jess Gilbert, Gwen Sharp, and M. Sindy Felin, "The Loss and Persistence of Black-Owned Farms and Farmland: A Review of the Research Literature and Its Implications," *Southern Rural Sociology* 18 (2002): 1–30. See also Michael D. Schulman, Patricia Garrett, Regina Luginbuhl, and Jody Greene, "Problems of Landownership and Inheritance among Black Smallholders," *Agriculture and Human Values* 2 (1985): 40–44; and Harold A. McDougall, "Black Landowners Beware: A Proposal for Statutory Reform," *New York University Review of Law and Social Change* 9 (1979–80): 127–61.

19. Ward Sinclair, "Agriculture Agency's Rights Plan Under Study," *Washington Post*, April 4, 1986, A17; Sinclair, USDA Fires Black Employee after Criticism of Agency," *Washington Post*, May 21, 1986, A4; "Aide Settles U.S. Bias Suit," *Washington Post*, November 15, 1986, A1. See also Sinclair, "Panel Asks Probe of USDA Race Bias Charges," *Washington Post*, May 31, 1986, A6; "A Racial Problem at USDA," *Washington Post*, June 4, 1986, A22; and "Uprooting Bias at Agriculture," *New York Times*, July 4, 1986, A26.

20. U.S. Congress, House, Committee on the Judiciary, *Hearings before the Subcommittee on Civil and Constitutional Rights on Equal Employment Opportunities at the Department of Agriculture*, 100th Cong., 1st sess. (Washington, 1987), November 5, 1987, testimony of Peter C. Meyers, 5–7, 12, 15–17.

21. Ibid., testimony of June K. W. Kalijarvi, first quote on 37, second quote on 64, third quote on 39.

22. Ward Sinclair, "FmHA Policies Harm Minorities, Groups Say," *Washington Post*, November 27, 1987, A10. On the larger rural crisis, see Miller, "Role of the Farmers Home Administration," 587–623.

23. Ward Sinclair, "At USDA, a 5-Year Plan to Fight Discrimination," *Washington Post*, April 20, 1988, A19; Sinclair, "Agriculture Cracks Down on FmHA Unit for Rights Violations," *Washington Post*, August 5, 1988, A15.

24. U.S. Congress, House, Committee on Government Operations, *Hearings before the Subcommittee on Government Information, Justice, and Agriculture on Decline of Minority Farming in the United States*, 101st Cong., 2nd sess. (Washington, 1990), July 25, 1990, statement of Mike Espy, 6; testimony of David H. Harris Jr., 19 (quotes), 26.

25. Ibid., statement of David H. Harris Jr., 31–34, quote on 45.

26. Ibid., testimony of Randi Ilyse Roth, first two quotes on 59, third quote on 63.

27. Ibid., affidavit of Betty Puckett, July 10, 1990, 88; affidavit of Olly Neal, July 19, 1990, 89–91, first quote on 90, second quote on 91. See also ibid., affidavit of Martha A. Miller, 92–95.

28. Ibid., statement of Ben Burkett, first quote on 152, second quote on 153; testimony of Burkett, 159. See also Lu Ann Jones, interview with Ben Burkett, Hattiesburg, Miss., October 26, 1987, OHSA.

29. Spencer Rich, "Local Officials Getting Above-Average Subsidies," *Washington Post*, June 29, 1995, A19.

30. "Civil Rights at the United States Department of Agriculture: A Report by the Civil Rights Action Team," (Washington, 1997), 3.

31. Ibid., 3–8.

32. Ibid., 18, 20, 22–23, 25–26. See also U.S. Commission on Civil Rights, "Ten-Year Check-up: Have Federal Agencies Responded to Civil Rights Recommendations?" (staff draft), June 12, 2003.

33. *Pigford v. Glickman*, 185 F.R.D. 82 (D.D.C. 1999), 85, 87, 89, 92–95. See also "Ten-Year Check-up."

34. *Pigford v. Glickman*, 185 F.R.D. 82 (D.D.C. 1999), 95–101. The case produced a mountain of documentation. Alexander Pires, the prosecution attorney, stated: "I have an office full of admissions. I have tape recordings of Mr. Glickman. I have tape recordings of Government officials. I've interviewed everybody there is to interview. I have documents. I have the CRAT Report annotated" (100). Carrie Johnson, "$1.25 Billion Settlement Reached with Black Farmers," *Washington Post*, February 19, 2010, A1, A10.

35. Dan Morgan, Gilbert M. Gaul, and Sarah Cohen, "Farm Program Pays $1.3 Billion to People Who Don't Farm," *Washington Post*, July 2, 2006, A1, A13; Morgan, Gaul, and Cohen, "Growers Reap Benefits Even in Good Years," *Washington Post*, July 3, 2006, A1, A8–9; Morgan, Gaul, and Cohen, "No Drought Required for Federal Drought Aid," *Washington Post*, July 18, 2006, A1, A11; Morgan, Gaul, and Cohen, "Aid Is a Bumper Crop for Farmers," *Washington Post*, October 15, 2006, A1, A12; Morgan, Gaul, and Cohen, "Crop Insurers Piling Up Record Profits," *Washington Post*, October 16, 2006, A1, A12; Cohen, "Deceased Farmers Got USDA Payments," *Washington Post*, July 23, 2007, A1, A4.

36. Darryl Fears, "USDA Is Called Lax on Bias," *Washington Post*, May 18, 2008, A6; Krissah Thompson, "USDA Chief Details Agency Efforts to Improve Record on Civil Rights," *Washington Post*, February 16, 2010, A11; Thompson, "Vilsack Vows to Implement Fixes over Discrimination at USDA," *Washington Post*, May 11, 2011, A13.

37. *Keepseagle v. Vilsack*, Civil Action no. 1: 99CV03119, District Court, D.C., Judge Emmet G. Sullivan, "Memorandum of Points and Authorities in Support of the Motion for Certification of Damages Claims under Federal Rule of Civil Procedure 23(b)(3)," quotes on 11–12 (nn. 9–12), 18 (nn. 16–17). For examples of discrimination, see

Krissah Thompson, "USDA Plaintiffs Celebrate Settlement," *Washington Post*, October 21, 2010, A19.

38. *Keepseagle v. Vilsack*, Civil Action no. 1: 99CV03119, "Memorandum of Law in Support of Motion for Preliminary Approval of Settlement, and an Order Certifying Settlement Class and Approving Certain Provisions in Settlement Agreement"; Krissah Thompson and Spencer S. Hsu, "Bias Case Settled with Native American Farmers," *Washington Post*, October 20, 2010, A3.

39. Jody Feder and Tadlock Cowan, "*Garcia v. Vilsack*: A Policy and Legal Analysis of a USDA Discrimination Case," Congressional Reference Service, December 17, 2010; Mary Clare Jalonick, "Offer Made to Hispanic, Female Farmers to Settle Bias Claims," *Washington Post*, February 26, 2011, A2.

40. Ann Valk and Leslie Brown, *Living with Jim Crow: African American Women and Memories of the Segregated South* (New York, 2010), 164–70, quote on 169. This case is amply documented in newspapers, in magazines, on television, and on the web.

# ACKNOWLEDGMENTS

In a draft of this book completed just after reading Keith Richards's *Life*, I included several pages of less than relevant observations on southern music and its international influence. Fortunately, both John Dittmer and Anthony Badger, who read the manuscript for the University of North Carolina Press, kindly suggested cuts in music and offered sage advice on improving the manuscript. Lu Ann Jones, a keen student of rural life who has interviewed nearly 200 southern farmers, read an earlier draft and offered invaluable suggestions. Susan Youngblood Ashmore, who has written eloquently of civil rights in Alabama, read two chapters and not only furnished me with wise advice but also suggested pertinent sources. Monica Gisolfi read portions of the manuscript and offered ideas on improving it. Elaine De-Lott Baker has been incredibly helpful, not only reading several chapters but also enduring an interview, sending me documents from her time in SNCC and COFO, and providing photographs that she took in Panola County, Mississippi, in 1965. Constance Curry, who was at SNCC's birth and is still actively writing and gathering documentation, has supported this project and offered enduring friendship.

One of the delights of historical research is relying on archivists. Joe Schwarz at the National Archives and Records Administration has over the years patiently guided me to both obvious and obscure sources. He good-naturedly shrugged his shoulders and smiled when I sent numerous Smithsonian fellows to him, and he gave them all careful attention. At Auburn University, archivist Dwayne Cox steered me through collections, uncovered crucial sources, and announced that Willie Strain lived in nearby Tuskegee, as well as sharing several meals. Mattie Abraham, coordinator of Special Collections at Mississippi State University, and her staff have over the years welcomed me and patiently brought out piles of documents. Vanessa Williamson at the U.S. Commission on Civil Rights guided me to pertinent hearings and photographs. I also received assistance at the Library of Congress's Manuscript Division, the Hargrett Rare Book and Manuscript Library at the University of Georgia, Fort Valley State University, the University of Mississippi, Southern University, the University of North Carolina, the South Caroliniana Library, the Amistad Collection at Tulane University, the Walter P. Reuther Library at Wayne State University, and the State Historical Society of Wisconsin. Special thanks to Kathleen Below, who diligently searched the SNCC papers for relevant documents. Thanks also to Grace Palladino for her friendship and her matchless annotation work.

Kate Torrey at the University of North Carolina Press lent encouragement and offered valuable suggestions on improving the manuscript. A special thanks to Paula Wald, who copyedited the manuscript with impressive skill and engagement.

# INDEX

Note: Page numbers in italics indicate illustrations.